SELFORGANIZATION

Library of Congress Cataloging in Publication Data

Selforganization portrait of a scientific revolution / edited by
Wolfgang Krohn, Günter Küpper, Helga Nowotny.
p. cm.
ISBN 978-90-481-4073-2 ISBN 978-94-017-2975-8 (eBook)
DOI 10.1007/978-94-017-2975-8
1. Self-organizing systems. 2. Science--Social aspects.
I. Krohn, Wolfgang. II. Küppers, Günter. III. Nowotny, Helga.
Q325.S475 1990
003'.7--dc20 90-38777

Printed on acid-free paper

Table of Contents

WOLFGANG KROHN, GÜNTER KÜPPERS AND HELGA NOWOTNY / Selforganization - the Convergence of Ideas. An Introduction 1

I. Epistemological Foundations

HUMBERTO R. MATURANA / Science and Daily Life: The Ontology of Scientific Explanations 12

GERHARD ROTH AND HELMUT SCHWEGLER / Self-Organization, Emergent Properties and the Unity of the World 36

BERND-OLAF KÜPPERS / On a Fundamental Paradigm Shift in the Natural Sciences 51

NIKLAS LUHMANN / The Cognitive Program of Constructivism and a Reality that Remains Unknown 64

II. Selfreference and Selfregulation in Social Systems

GUNTHER TEUBNER / How the Law Thinks: Toward a Constructivist Epistemology of Law 87

PETER M. HEJL / Self-Regulation in Social Systems 114

KURT LUDEWIG / Systemic Therapy - A Particular Drift Between Systems Theory and Psychotherapy 128

SIEGFRIED J. SCHMIDT / Literary Systems as Self-organizing Systems 143

CHRISTOPHER S. BRAIDER / Chekhov's Letter: Linguistic System and its Discontents 154

III. The Appearance of Structure

MICHAEL HEIDELBERGER / Concepts of Self-Organization in the 19th Century 170

MICHAEL STADLER AND PETER KRUSE / Cognitive Systems as Self-Organizing Systems 181

IV. The Selforganization of Science

RUDOLF STICHWEH / Self-Organization and Autopoiesis in the Development of Modern Science 195

WOLFGANG KROHN AND GÜNTER KÜPPERS / The Selforganization of Science - Outline of a Theoretical Model 208

HELGA NOWOTNY / Actor-Networks versus Science as Self-Organizing System: A Comperative View of two Constructivist Approaches 223

RAINER PASLACK / Self-Organization and New Social Movements 240

Person Index 255

Subject Index 257

Selforganization - the Convergence of Ideas
An Introduction

WOLFGANG KROHN*, GÜNTER KÜPPERS*, HELGA NOWOTNY**

* *University of Bielefeld,* ** *University of Vienna*

This volume unites authors who are participating in a fundamental change of their discipline. Some of them even speak of a scientific revolution. Observing a present-day scientific revolution encompassing many fields of the natural and social sciences as well as of the humanities that touches epistemology and the self-understanding of social groups in terms of the »Weltanschauung«, is a rare opportunity indeed. Thus, the main objective of this volume consists in opening the view on this change from different perspectives.

Whether this revolution will be victorious or will later be referred to by historians only as a strange passing fashion cannot be answered here. Nor can it be the task of a volume like this to argue in a coherent way for the significance of such a revolution. Finally, the decision to have experts from various fields express their opinion is to be viewed as a decision in favor of the subject's topicality. It is not meant to function as a critical assessment of controversial opinions.

The focus of the claimed scientific revolution is the concept of self-organization that has come into use as the generic term for a category of related, yet nonidentical concepts that are being applied within the context of various fields of research: autopoiesis, synergetics, dissipative structures, self-referential systems, and deterministic chaos. The purpose of this introduction is a) to give a short overview on the most important fields in which these concepts have been developed and defined, b) to show how they came into contact with one another, interconnected, and became part of a research programme; and c) to set forth some categories by means of which the basic approach and its potential importance for the sciences becomes clearly visible; and d) to provide the opportunity for discussing some specific issues on the use of the concept in the field of social sciences.

The revolutionary character of the majority of revolutions in the history of science is not due to their introducing entirely new ideas into the debate but rather to their offering surprising solutions to problems that have been the object of discussions over a long period. This holds true for the present example. The concept of self-organization in its modern sense may be traced back to I. Kant who, in his »Critique of Judgement«, grapples with the inner purposiveness of nature. According to Kant and other thinkers of his time, the problem of thought consisted in the fact that the purposes as such failed to provide an explanation for purposiveness. The purposes operate and are caused in a way that is no less »blind« than all causal, i.e., mechanical, processes. The conceptual solution offered by Kant is closely linked with the theories of today:
»In such a natural product as this every part is thought as owing its presence to the agency of all the remaining parts, and also as existing for the sake of the others and of the whole, that is as an instrument, or organ. But this is not enough - for it might be an

W. Krohn et al. (eds.), Selforganization. Portrait of a Scientific Revolution, 1–10.

instrument of art, and thus have no more than its general possibility referred to an end. On the contrary the part must be an organ producing the other parts - each, consequently, reciprocally producing the others. No instrument of art can answer to this description, but only the instrument of that nature from whose resources the materials of every instrument are drawn - even the materials for instruments of art. Only under these conditions and upon these terms can such a product be an organized and self-organized being, and, as such, be called a physical end«.[1]

The problem basically consisted in understanding how to maintain and increase a sublime whole. The close relationship with the basic problem of bourgeois society may be seen instantaneously: how can the particular selfishness of individuals ensure society's progress and solidarity? In his »Wealth of Nations«, Adam Smith formulated a general model according to which the balance of the whole occurs just at the point when nobody cares about it. The specific regulating mechanism of the feedback Smith had propounded may be found in the new power machines that had a great impact on further development. In 1784 James Watt was granted a patent for his centrifugal regulator of the steam engine's speed. The temporal simultaneity in articulating the idea of self-organization in nature, economy, and technology is astounding - no less than in the parallel events occurring in our own time to which this book is devoted.[2]

It would certainly also be possible to stress the historical continuity of the problem awareness. From pre-Socratic philosophy until the present, one would presumably recognize in the fluctuations between teleologic-holistic and mechanistic-reductionist organization concepts of world and society the variation of one of those long-term »themata« of cognition as expounded by Holton.[3]

By presenting some positions that are typical of the post World War II period, in particular positions of the 1970s, emphasis should be placed not only on those continuities with an old topic but much more on the revolutionary innovations that emanated from some paradigmatic solutions to specific problems for the general problem of the origin of order in the formation of complex self-organization.

Heinz von Foerster was among the first to attempt a description in a physically exact way of the phenomenon of self-organization as an increment of order. In his work »On Self-Organizing Systems and their Environment«[4], published in 1960, he develops, following Shannon's Information theory, a measure for order by formulating the conditions for its increase: a system can achieve a state of higher order if and only if the relative increase of the maximally possible disorder within the system is greater than the presently prevailing actual disorder, since under such conditions the relative order increases automatically. This may occur in two ways that are independent of each other and cooperate in real systems. In the first process, environmental perturbations (undirected energy) are selected by the system to be subsequently incorporated into the structure of the system. In the second case, the system expands. This expansion does not occur arbitrarily but is due to an import of »building blocks« that are compatible with the system's structure. Heinz von Foerster called the first of these principles »order from noise«, while the second, according to Erwin Schroedinger, is »order from order«.

Not only was Heinz von Foerster one of the founders of modern self-organization research, but he also created an institutional framework for his ideas. At the »Biological Computer Laboratory« (BCL) of the University of Illinois (Urbana) he convened and encouraged an interdisciplinary group of researchers to tackle the task of elaborating a theory of self-organizing systems by drawing from the findings of biological systems theory, cybernetic learning theory, and automata theory.[5]

Ilya Prigogine is a physical chemist. Since the mid-1940s, he has been interested in and concerned with the dynamics of chemical reactions and, in this context, with the thermodynamics of irreversible processes. From 1967 onward, the initial focus on questions of the stability of thermodynamic systems shifted to the question: »How may dissipation generate ›order‹, both in time and space«.[6] In the beginning, questions centered around processes in the proximity of an equilibrium, but soon it became evident that structure formation occurred at quite some distance from thermodynamic equilibrium. While in classical thermodynamics, oscillations around the state of equilibrium are balanced out in the so-called stationary state[7] remote from the thermodynamic equilibrium, oscillations may increase. With the help of fluctuations, the stationary state constantly examines its stability. In the case of instability, critical perturbations continue to increase until they reach macroscopic dimensions. In this way, fluctuations - environmental perturbations or eigenfluctuations - may drive the system into a completely new state and thus become the motor of system development (»order through fluctuation«). Instabilities at the same time can break symmetry, i.e., bifurcations occur: the system may choose among two states, though determined by causality. The »selection« of the future path of development is therefore absolutely unpredictable. At the same time, the system becomes dependent on its own system history.

Also in the early 1960s, Hermann Haken began to elaborate a theory of the laser. He demonstrated how the interaction between many independent laser atoms and the photoelectric-field they have generated under certain constraints leads to a coordination of the laser atoms' activities and to a practically endless, monochromatic light beam, i.e. laser light. The laser atoms are »enslaved« by their own light field and forced to move rhythmically in time. It was on these grounds that Haken called his new theory synergetics or the theory of cooperation. Today, the laser is a paradigmatic example for self-organization.

At the same time as Haken was working on his laser theory, Manfred Eigen succeeded in demonstrating that the genesis of life is the result of a selection process at the molecular level and that Darwin's ideas are not only limited to the origin of the species. Eigen was able to show how complex organic structures with environmentally adapted characteristics emerge through an evolutionary process of selection in which adaptation is optimized by the structures themselves. The prerequisites for this self-organizing performance of the macromolecules are openness for the transfer of matter and energy, self-reproduction (autocatalysis), and mutation.

Chaos research also started in the early 1960s. At that time, E.N. Lorenz developed a computer-based global weather model and made a surprising discovery. He realized that though his simulations produced certain basic patterns, they never repeated themselves in exactly the same way. Two simulations starting from an almost identical point thad entirely different results. The model obviously was »sensitively dependent«[8] on the initial conditions; its development was unpredictable. A simple, deterministic system of equation showed a chaotic behavior. But the chaos possessed a kind of internal order; a kind of regular irregularity prevailed whose legitimacy and inherent laws are being investigated by the current program of chaos research.

The spectrum of the successful investigations initiated in the 1960s mainly encompassed various fields of the natural sciences and engineering. The interrelation of the methods adopted was very limited.[9] Viewed from this initial phase, it seemed improbable that in the 1970s a comprehensive research program would become established within the fields of mathematics, physics, biology, cybernetics, and, shortly afterwards, also in socialpsychology, gestalt theory, and family therapy. In addition, it was expected that the various approaches would continue to retain their disciplinary orientation in order to get a firm footing there. Interviews that we conducted with the founders of modern self-organization research confirmed this.[10] Von Foerster, who saw himself fully bound up with the tradition of cybernetics, posed the question: »How can one formulate problems of organization? Is it, for instance, possible to use information theories for organization?« Eigen's research work was initially still centered around his work on reaction kinetics. »Formerly we did much more research on the kinetics of instant reactions by first developing it methodically before passing through various phases. At first we examined the very quick protolytic reactions and then the metal complexes. Subsequently it was the turn of organic chemists and finally that of the biochemists. Thereafter we started to examine the reactions of enzymes.« When Haken was with Bell Telephone Laboratories in 1950, he heard, »that over there a new light source was envisaged to be built, and in the years to follow, i.e., in 1961, 1962, I set out to develop the fundamentals of the laser theory, initially a semi-classical theory«, closely linked with the quantum field theory of solids. Prigogine's work also concentrated initially on research within his special field. »It was all physical chemistry.«

The general significance of these initial approaches was not recognized in the 1970s. At that time one could not as yet speak of a general theory of self-organization. Nevertheless, these first approaches already contained a »revolutionary« core whose central function for a theoretically new approach within the rather pragmatic relation with the technically specific context remained hidden. At least this is what the reconstruction of the actors conveys. A clear explanation for this is contained in the words of Haken: »With hindsight this is (i.e., the laser, the authors), of course, a good example for self-organization; only I was not aware of it then.«

Embracing and making the various developments an overlapping disciplinary research program proceeded hand in hand with the discovery of surprising analogies of the various concepts (up about 1975). Despite phenomenologically heterogeneous fields, type related equations were used (e.g. between Haken and Eigen) as well as analogous theoretical concepts such as stability, fluctuation, bifurcation, selection, equilibrium, etc. However, this alone would not suffice to explain why questions have arisen that are beyond the individual problems posed in the various fields of research.

The context of the discovery of this common feature was the intellectual and social climate of the 1970s. In a phase of criticism, the search for fundamentally new scientific approaches and for new forms of social organization was supported. The chances of self-organization were propitious: they consisted in scientific outsider positions that did not fit into the obsolete cliches of »right« and »left«. The spirit of that time becomes tangible in the sentences written by Gregory Bateson (1978) in a manifesto for the university of Berkeley: »It is a matter of *obsolescence*. While much that universities teach today is new and up to date, the presupposition or premises of thought upon which all our teaching is based are ancient, and, I assert, *obsolete*. I refer to such notions as: a) The Cartesian dualism separating ›mind‹ and ›matter‹. b) The strange physicalism of the metaphors which we use to describe and explain mental phenomena - ›power‹, ›tension‹, ›energy‹, ›social forces‹, etc....

Pragmatically, it is clear that these premises and their corollaries lead to greed, monstrous over-growth, war, tyranny, and pollution. In this sense, OUR premises are dayly demonstrated false, and the students are half aware of this.

Intellectually, the premises are obsolete in that systems theory, cybernetics, holistic medicine, ecology, and gestalt psychology offer demonstrably better ways of understanding the world of biology and behavior.«[11] Bateson frequently visited BCS and was the founder of a theory of mental diseases based on the principle of self-organization.

In his preface to the book »Das Spiel«, written in conjunction with Ruthild Winkler, Eigen argues: »Starting points of our considerations have been the molecular theory of evolution, which had been worked out some years ago, and certain related game theory models, which were developed in order to simulate natural phenomena such as equilibrium, selection and growth. Though we frequently refer to examples from biology in this book, the scope is much wider and comprises scientific, philosophical, sociological and esthetic points of view.«[12] In the preface to his book »The Tao of Physics«, published in 1974, Fritjof Capra writes: »Five years ago, I had a wonderful experience. It stimulated me to write this book. One afternoon in late summer I sat on the seashore watching the waves rolling towards me and feeling the rhythm of my breath when all of a sudden I became aware of my environment as a part of a gigantic cosmic dance«.[13] In the preface to his book »From Being to Becoming« Ilya Prigogine refers to a trend in Europe and the United States that favors a closer connection of »philosophical and scientific topics«. He dedicates his book »Order out of Chaos« to, among others, Erich Jantsch, who in his book »The Self-Organization of the Universe« later on tried to establish a new basis for understanding the sciences.[14] The relevance of the intellectual and social climate of the 1970s, which was characterized by students riots, the Vietnam crisis, and the zodiacal sign of Aquarius for a New Age, is evident. Alan Watts and John C. Lilly suggested, for instance, organizing a conference on Spencer-Brown's book »Laws of Forms«. It took place at the Esalen Institute (Big Sur) in 1973. Among the participants were Bateson, von Foerster, Pribram, Brand, and Tart.[15] Similar meetings took place with Ivan Illich in Cuernavaca (Mexico). The »summer camps for intellectuals« in the alternative »Lindisfarne Association« (USA) attracted representatives from artistic, esoteric-psychological, ecological, alternative, and techno-scientific circles. Among them were Bateson, E.F. Schuhmacher, J. Salk, W.J. Thompson, P. Saleri, S. Mendlovitz, and F. Varela.[16]

Thus the initial search for common principles connecting the various fields of research was »externally« induced and the first work that tried to do justice to this task was the popular-science book by *Jantsch*.[17] The result was a semantic gestalt switch. The development of order and the evolution of complex systems was no longer perceived as a special problem of various fields but rather as a central problem of universal scope. Theories that had arisen independently of each other gave rise to the development of a new program of research.

The generalization aimed at transferring the findings to all fields of research so far, had either neglected the phenomena of self-organization or not perceived them as such. This is proven by some »basic books«, written by authors who were central figures. *Haken* tried[18] to integrate his concept of synergetics for specific phenomena of biological evolution, ontogenesis, the economy, political science, and the evolution of science. In an enlarged perspective, *Eigen* is concerned with economic, ecological, and even aesthetical problems.[19] *Prigogine*, in demonstrating the general validity of self-organization theory quotes examples from the fields of ecology, economy, and biology.[20] The working group convened by Prigogine in Brussels applied the concept of self-organization to a number of different phenomena, such as, for example, the stock-market, traffic flow, urban developments (»Brussaville«-project), the social behavior of termites, weather conditions, the heart, and the dynamics of tumors.[21] This, however, was frequently done in a metaphorical way only. The means of popularization as well as the emphasis of the new world outlook made the task of an exact modelling recede into the background. It was this circumstance, however, that also attracted scientists who were not directly involved in formulating the new theory, yet who, in the concepts of self-organization, discovered possibilities that enabled them to address unsolved problems of their special discipline in a new way. This turn is characteristic for the current state of research: the competent application of concepts to special branches of science in competitive discussions with colleagues. The authors of this volume predominantly belong to this generation of research. They are wary of global assertions. They recognize the dangers of an inflatory use of the central concept for empirical research and believe that the further success of the concept depends on the development of efficient models for the solution to specific problems. As may be seen by some contributions in this volume, they also realize that in order to prove their epistemological claims unfounded assertions have to be withdrawn.

This is in bold outline the structure of the purported scientific revolution addressed by this book. From heterogeneous origins, a research program has developed whose coherence is expressed by some central concepts that mark the »gestalt switch« from traditional concepts:

(1) Complexity is recognized as a genuine phenomenon that does not lend itself to being reduced to something simple or to being explained by the »overlapping« of simple elements. The relevant presupposition is the understanding that even something simple

may be complex without being able to be replaced by something »still more simple«. This became evident with the help of computer models of deterministic-recursive systems in which simple mathematical equation systems provide an extremely complex behavior.

(2) Irregularity of nature is not treated as an anomaly but becomes the focus of research and thus is declared to be normal. One looks for regularity within irregularity. Non-equilibrium processes are recognized as the source of order and the search for equilibrium is replaced by the search for the dynamics of processes.

(3) The classical system-environment model, according to which the adaptation of a system to its environment is controlled externally and according to which the adaptation of the system occurs in the course of a learning process, is replaced by a model of systemic closure. This closure is operational in so far as the effects produced by the system are the causes for the maintenance of systemic organization. If there is sufficient complexity, the systems perform internal self-observation and exert self-control (»Cognition« as understood by Maturana as self-perception and self-limitation, e.g., that of a cell vis-à-vis its environment).[22] But any information a system provides on its environment is a system-internal construct. The »reference to the other« is merely a special case of »self-reference«.

The social sciences frequently have suffered from the careless way in which scientific ideas and models have been transferred. One may thus be inclined to believe that we are advocating another inappropriate transfer. A number of discussions reveal that it is the concept of »closure« that is at fault. On the one hand, it is essential for the definition of self-organization, on the other hand, it is counter-intuitive if social systems are considered.[23]

It is to be hoped that the contributions of this book will defend their cause through content, and that they will make plausible the assumption that no epistemic rupture or a principle limitation of the conceptual transfer between the natural sciences and the social sciences exist. Briefly stated, operational »closure« may initially be applied only to the axiom according to which systems exist merely because of their recursive organization. This organization secures the continued existence of the system. The examination of concrete systems must focus on the discovery of mechanisms that are responsible for self-conservation. These mechanisms determine the concrete mode of operation of a system, irrespective of whether the system is a social group or the brain of a vertebrate, a system of law or the climate of the earth. Notwithstanding this self-preservation, systems may have an unlimited number of contacts with their environment; they even are obliged to have them, as only in this way may they ensure the influx of their resources. If social systems are involved, these resources comprise, in addition to money, qualified manpower and legitimation.

Whether the analysis of systems focuses more on interactions with the environment or on the modes of »closure« is therefore not a question of principle but one of interest. However, if one wants to determine a system empirically, one cannot avoid indicating its operational »closure«.

The traditional viewpoint of system theory certainly was that social systems are to be defined according to their external functions and performances, i.e., via input and output. This approach was thwarted by the fact that systems empirically always behave differently, i.e., not in conformity with what the functions prescribe. This is due to the circumstance that they primarily orient themselves towards the conditions dictated by their mechanisms of self-preservation and that they repudiate functional interference as perturbation or alternatively they transform them into system structures that are unpredictable.

It follows that one may admit what in any case has become common knowledge, namely that social institutions leave no stone unturned in order to ensure their continued existence. Functionalism took this seriously only in a moral way. Here, however, the fact that the system does not do what it is expected to do, is theoretically treated as a process of creating innovative structures.

The contributions of this book address the subject of self-organization at two levels. While the first level is limited to an epistemological and metatheoretical analysis, the second embraces conceptual and historical analyses of the use of self-organization in special fields of research. The spectrum of approaches at either level has been a wide one. It is our expectation that this variety reveals the progressive problem shift that underlies the new constructivist approach.

Acknowledgement: Most of these papers were presented at a conference on the topic of self-organization from 7th - 10th September 1988 in Bielefeld. This conference was financed by the University of Bielefeld, the Volkswagen-Stiftung and the Schweisfurth-Stiftung. The editors wish to thank the board of the Yearbook series for helpful comments on draft versions of the papers. They are also grateful to several native English speakers, Peter Germain, Berlin, Jonathan Harrow, Bielefeld and John Bednarz, Bielefeld, who either translated papers or assisted in rendering translations. Last not least we acknowledge the difficult job of bringing the manuscripts into the final version; this was done by Heidi Bhatti-Küppers and Michael Spehr.

Notes and References:

1. I. Kant, *The Critique of Judgement*, Oxford: Clarendon Press, 1980, Part II, § 4 (65).

2. On the contemporaneity and structural similarity of the Smith's and Watt's solutions, see O. Mayr, »Adam Smith and the Concept of the Feedback System, Economic Thought and Technology in 18th Century Britain«, in *Technology and Culture* **12** (1971), 1-22.

3. See G. Holton, *Thematic Origins of Scientific Thought*, Cambridge, Massachusetts: Harvard University Press, 1973.

4. H. v. Foerster, »On Self-Organizing Systems and their Environments«, in M.C. Yovits, S. Cameron (eds.), *Self-Organizing Systems*, Oxford, London, New York: Pergamon Press, 1960, p. 31-50.

5. Sometime members or guests at the BCL included W.R. Ashby, M.R. Gardner, G. Günther, A. Inselberg, H.R. Maturana, W.S. McCulloch, G. Pask, S. Umpleby, and F.J. Varela. See *The Collected Works of the Biological Computer Laboratory*, Urbana: University of Illinois, 1976.

6. I. Prigogine, *Introduction to Thermodynamics*, 3rd ed., New York: Interscience, 1967, p. XII.

7. A stationary state is like a static equilibrium not explicitly time dependent but is characterized by processes that are independent of time. A stationary state is therefore dynamic.

8. E.N. Lorenz, »On the Prevalence of Aperiodicity in Simple Systems«, in M. Grmela, J.E. Marsden (eds.), *Global Analysis*, Berlin, Heidelberg, New York: Springer, 1979, p. 53-75.

9. And the list is incomplete. It could be expanded through ecosystem research, the theory of nonlinear oscillations, nonlinear codes in computer simulation (hydrodynamics and plasma physics), and psychiatric research (double bind).

10. These interviews are part of a program on the history of self-organization that is being carried out by the USP Wissenschaftsforschung at the University of Bielefeld with the aid of a grant from the Volkswagen Foundation.

11. G. Bateson, *Mind and Nature: A Necessarity Unity*, New York: Dutton, 1979, p. 217-218.

12. See M.Eigen, R. Winkler, *Das Spiel: Naturgesetze steuern den Zufall*, Munich: Piper, 1975.

13. See F. Capra, *The Tao of Physics*, New York: Random House, 1975.

14. I. Prigogine, *From Being to Becoming: Time and Complexity in the Physical Sciences*, New York: Freeman, 1980, p. XVI. I. Prigogine, I. Stengers, *Order out of Chaos. Man's Dialogue with Nature*, London: Heinemann, 1984. E. Jantsch, *Die Selbstorganisation des Universums*, Munich: Hanser, 1979.

15. See J. and A. Lilly, *The Dyadic Cyclone*, Malibu: Human Software Ing., 1976. J. Brockman (ed.), *About Bateson*, London: Wildwood House, 1978.

16. One of the sources for this information is the Lindisfarne book *Earth's Answer*, New York: Harper & Row, 1977. We wish to thank Rainer Paslack for this information.

17. The many facets of the 1970s are linked together in the person of E. Jantsch: a physicist, an OECD advisor with critical attitudes towards linear planning, and a person living in California exposed to the ideas of the New Age movement. His book covers cosmology, thermodynamic chemistry, biology, evolution, society, and ethics. Although many of its conclusions were rather hasty, the book as a whole is a keystone of this phase of development. See Jantsch, *op.cit.*, 1975, note 14.

18. H. Haken, *Erfolgsgeheimnisse der Natur. Synergetik: die Lehre vom Zusammenwirken*, Stuttgart: Deutsche Verlagsanstalt, 1981.

19. Eigen, Winkler, *op.cit.*, 1975, note 12.

20. Prigogine, *op.cit.*, 1979, note 14.

21. See the articles from Prigogine, Nicolis, van den Broeck, et al., in *Nouvelles de la Science et des Technologies*, Vol. 2, No. 1-2, Brussels, March 1984.

22. H.R. Maturana, »Biologie der Kognition«, in H.R. Maturana, *Erkennen: Die Organisation und Verkörperung von Wirklichkeit*, Braunschweig, Wiesbaden: Vieweg, 1982, p. 22.

23. In commenting on one of the contributions to this book, the distinguished editor of the Yearbook series, R. Whitley remarked: »I find it rather strange that the author asserts that the legal system is a totally closed one reproducing itself with apparent ease and yet in his discussion towards the end of the paper he talks about development and change and the relationship between the law and other institutions. In general I can see no justification for asserting that the law, or any other major social institution, is a totally closed system. It is also not clear to me how you can talk about law as a communication system without including some discussion of the material necessities for reproducing that system. It seems to

me that this sort of discourse runs the risk of retreating to conceptual idealism. Anyway, no doubt the bulk of the papers in the book take similar point of view...«

I. Epistemological Foundations

Science and Daily Life: The Ontology of Scientific Explanations

HUMBERTO R. MATURANA
University of Chile - Santiago, Chile

Introduction

Although according to its etymology the word science means the same as the word knowledge, it has been used in the history of Western thinking to refer to any knowledge whose validity can be defended on methodological grounds, regardless of the phenomenal domain in which it is claimed. In modern times, however, this has progressively changed, and the word science is now most frequently used to refer only to a knowledge validated through a particular method, namely, the scientific method. This progressive emphasis on the scientific method has arisen under two general implicit or explicit assumptions of scientists and philosophers of science alike, namely: a) that the scientific method, either through verification, through corroboration, or through the denial of falsification, reveals, or at least connotes, an objective reality that exists independently of what the observers do or desire, even if it cannot be fully known; and b) that the validity of scientific explanations and statements rests on their connection with such objective reality. It is of this kind of knowledge that I shall speak in this article when speaking of science, and in the process I shall implicitly or explicitly disagree, without giving a full philosophical justification, with one aspect or another of what many classic thinkers of the philosophy of science who discuss these matters in depth have said.[1] And I shall do so because I shall speak as a biologist, not as a philosopher, reflecting about science as a cognitive domain generated as a human biological activity. Furthermore, I shall do these reflections attending to what I see that we modern natural scientists do in the praxis of science in order to claim the scientific validity of our statements and explanations, and I shall show how that which we do as scientists relates to what we do as we live our daily lives revealing the epistemological and ontological status of that which we call science.

The Observer and Observing

We scientists do science as observers explaining what they observe. As observers we are human beings. We human beings already find ourselves in the situation of observers observing when we begin to observe our observing in our attempt to describe and explain what we do. That is, we find ourselves already in language making distinctions in language when we begin to reflect in language upon what we do and how we do what we do as we operate as languaging animals. In other words, it happens to us that we are already languaging living systems doing what we do, including our explaining, when we begin to explain what we do, and that we are already in the experience of observing when we begin to observe our observing. Observing is that which we observers do as we distinguish in language the different kinds of entities that we bring forth as objects of our descriptions, explanations and reflections in the course of our participation in the different conversations in which we are involved along our daily lives, regardless of the

W. Krohn et al. (eds.), Selforganization. Portrait of a Scientific Revolution, 12–35.

operational domain in which these take place. The observer happens in observing, and, when the human being that the observer is dies, the observer and observing come to an end. In these circumstances, when reflecting upon what the observer does, the cognitive abilities of the observer must either be taken as given, unexplainable properties, or they must be explained by showing how they arise as a result of the biology of the observer as a human being. I have done the latter in several articles that I invite the reader to read.[2] Now, however, I shall proceed on the assumption that the reader accepts that his or her properties and abilities as an observer result from his or her operation as a living system, even though the experience of being an observer just happens to him or her as a matter of course while he or she remains experientially blind to its origin.

Cognition

Living systems are structure-determined systems. As such they do not admit instructive interactions, and everything that happens in them happens as a structural change determined at any instant in their structure either in the course of their own internal dynamics or triggered but not specified by the circumstances of their interactions. In other words, nothing external to a living system can specify in him or her what happens in him or her, and, since the observer is a living system, nothing external to the observer can specify in him or her what happens in him or her. It follows, then, that the observer as a living system constitutively cannot make explanations or statements that reveal or connote anything independent from the operations through which he or she generates his or her explanations and statements. Therefore, that which we observers connote or imply with the word cognition, as we use it in daily life in our interpersonal coordinations of actions and relations when we answer questions in the domain of knowing, must reveal what we do or how we operate in those coordinations of actions and relations as we generate our cognitive statements. That in daily life we act under the implicit understanding that cognition has to do with our interpersonal relations and coordinations of actions, is apparent in that we claim cognition in others and in ourselves only when we accept the actions of others or our own actions as adequate because they satisfy the particular criterion of acceptability that we accept for what constitutes an adequate action in the domain of actions entailed in the question. Accordingly, that which we as observers connote when we speak of knowledge in any particular domain is constitutively what we consider as adequate actions (distinctions, operations, behaviors, thoughts, or reflections) in that domain, assessed according to our own criterion of acceptability for what constitutes an adequate action in it. In other words, knowledge is constituted by an observer as an operational capacity that he or she attributes to another living system, which could be him or herself, by accepting his or her actions as adequate in some operational domain specified in that attribution. Therefore, there are as many cognitive domains as there are domains of adequate actions (distinctions, operations, behaviors, thoughts, or reflections) that the observers accept, and each of them is operationally constituted and operationally defined in the experiential domain of the observer by the criterion that he or she uses for accepting as adequate actions (distinctions, operations, behaviors, thoughts, or reflections) the actions that he or she accepts as proper to it.[3] I call the criterion that an observer uses to accept certain actions as the actions that define and constitute a cognitive domain the criterion of acceptability that defines and constitutes that cognitive domain. Science as a cognitive domain is not an exception to this manner of constitution, and I call the criterion of acceptability that defines and constitutes science as a cognitive domain, and that simultaneously constitutes as a scientist the person who applies it, the criterion of validation of scientific explanations. It is this criterion of acceptability that constitutes science as a cognitive domain that I shall be discussing below when I speak of science.

Acts

We usually speak of acts as external operations of our bodies in some environment. In this article I am speaking of acts in a more general and fundamental manner that includes our external bodily operation as a particular case. I am calling acts all that we do in any operational domain that we bring forth in our discourse, however abstract it may seem. Thus, to think is to act in a domain of thinking, to walk is to act in a domain of walking, to reflect is to act in a domain of reflecting, to talk is to act in a domain of talking, to strike is to act in a domain of striking, and so forth, and to explain scientifically is to act in the domain of scientific explaining. All acts as operations of a living system take place as part of its dynamics of states regardless of whether it has a nervous system or not, but include the dynamics of the nervous system if there is one. Therefore, as operations in the dynamics of states of a living system, all acts are phenomena of the same kind, regardless of the domain in which an observer claims that they take place when he or she considers them while beholding the living system in relation to an environment. Furthermore, all the acts of a living system that involve in it the operation of a nervous system take place in it as dynamic configurations of changes of relations of activity in its nervous system as a closed network of changing relations of activity between its neuronal components.[4] Thus, to think, to walk, to talk, to have a spiritual experience, and so forth, are all phenomena of the same kind as operations of the internal dynamics of the organism (nervous system included), but are all phenomena of different kinds in the relational domain of the organism in which they are brought forth by the distinctions of the observer.

Emotions

In daily life we distinguish different emotions in ourselves, in other human beings, and in other animals by observing the different domains of actions in which we and they operate at any instant. Thus we may say, »Do not speak with so-and-so now because he is angry, and he will not listen to you or do what you ask«. Emotions are dynamic body dispositions that specify the domains of actions in which animals in general, and we human beings in particular, operate at any instant. Accordingly, all animal acts arise and are realized in some emotional domain, and it is the emotion that defines the domain in which an act (a movement or an inner body posture) takes place, regardless of whether, for an observer who beholds the animal in an environment, it occurs as an abstract or as a concrete act or of what specifies that act (movement, or inner body posture) as an action of a particular kind. In fact, we know from our daily human life that as we move from one emotion to another we change our domain of actions, and that when we see someone change his or her domain of actions, what we see is a change of emotion. In other words, it is the emotion under which we act at any moment in any operational domain that defines what we do at that moment as an action of a particular kind in that operational domain. Therefore, if we wish to understand any human activity, we must attend to the emotion that defines the domain of actions in which that activity takes place as an action and, in the process, learn to see the actions desired in that emotion.

Language

If we see two persons at a distance too great to hear them and we want to be able to claim later whether they had been talking to each other or not we observe the course of their interactions looking in these for consensual coordinations of consensual coordinations of actions in forms that we may be able to recognize easily, like requests and promises, directions for actions in answer to questions, or complaints. In other words, when we attempt to determine whether two or more persons are interacting in language or not we do not only look for their consensual coordinations of actions, but we look for a dynamics of recursion in their consensual coordinations of actions. That is, we look for the occurrence of consensual coordinations of consensual coordinations of actions as operations in a learned and not in an instinctive domain of coordinations of actions. I maintain that this, our manner of knowing in daily life whether two or more persons interact in language or not, reveals what we do when we are languaging, whichever our domain of operations. In other words, I claim that language takes place when two or more persons in recurrent interactions operate through their interactions in a network of criss-crossing, recursive, consensual coordinations of consensual coordinations of actions, and that all that we human beings do, we do in our operation in such a network as different manners of operating in it. In other words, I claim that we human beings exist as such in language, and that all that we do as human beings we do as different manners of operating in language. Moreover, I also claim that that is what language is as a biological phenomenon in its phylogenic origin and in its ontogenic constitution: an operation in a domain of consensual coordinations of consensual coordinations of actions that arose as a result of intimate coexistence in coordinations of actions in the lineage of bipedal primates to which we belong and which has to be established anew in each child during its coontogeny with the adults with whom he or she grows.[5] That is, I maintain: a) that languaging is not a system of operation with abstract symbols in communication; b) that symbols do not preexist language but arise after it and in it as distinctions made by an observer of consensual relations of coordinations of actions in language; c) that although language takes place through the body interactions and the body changes involved in the consensual coordinations of consensual coordinations of actions of those in languaging, it does not take place in the body of the participants because it takes place in the flow of their consensual coordinations of consensual coordinations of actions; d) that although languaging does not take place in the bodies of those in language, the course of the body changes of those that interact in language braids with the course of their coordinations of actions, and the course of their coordinations of actions braids with the course of their body changes as these arise in the flow of their languaging;[6] and e) that that which an observer sees as the contents of a process of languaging is a distinction in language that an observer makes of the relations of a process of languaging in a network of languaging. The result of this condition of constitution of language is that we human beings exist as observers in language, and whatever we distinguish in language is operations in language according to circumstances that have arisen in us in language.

Conversations

As mammals we are animals that learn to coordinate the flow of our emotions and behaviors consensually as we live together. As languaging animals we learn to live in consensual coordinations of consensual coordinations of actions also through living together. As human beings we grow and live in consensual coordinations of emotions and consensual coordinations of consensual coordinations of actions that braid with each other and form closed networks of consensual coordinations of emotions and language. In these closed networks of consensual coordinations of emotions and language our actions, and the flow of our actions in language, change as our emotions change, and our emotions and the flow of our emotioning change as our coordinations of actions in language change. I call conversation our operation in this braided flow of consensual coordinations of languaging and emotioning, and I call conversations the different networks of braided, consensual coordinations of languaging and emotioning that we generate as we live together as human beings.[7] As languaging animals we exist in language, but as human beings we exist (we bring ourselves forth in our distinctions) in the flow of our conversations, and all our activities as such take place as different kinds of conversations. Accordingly, our different domains of actions (cognitive domains) as human beings (cultures, institutions, societies, clubs, games, etc.) are constituted as different networks of conversations, each defined by a particular explicit or implicit criterion of validation that defines and constitutes what pertains to it. Science as a cognitive domain is a domain of actions, and as such it is a network of conversations that entail statements and explanations validated by the criterion of validation of scientific explanations under the passion for explaining.

The Scientist

Science is a human activity. Therefore, whatever we scientists do as we do science has validity and meaning, as any other human activity does, only in the context of human coexistence in which it arises. All human activities are operations in language, and as such they occur as coordinations of coordinations of consensual actions in conversations that take place in domains of actions specified and defined by some fundamental emotion.[8] The fundamental emotion that specifies the domain of actions in which science takes place as a human activity is curiosity under the form of the desire or passion for explaining. Furthermore, that which constitutes science as a particular kind of explaining is the criterion of validation that we scientists use, explicitly or implicitly, to accept our explanations as scientific explanations as we practice science under the passion for explaining. I call this criterion of validation of explanations that we scientists use, and which I shall describe further below, the criterion of validation of scientific explanations. It follows from what I have said that we scientists become scientists, while operating under the passion for explaining, when we constitute science as a particular domain of explanations by being rigorous in our endeavor to be always impeccable in the application of the criterion of validation of scientific explanations as we generate explanations that we call scientific explanations. It is this manner of constitution of science and the scientist that gives the use of science its particular operational effectiveness in the worlds in which we modern human beings live.

Scientific Explanations

When in ordinary daily life we answer to ourselves or to somebody else a question that demands from us an explanation of a particular experience (situation or phenomenon), we always answer it by proposing a reformulation of that experience (situation or phenomenon) in terms of other experiences different from those used in the original formulation of the question. If the proposed reformulation is accepted as such by the person asking the question, it becomes ipso facto an explanation, and the question as well as the desire for asking it vanish. As this happens, the accepted explanation becomes an experience that can be used as such for other explanations. In other words, explanations are propositions presented as reformulations of experiences that are accepted as such by a listener in answer to a question that demands an explanation. That is, a proposition presented as a reformulation of an experience that is not accepted as such is not an explanation. Therefore, there are as many different kinds of explanation as there are different criteria that we explicitly or implicitly use for accepting the different kinds of reformulation of experiences that we accept as explanations in answer to our questions. At the same time, the different criteria of acceptability that we use in our listening for explanations define the different explanatory domains with which we operate in our daily lives. It is because explanatory domains are constituted in this manner that that which defines science as a particular explanatory domain is the criterion of validation of explanations that scientists use, and that that which defines a scientist as a particular kind of person under the passion for explaining is the use of the criterion of validation of explanations that constitutes science as an explanatory domain. Finally, as explanations are experiences of the observer that arise as he or she operates in his or her domain of experiences, all explanatory domains constitute expanding experiential domains in which the observer lives new experiences, asks new questions, and unavoidably generates new explanations in an unending, recursive manner, if he or she has the passion for explaining.

Criterion of Validation of Scientific Explanations

If we attend to what we do as scientists in the praxis of science when we propose a scientific explanation of any particular phenomenon (experience) that we want to explain, we may notice that we accept a given reformulation of the phenomenon to be explained as a scientific explanation only if it is presented as one of four interrelated operations that we must satisfactorily realize in our domain of experiences to validate our explanations as we operate as scientists. I call these four operations, which must be conjointly satisfied in order that a particular reformulation of experiences included among them may be accepted as a scientific explanation, the criterion of validation of scientific explanations; and I call the observer who performs them, and who accepts their conjoint satisfaction as the criterion of validation of his or her explanations, a standard observer or scientist. These four operations are the following:

i) The presentation of the experience (phenomenon) to be explained in terms of what a standard observer must do in his or her domain of experiences (praxis of living) to experience it.

ii) The reformulation of the experience (phenomenon) to be explained in the form of a generative mechanism that, if realized by a standard observer in his or her domain of experiences, would allow him or her, as a result or consequence of its operation, to have in his or her domain of experiences the experience to be explained as presented in point (i).

iii) The deduction from the operation of the generative mechanism proposed in (ii), as well as from all the operational coherences of the domain of experiences of a standard observer entailed by it, of other experiences that a standard observer should have through the application of those operational coherences and of the operations that he or she must realize in his or her domain of experiences to have them.

iv) The experience by a standard observer of the experiences (or phenomena) deduced in (iii) through his or her realization in his or her domain of experiences of the operations also deduced in (iii).

It is only when these four conditions are conjointly satisfied in the praxis of living of a standard observer while he or she is in the process of explaining a particular aspect (phenomenon) of his or her domain of experiences that: a) a standard observer can claim as a scientist that the generative mechanism that he or she proposes in (ii) is a scientific explanation of the experience that he or she presented in (i) as the phenomenon (experience) to be explained; b) that such an explanation is valid while these conditions apply; and c) that such an explanation is valid in the community of scientists as standard observers who accept that the criterion of validation of scientific explanations is satisfied. In other words, there is no single aspect or operation of the criterion of validation of scientific explanations that is by itself scientific, and, therefore, there are no operations such as scientific observations, deductions, corroborations, or predictions. There are only scientific explanations as propositions of generative mechanisms that are accepted as valid exclusively to the extent that they are part of the satisfaction of the criterion of validation of scientific explanations, and scientific statements as statements that are accepted as valid because they arise directly or indirectly as the result of the application of scientific explanations. Let us consider now some consequences and implications of understanding how we generate science as a cognitive domain through the application of the criterion of validation of scientific explanations for our understanding of what we do when we do science.

1. Explanations in general, as reformulations of experiences accepted as such by an observer, do not replace, and are not expected to replace, the experiences that they explain; they only present the operational conditions of the domain of experiences of the observer under which he or she claims that the explained experience arises. Scientific explanations are not an exception to this, and they also do not replace the experiences that they explain: a scientific explanation only says what happens in a particular part of the domain of experiences of the observer if the operational conditions of the criterion of validation of scientific explanations are satisfied in it.

2. Scientific explanations arise as actual human actions in the domain of experiences of individual standard observers and are valid as such in a community of them, even though each standard observer lives his or her experiences in the total loneliness of his or her structural determinism as a living system.[9] This situation does not constitute a contradiction because membership in the community of standard observers does not depend on the individual ability of making reference to an independent objective reality that the standard observer as a living system cannot do, but on the consensual participation in the domain of scientific explanations. Accordingly, only those observers who can participate with other observers, and to their complete satisfaction, in the realization of the criterion of validation of scientific explanations and, furthermore, accept this as their only criterion of validation for their explanations, are scientists (standard observers) and members of the community of scientists. Those observers who, for one reason or another, cannot or would not do this are either discarded as standard observers or scientists by the members of the community of scientists under the claim that they are bad observers, or they are not considered at all.

3. As generative mechanisms constituted and accepted in the context of the satisfaction of the four operations of the criterion of validation of scientific explanations, scientific explanations are constitutively mechanistic in the sense that they deal only with structure-determined systems and with the structural determinism entailed in the operational domain in which they are proposed. As such, scientific explanations take place in the domain of experiences of the standard observer, and the domain of structural determinism in which they happen belongs to the area of operational coherences of the praxis of living of the standard observer in which he or she proposes them: a scientific explanation operates only in the area of structural determinism in which it is proposed. Nevertheless, this situation is not a limitation of scientific explanations, on the contrary, it is their condition of possibility.

4. Contrary to a common implicit or explicit belief, scientific explanations, as propositions of generative mechanisms that give rise as a consequence or result of their operation to the experiences (phenomena) to be explained, constitutively do not and cannot operate as phenomenic reductions or give rise to them. This nonreductionist relation between the phenomenon to be explained and the mechanism that generates it is operationally the case because the actual result of a process and the operations in the process that give rise to it in a generative relation intrinsically take place in independent and nonintersecting phenomenal domains. This situation is the reverse of reductionism; scientific explanations as generative propositions constitute or bring forth a generative relation between otherwise independent and nonintersecting phenomenal domains, which they thus de facto validate. The liberation from the belief that scientific explanations are or constitute reductionist propositions that the understanding of the criterion of validation of scientific explanations carries, permits us to see, particularly in the domain of biology, that there are phenomena like language, mind, or consciousness that require an interplay of bodies as a generative structure but do not take place in any of them. In this sense, science and the understanding of science lead us away from transcendental dualism.

5. The fact that in a scientific explanation the phenomenon to be explained should arise in a different phenomenal domain than the one in which the generative mechanism that would give origin to it as a result of its operation takes place, constitutes the phenomenon to be explained as a phenomenon in an abstract relational phenomenal domain with respect to that in which its generative mechanism occurs. There are two basic consequences of this situation: a) there is in principle no restriction about the kind of phenomena that can be explained scientifically, however abstract they may seem, because the phenomenon explained constitutively takes place in an abstract relational domain with respect to the mechanism that gives rise to it; and b) the fact that scientific explanations are mechanistic propositions does not restrict the possibility of using them to explain what seem to be nonmechanistic phenomena such as self-awareness or spiritual experiences.

6. Since the criterion of validation of scientific explanations is defined and constituted only in terms of the operational coherences of the domain of experiences of the standard observers, it does not entail any supposition about an objective independent reality. Accordingly, a standard observer can use scientific explanations only to explain his or her experiences by reformulating them with other experiences, in the application of the operational coherences that these entail through the satisfaction of the criterion of validation of scientific explanations, and not use them to reveal or connote anything deemed to be independent of what he or she does. In fact, the contrary happens, because for these same reasons scientific explanations enter into the constitution of the world or worlds that we standard observers live through the transformation and expansion of our domain of experiences that they bring forth as we operate with the consequences of them in our experiential domain. In these circumstances, the claim that scientists make about the universal validity of scientific explanations and statements does not refer to a pretended revelation through them of an objective, independent, and, hence, universal

reality, but to their validity through the application of the operational coherences that they entail in the world or worlds brought forth through the application of the criterion of validation that constitutes them.

7. Scientific explanations, as reformulations of experiences with other experience in the domain of experiences of the standard observer, pertain to the praxis of living in daily life of the standard observer as a human being. Furthermore, the manner in which we human beings de facto validate our actions in daily life within any operational domain entails the same operational coherences as the criterion of validation of scientific explanations. The difference between our operation in daily life as scientists and as nonscientists rests on our different emotions, on our different desires for consistency and impeccability in our actions, and on our different desires for reflection about what we do. We human beings are multidimensional beings in our desires, in our concerns, and in our enjoyments; and due to this we realize in our daily living many different kinds of beings through different conversations that intersect in our bodies, each grounded on a particular emotion. Thus, as scientists, we are under the passion of explaining, and every doubt, every question, is in us, in principle, an acceptable and desired occasion for our realization as such. Furthermore, we as scientists are also, at least in principle, careful not to confuse experiential or phenomenal domains in our explanatory propositions as we validate them with the criterion of validation of scientific explanations. Finally, we as scientists commit ourselves to use only the criterion of validation of scientific explanations in our explaining. As nonscientists, in daily life we are not careful in this manner, and we use successively many different criteria to validate our explaining and our assertions as we spontaneously change phenomenal domains in our discourse, frequently without being aware that we do so, and we are not interested in applying the conceptual rigor of scientific explanations. Yet, as a result of our structural determinism as living systems, we intrinsically operate in the experience of daily life according to the operational coherences entailed in the criterion of validation of scientific explanations. Or, said in a more fundamental manner, the criterion of validation of scientific explanations is a formalization of the operational validation of the flow of the praxis of living of living systems.

8. Einstein said, and many other scientists have agreed with him, that scientific theories are free creations of the human mind, and he marvelled that through them one could understand the universe. The criterion of validation of scientific explanations as operations in the praxis of living of the observer, however, permits us to see how it is that the first reflection of Einstein is valid, and how it is that there is nothing marvelous in that it is so. Let us reflect upon this. The statement and marvelling of Einstein arise from an implicit supposition that there is an independent reality that the observer finds and explains with explanatory propositions that Einstein sees as arising independently from any direct observation or experimentation with that objective reality. If Einstein's implicit supposition were right, then his marvelling would be right too. The criterion of validation of scientific explanations tells us, however, that to the extent that it does not require any supposition about an objective and independent reality, all that a standard observer does in the generation of a scientific explanation arises in him or her as an

expression of his or her experiential dynamics without any reference to that supposed objective independent reality. Accordingly, points (i) and (ii) in the criterion of validation of scientific explanations pertain totally to the arbitrariness of the observer's mind in the sense that they arise fully in the spontaneity of the reflections of the standard observer as these arise in the flow of his or her structural determinism. The observer does not find a problem or phenomenon to be explained outside him or herself, but on the contrary, he or she constitutes one in his or her domain of experiences as he or she finds him or herself in a question that he or she desires to answer. Furthermore, the observer brings forth the generative mechanism that he or she proposes in the attempt to explain the phenomenon that he or she wants to explain as an ad hoc proposition specifically designed with elements of his or her experience to generate it as a result of its operation, and with no other necessary justification than that. Strictly speaking, here is where the poetry of doing science resides. The other two points, (iii) and (iv), of the criterion of validation of scientific explanations entail operations of a different kind. They arise as deductions, from the operational coherences entailed by point (ii) in the domain of experiences of the observer, of other experiences that he or she has lived, lives, or could live in the form of actual operations in that domain. As such, points (iii) and (iv) are totally subordinated in their generation to points (i) and (ii) that specify when and how they would take place. Finally, since to understand a given experience is to operate in awareness of the circumstances that generate it, and since everything in a scientific explanation occurs in the domain of experiences of the standard observer as his or her actual operation in it, scientific theories cannot but arise as free creations of our operation as standard observers, and cannot but take the form of reformulations of our experiences with elements of our experiences that constitute de facto our understanding of our domain of experiences as we live it through our scientific explanations of it.

9. Under the implicit or explicit assumption that science has to do with the revelation of the properties of an objective (ontologically independent) reality, it is frequently believed, even by scientists, that for any theory or explanation to be scientific, it must entail quantifications and predictions. Thus, one frequently hears, at least colloquially, a rather satirical distinction between hard and soft sciences according to whether they are quantitative or not, implying that the soft ones are not really sciences, or that one cannot develop a scientific theory in any particular domain if one does not or cannot make measurements and propose predictions in it with a validity that can be asserted with objective quantitative observations. These beliefs are both misleading and inadequate because they obscure our direct view of the operations through which the standard observers constitute science as a cognitive domain. What makes a scientific explanation or theory scientific is not quantification or the possibility that it creates for an observer to predict some of his or her future experiences with it, but that it is validated as it arises through the application of the criterion of validation of scientific explanations without reference to quantification or any restriction of domain. A standard observer can generate a scientific explanation or theory in any domain in which he or she can apply the criterion of validation of scientific explanations. What happens with quantifications and predictions is something different. To quantify or to measure is to

apply a metric, defined as a system of repeatable comparisons that an observer makes between two areas of his or her domain of experiences, to another area in it. As such, a measurement or quantification does not constitute an independent or objective validation of any statement that the observer makes, but, if it is properly done, it facilitates or makes possible his or her deductions in the area of operational coherences of his or her domain of experiences in which it applies. Similarly, a prediction, as a tentative computation of a change of state in a not fully described structure-determined system brought forth by the observer as an abstraction of his or her operational coherences in his or her domain of experiences, is also an operation that, if successful, cannot by itself constitute an objective validation (a validation independent of what the observer does) of the structure-determined system in which it takes place. Quantifications (or measurements) and predictions can be used in the generation of a scientific explanation but do not constitute the source of its validity.

10. The notions of falsifiability[10], verifiability, or confirmation would apply to the validation of scientific knowledge only if this were a cognitive domain that revealed, directly or indirectly, by denotation or connotation, a transcendent reality independent of what the observer does, and if point (ii) in the criterion of validation of scientific explanations were a model of that transcendent reality rather than a generative mechanism that gives rise to the experience to be explained as it is presented in point (i). However, since according to the criterion of validation of scientific explanations, scientific explanations and scientific statements do not refer to an independent reality and do not pretend to do so, such notions do not apply in the domain of science, even though they may have a metaphorical use. No doubt, points (iii) and (iv) of the criterion of validation of scientific explanations offer to an observer the operational possibility of believing that he or she realizes the falsification, verification, or confirmation of what is proposed in (ii), if he or she believes that in this proposition he or she has a model of how the objective reality, which he or she intends to know while assuming that it exists independently of his or her doings, is. Yet, such a belief, as it is apparent from all that I have said, is a source of blindness for our understanding of what we do as we do science, leading us not to see that the validity of what we do in science rests exclusively in the operational consensuality in which it arises as a manner of human coexistence under conditions in which the notions of falsification, verification, or confirmation do not and cannot apply. Since the implicit belief in an objective, independent reality as a source of universal validity of knowledge is a basic part of our Western culture in which science arose and we usually believe also that the strength of science rests on it, let me again repeat some of the arguments that indicate that science and its operational validity and effectiveness do not depend on it. a) Scientific explanations arise operationally as generative mechanisms accepted by us as scientists through operations that do not entail or imply any supposition about an independent reality, so that in fact there is no confrontation with one, nor is it necessary to have one, even if we believe that we can have one. b) If we listen to what I said above in points (iii) and (iv) of the criterion of validation of scientific explanations, we cannot but notice that these points do not entail any supposition about an independent reality, and that they refer only to the operational

coherences of the domain of experiences of the standard observer, regardless of what he or she may think about what reality is. c) The standard observer as a living system is a structure-determined system, and as such he or she cannot operationally make a distinction that can in any way be claimed to be a distinction of something independent of his or her doing, and, due to this, he or she cannot have any operational conception of anything that does not pertain to his or her domain of experiences.[11] d) No matter what we may say, we scientists proceed in our research under the inner body disposition (emotion) of following the path of validation of our explanatory propositions, not of finding the conditions of their falsification.

Science as a Cognitive Domain

The use of the criterion of validation of scientific explanations defines and constitutes scientific explanations. The use of scientific explanations to validate a statement makes that statement a scientific statement. The use of scientific explanations by the members of a community of standard observers to directly or indirectly validate all their statements defines and constitutes science as a cognitive domain that defines as a scientific community the community of those observers that use it. Therefore, ontologically, in its manner of constitution as a cognitive domain, science is no different from other cognitive domains because it is defined and constituted as all cognitive domains are, namely, as a domain of actions defined by a criterion of validation or acceptability used by an observer or by the members of a community of observers to accept those actions as valid in a domain of actions defined by that very same criterion of acceptability.[12] Let us now see some implications that this manner of constitution has for our understanding and use of science as a cognitive domain.

1. It is frequently claimed that scientific knowledge has to be accepted as universally valid because the explanations and statements that belong to it are validated through their continuous confrontation with an independent objective reality. The manner of constitution of science as a cognitive domain, however, indicates that such a notion cannot be sustained, and that scientific explanations and statements are universally valid only in the community of those observers who accept the criterion of validation of scientific explanations as the criterion of validation of their explanations. In this respect, science is no different from any other cognitive domain, since all cognitive domains are by constitution valid only in the particular community of those observers who accept the criterion of validation of adequate actions that defines such a community. It is also often claimed that it is the universality and objectivity of scientific knowledge that gives scientific rational arguments their compelling power and scientific statements their convincing character. In this respect, however, science is no different from other cognitive domains either, because all properly rational arguments (arguments without misapplied operational coherences) are by constitution valid only in the cognitive domain in which they belong. Finally, science as a domain of adequate actions in the domain of experiences of an observer is also no different from any other cognitive domain that an

observer may claim, because all cognitive domains are domains of adequate actions of an observer in his or her domain of experiences.

2. The peculiarities of science as a cognitive domain arise from its manner of constitution through the application of the criterion of validation of scientific explanations. Let us see some of them. a) The criterion of validation of scientific explanations describes what we modern natural scientists explicitly or implicitly do in the praxis of scientific research or what we expect should have been done by someone who claims that he or she is proposing a scientific explanation or making a scientific statement. b) Since the criterion of validation of scientific explanations consists by constitution in the rigorous application (without confusion of domains) by a standard observer of the operational daily validation of his or her praxis of living, all scientific explanations and statements pertain to the praxis of living of the standard observer, and the standard observer cannot generate through science statements or explanations that are not de facto constituted in the operational coherences of his or her praxis of living. c) Since the criterion of validation of scientific explanations allows a standard observer to validate generative relations in nonintersecting phenomenal domains, science is by constitution a multidimensional cognitive domain, and a standard observer can expand, through the generation of scientific explanations and scientific statements, all the dimensions of human experience in many nonintersecting phenomenal domains. Finally, d) since the criterion of validation of scientific explanations entails the operationality of reflection, science as a cognitive domain is by constitution a domain in which a standard observer can recursively treat any experience in it as an object of reflection in it without falling out of the operational coherences of his or her praxis of living.

3. We scientists claim that our emotions do not participate in the generation of our scientific statements and explanations. We claim this because the criterion of validation of scientific explanations fully specifies in a manner that is constitutively independent of our emotioning the operations that we have to perform as standard observers to generate a scientific explanation. In addition, as we learn to be scientists, we learn to be careful in not letting our preferences and desires to distort, and thus to invalidate, our application of the criterion of validation of scientific explanations, and we also learn to recognize that when we allow this to happen we commit a serious blunder. Our emotions, however, enter legitimately and constitutively into what we scientists do in the foundation of the circumstances of our scientific explaining, because they specify at every moment the domain of actions in which we operate as we generate our questions. We human beings bring forth with our actions in our domain of experiences the worlds that we live as we live them in our domain of experiences as human beings, and we move in the worlds that we bring forth changing our concerns and our questions in the flow of our emotioning. In other words, the poetry of science is grounded on our desires and concerns, and the course followed by science in the worlds that we live is guided by our emotions, not by our reason, as our desires and concerns constitute the questions that we ask as we do science. It is the conversations in which we are immersed as we do science that determine the course of science. And this cannot be otherwise because whatever we human beings do arises in our operation as such in our domain of experi-

ences through the continuous braiding of our languaging and our emotioning that is all that we human beings do.[13] Therefore, we do not find problems or questions to be studied and explained scientifically outside ourselves in an independent world. We constitute our problems and questions as we flow in our praxis of living, and we ask the questions that we in our emotioning desire to ask. Our emotions do not enter in the validation of our scientific explanations, but what we explain arises through our emotioning as a concern that we do not want to ignore, and we explain what we wish to explain, and we explain it scientifically, because we like to explain in that way. So, science as a cognitive domain exists and grows as such always expressing the concerns, desires, ambitions, aspirations, and fantasies of the scientists, regardless of their claims to objectivity and emotional independency.

4. The fact that science as a cognitive domain is constituted and validated in the operational coherences of the praxis of living of the standard observers as they operate in their experiential domains without reference to an independent reality, does not make scientific statements subjective. The dichotomy of objective - subjective pertains to a cognitive domain in which the objective is an explanatory proposition that asserts, directly or indirectly, the operational possibility of pointing to an independent reality. Science does not, and cannot, do that. The fact that the criterion of validation of scientific explanations constitutes science as an explanatory domain that arises in the recursive operation of the standard observer within the operational coherences of his or her domain of experiences, makes the scientific reference to anything conceived as an objective, independent reality operationally impossible. Science as a cognitive domain is realized in the operational coherences of the domain of experiences of the standard observer in his or her praxis of living as a human being, and as such the observer lives it as a domain of constitution of experiences in his or her domain of existence.[14]

5. The notions of objectivity and universality in science can be used in different and at the same time more relevant manners for the praxis of science than those to which I have objected above. Thus, we can listen to the claim of objectivity in the praxis of science as the commitment of the standard observer not to let his or her desires or preferences distort or interfere with his or her application of the criterion of validation of scientific explanations. Similarly, we can listen to the claim of the universality of science as claiming that because science as a cognitive domain takes place in the praxis of living of the standard observer as a human being, every human being can, in principle, operate as a standard observer if he or she wishes to do so. Therefore, the claims of objectivity and universality in science are moral, not ontological claims. The legitimacy of these claims in the manner just mentioned above, even if we do not usually understand them like this, is one of the conditions that gives to scientists both the possibility of escaping the always present temptation of fanaticism and the possibility of keeping science as a cognitive domain always open to the understanding and practice of every human being.

6. Since the criterion of validation of scientific explanations can be made fully explicit operationally and does not entail any direct or indirect reference to an independent reality or objective world, it can be learned, used, and applied with total

independency of the beliefs of the standard observer about reality, values, and spiritual life. This is why we scientists are not always wise. The path of modern science is not the path of wisdom in living the worlds we human beings live, even though it does not necessarily contradict it. The path of wisdom, as the path of understanding, care, and responsibility for the consequences of our actions in the worlds that we human beings bring about and live, must be specially cultivated if it is to have any presence at all in our daily lives. Spiritual experience is an experience of belonging in a community or cosmic realm that happens to us human beings in one way or another along the paths of our lives, and it has frequently been of basic significance for the harmony and health of our human social life. As such, spiritual experiences cannot be denied, and science does not deny them. Indeed, experiences are never a problem in the domain of human coexistence, it is in the domain of explanations and of use of our experiences where we may enter into deadly disputes. It is the fanaticism that may arise around the explanation of experience when someone claims to have access to some transcendent truth that constitutes a source of strife and suffering in human coexistence. To the extent that science does not stand on any system of beliefs because it belongs exclusively to the domain of the operational coherences of the praxis of living of the standard observers as a human beings, science constitutes for us the operational possibility of being responsible for our actions by giving us the possibility of becoming aware of our emotions and of knowingly acting according to our desires.

7. Notions of progress, of social responsibility, and of ethics do not apply to science as a cognitive domain. Indeed, science, as any other cognitive domain, is operational in its constitution, and as such it is value-free. Notions of progress, social responsibility, and ethics apply to human actions and, therefore, to what we scientists, artists, technologists, or whatever kind of person we may be, desire and do as human beings. We human beings operate and exist as an intersection of our conditions of observers (in conversations) and living systems, and as such we are multidimensional beings, actual bodynodes of a dynamic intercrossing network of discourses and emotions that continuously move us from one domain of actions to another in a continuous flow of many changing conversations. Accordingly, we scientists practice science as a manner of living under one of the numerous emotions that constitute us in our living as normal emotional human beings, namely, under the passion or desire for explaining. Nevertheless, which area of our domain of experiences we chose at any moment to investigate, to explain, or to analyze as scientists, or to what uses we choose to put the results of our scientific activities, depends on other emotions that show up in us in the course of our lives. Once scientific knowledge is there, we can use it for whatever purpose we may wish as our different desires, passions, fears, or aims arise in us with the flow of our languaging and emotioning. As this happens, we begin to act, let us say, as technologists, entrepreneurs, artists, cheaters, politicians and so forth, as some of our other human concerns show up as we become involved with technical, productive, aesthetical, personally advantageous, or social actions. It is in this context that notions of progress, ethics, and social responsibility acquire their presence. Thus, the notion of progress has to do with what we consider better or desire to be the case in human life;

the notion of social responsibility has to do with our awareness of whether we do or do not want the consequences of our actions; and the notion of ethics has to do with our concern for the consequences of our actions on the life of other human beings that we accept in coexistence with us. Yet, the notions of progress, ethics, and responsibility do not pertain to science as a cognitive domain, but they apply to us scientists as human beings because every thing that we do involves us and the human and nonhuman communities of living beings to which we belong and who support us in our endeavors.

8. I have not spoken of truth or naturál law; indeed this is not an accident. We usually use notions of truth and nature to explicitly or implicitly connote a reality independent of what we do as observers, either as a reference through which we may validate our statements or as an argument that gives them universality. From all that I have said, however, it is apparent that science has nothing to do with a notion of truth that is independent of the criterion of validation that constitutes a scientific statement or explanation as valid in the cognitive domain constituted by the criterion of validation of scientific explanations. At the same time, the notion of relative truth does not apply to science either, because such a notion is valid only in reference to an absolute truth. In other words, the dichotomy between absolute and relative truths does not apply in science because all that one may be saying in science when claiming that a statement is scientifically true is that it is a scientific statement. Something similar happens with the notion of nature. Since the observer cannot make any cognitive statement about anything independent of his or her operation as a living system, the notion of nature constitutively can refer only to what the observer does (in language) as a human being explaining his or her experiences as such, and, hence, it cannot refer to anything deemed independent of what the observer does.[15] Nature is an explanatory proposition of our experience with elements of our experience. Indeed, we human beings constitute nature with our explaining, and with our scientific explaining we constitute nature as the domain in which we exist as human beings (or languaging living systems). Scientific explanations and statements are not validated through a reference to nature, but nature is operationally constituted (known) and expanded as we constitute it as our domain of experience as living systems through our scientific explanation of our experience with elements of our experience. To act according to our awareness of our continuous constitution of nature through our explanations of our experience being aware of the constitutive nature of nature, is our greatest responsibility as human beings in general, and as scientists in particular.

9. Science, as a cognitive domain, changes as the questions that the standard observer asks and the explanations that he or she accepts change, and these change as his or her domain of experiences changes in the flow of his or her praxis of living. Since language takes place in the domain of consensual coordinations of consensual coordinations of actions, and these take place through the body encounters of the participants in it, the bodyhood of those in language changes according to the flow of their languaging, and the flow of their languaging changes contingently to the changes of their bodyhood. Due to this recursive braiding of bodyhood changes and consensual coordinations of actions in language, everything that the observer does as a human being

takes place at the level of his or her operational realization in his or her bodyhood in one and the same domain, namely, in the domain of sensorimotor correlations through which we do all that we human beings do. In these circumstances, human activities that are totally different in the conversational domains in which they are distinguished as human activities, such as theoretical and practical actions, do not differ in their actual realization through the bodyhood of the acting human beings.[16] In other words, the result of the braiding of bodyhood and consensual coordinations of actions is that the practical and the theoretical cognitive domains arise as two different but mutually entailing aspects of the consensual coordinations of actions of the observers, and constitutively what arises in the theoretical domain gives rise to adequate actions in the practical domain that it entails, and vice versa. Our only difficulty as observers with the theoretical and the practical is that we are not always aware in what area of our domains of experience the mutually entailing practical and theoretical domains take place in each particular case when we begin to consider some practical or theoretical action. As a result, as observers of our experiences in our reflections in language, we may expect experiences in one area of our experiential domain that pertain to another.

10. Creativity is an assessment made by an observer that claims novelty according to his or her listening in the activities, operations, or distinctions of another observer, that could also be him or herself, when these activities, distinctions, or operations appear to him or her as unexpected. The same applies when we speak of creativity or novelty in any domain of science. Everything happens to us as experiences that just occur to us without effort as a matter of course, and this is why we are blind to their origin and have to invent explanations when we want to account for them. Furthermore, everything happens to us as distinctions in language that we make in the conversations that we live, blind to the dynamics of states of our bodyhood through which they arise. As a result, the more complex and rich our dynamics of state (including the dynamics of states of our nervous system, of course), the more unexpected will be our participation in the different conversations in which we are involved, and the more creative we shall be in the eyes of the surprised beholder. And the more complex and multidimensional our life, the more novel, strange, or unexpected will appear our actions and distinctions in language to those with whom we live without sharing all our conversations. The same happens in the practice of scientific conversations. Novelties, changes, or conceptual revolutions occur in science when a standard observer, as a result of his or her recursive operation in his or her domain of experiences through conversations outside the accepted domain of scientific reflections, and in the context of the continuous structural change that he or she necessarily undergoes in them, brings forth, simply as a matter of course, some unexpected configuration of operational coherences acceptable in the scientific community. Novelties in science, therefore, constitute new dimensions of operational coherences in the domain of experiences of the standard observers, but they do not disclose any hidden independent reality.

11. Due to its manner of constitution, science as a cognitive domain is a domain in which the observer brings forth existence in its own domain of existence as the domain in which he or she distinguishes him or herself as a living system. In this sense science

is a domain of ultimate explanations, not because scientific explanations do not change or are objective or true in a transcendent sense, but because they arise and remain in our domain of constitution as living systems.

Conclusion

Science is the domain of scientific explanations and statements that we scientists generate through the application of the criterion of validation of scientific explanations. As such, we scientists deal in science with the explanation and understanding of our human experience (human life) and not with the explanation and understanding of nature or reality as if these were objective domains of existence independent of what we do. Human experience is content- free. In our experience we do not encounter things, objects, or nature as independent entities as it seems to us in the naivety of daily life; we live in experience, in the praxis of living of human beings in the flow of being living systems in language as something that happens in us and to us as we language it. This is why, as we as scientists explain our experience as human beings by reformulating it with elements of it through the application of the criterion of validation of scientific explanations, we find ourselves generating science as a cognitive domain that does not take us out of experience and keeps us in language. We human beings exist in language, and our experience as human beings takes place in language in a flow of consensual coordinations of consensual coordinations of actions that we bring forth in language. Objects, awareness, self-reflection, self, nature, reality, and so forth, all that we human beings do and are, takes place in language as distinctions or as explanations in language of our being in language. So experience takes place in language, science takes place in language as we use language to generate it, but not as an abstraction or as a mere discourse, but as something as concrete as any operation in the flow of consensual coordinations of actions in which we arise and exist. That we exist in language, and constitutively cannot exist outside of it because we are constituted in it, and that by being in language we only generate experiences in language, is not a limitation in us, but, on the contrary, it is the condition that makes possible science as an explanatory domain such that whatever we bring forth in it becomes part of our existing as human beings. Indeed, that by existing in language our experiential domain should be a closed domain from which we do not and cannot come out, appears as a limitation only if we think that we should be able to refer to an independent reality.

I have said above that a central aspect of doing science has to do with our quest for understanding our experience as human beings. And I mean by understanding the experience of adopting an operationality of reflection in language in which we can know what we know in the circumstances of constitution in language. In order to reflect, we need to release that which we constitute in our distinctions as an object of our reflection so that we may contemplate it, but we must do so without fear of losing that which we release, so that we may look at it without obscuring it through our attachment to it. Moreover, we must possess a procedure that, if properly applied, allows us to operate in our reflections as if we were indeed dealing with something that exists independently from what we do, so that we may realize the operation of contemplation. The criterion of validation of scientific explanations provides us with this procedure in a manner that no other methodological approach could do because it does not require any assumption about the origin of our abilities as observers, so that these can also be the subject of our inquiries with it. In other words, it is the fact that the criterion of validation of scientific explanations takes place as a system of operations of the standard observer in his or her domain of experiences that gives rise in him or her to further operations in it without entailing any supposition about the origin of his or her abilities that permits the standard observer to treat any aspect of his or her domain of experiences, including his or her abilities and capacities as an observer, as an object of his or her scientific inquiry.[17]

In a scientific explanation, a standard observer proposes an ad hoc mechanism that would give rise as a consequence of its operation to the experience that he or she wants to explain. In other words, the generative mechanism proposed in a scientific explanation is arbitrary and can be any, providing that it operates as such with respect to the experience to be explained. In these circumstances, all that makes scientific explanations operationally effective in our human praxis of living is that they arise as operations in it that give rise to further operations in it and not an impossible reference to anything like an independent objective domain of reality. At the same time, it is because of this that science is an operational domain in which the standard observer recursively creates knowledge in his or her praxis of living. Indeed, it does not matter how strange or crazy any particular generative mechanism proposed as a reformulation of our experiences may initially seem to us. If such a mechanism becomes validated through the criterion of validation of scientific explanations, it becomes so in our recursive operation in language with the operational coherences of our domain of experiences, and becomes as such for us a source of adequate actions in our praxis of living in the area of our domain of experiences in which it is valid. Our operational creativity through the use of science needs no other explanation, it occurs to us in our doings as standard observers; yet, that this should be so makes us scientists fully accountable for what we bring forth through the practice of science. As we explain our experience scientifically, thus it becomes the world we live. We cannot claim innocence anymore.

In our modern Western culture we speak of science and technology as sources of human well-being. However, usually it is not human well-being that moves us to value science and technology but rather the possibilities of domination, of control over nature, and of unlimited wealth that they seem to offer. We fight a hostile nature, we say, and we look for scientific knowledge as in search of an instrument that would permit us to control and to manipulate it rather than to understand it. We also speak of progress in science and technology in terms of control and domination and not in terms of understanding and responsible coexistence. The notions of control and of domination entail the negation of that which is controlled or dominated as well as positing it as something different and independent from oneself. I think that it is this basic attitude that limits our possibility of grasping without anxiety our constitutive participation in bringing forth the world that we live through science. Whatever we distinguish we distinguish it in language, whatever we experience we experience it as we distinguish our experience in language. That this should be so, however, is not a problem for us in our living, it just happens in us and to us as we distinguish our experiences. Experience, the happening of living, is not a problem for us, our problems arise with our explanations of our experiences and the demands that they impose upon us and those other human beings with whom we coexist. We human beings live a world of explanations and descriptions in language of our experiences as we bring them forth in language, and we even kill each other defending our explanations when we are in discordance about them. Furthermore, since explanations are reformulations of experiences with elements of experience in the operational coherences of experiences, we live different worlds as we bring forth in our praxis of living different operational coherences as we adopt different systems of explanations in our living. This is not a vacuous statement because language is constituted as a domain of consensual coordinations of actions in the domains of operational coherences of the observers, and if two observers accept different explanations they live different areas of operational coherences in their domains of experiences. Life happens to us, experience happens to us, the worlds we live happen to us as we bring them forth in our explanations. Moreover, since every system or mechanism operates only if the operational coherences that it entails are satisfied, life and experience happen to us as they happen to us only to the extent that the operational coherences that constitute them are satisfied. Finally, as we human beings exist in language, our bodyhood are nodes of operational intersection of all the operational coherences that we bring forth as observers in our explanation of our operation, and we live them through mapping them in our discourse as if they were taking place in a single operational domain. As long as we are not aware of this, we confuse operational domains and expect that a particular system or mechanism should operate in a different domain of operation than the one in which it takes place. Also, if we are not aware of this mapping of our discourse in our discourse, we shall not be able see that the value of science for human life lies in the possibilities that it opens for understanding it by permitting us to enter in recursive reflections on our domain of experiences through the application of the operationality of the criterion of validation of scientific explanations as part of our daily operations in our domain of experiences.

We scientists frequently demand a special status for science as a cognitive domain and require a special freedom and respect for our pursuit of scientific research, claiming that scientific knowledge deserves a special respect on account of its objective, factual nature. We scientists are concerned with objective facts, we say, and because of that our statements and explanations are free from fanaticism and prejudices. As we have seen, this is only partly true. The value and operational effectiveness of scientific statements and explanations rests on the fact that they pertain to the operational coherences of our praxis of living as human beings, and for this science does not deserve a special status. Yet, science deserves a special status because the criterion of validation that constitutes it entails the operationality of a reflective dynamics which, if properly learned and practiced by the standard observer, allows him or her to remain an observer of all his or her circumstances without attaching to them. The latter, however, does not always happen. What science and the training to be a scientist does not provide us with is wisdom. Modern science has arisen in a culture that values appropriation and wealth, that treats knowledge as a source of power, that prizes growth and control, that respects hierarchies of domination, that values appearances and success, and that has lost sight of wisdom and does not know how to cultivate it. We scientists in our efforts to do what we like most, namely, scientific inquiry, frequently fall prey to the passions, desires, and aims of our culture and think that the expansion of science justifies everything, becoming blind to wisdom and how it is learned. Wisdom breeds in the respect for the others, in the recognition that power arises through submission and loss of dignity, in the recognition that love is the emotion that constitutes social coexistence, honesty, and trustfulness, and in the recognition that the world that we live is always, and unavoidably so, our doing. But, if science and scientific knowledge do not provide us with wisdom, at least they do not deny it, and the awareness of this opens the possibility of learning it by living in it to those who escape the ambition for success and the desire for control and manipulation.

Overview

By accepting that science as a cognitive domain constituted through the application of the criterion of validation of scientific explanations does not deal with truth or reality in a transcendent sense but only with the explanation of human experience in the domain of human experiences, many questions disappear or fully change their character, and new insights are possible. Let us see some of them.

Truth stops being an argument that can be used without specifying its conditions of constitution and validation, and the observer has the possibility of relinquishing his or her claim to be the owner of it.

Nature stops being an autonomous domain of independent entities that can be used as a confrontational reference for accepting or rejecting any particular explanation as a scientific explanation to become something that arises through the operation of an observer in language in his or her domain of experiences.

Learning to be a scientist becomes, for the observer who desires to do so, a process of learning how to apply the criterion of validation of scientific explanations, of expanding some particular area of his or her domain of experiences, and of reaching a total personal commitment to act with honesty and operational impeccability in the generation of scientific explanations and statements, and, in the long run, in his or her daily life.

Emotions become respectable and apparent, and it becomes possible to see that for an observer to act responsibly is to act with awareness of whether he or she wants or does not want the consequences of his or her actions.

The application of the criterion of validation of scientific explanations as a manner of living enables the observer to operate systematically in a recursive reflection upon his or her circumstances and become responsible with respect to his or her actions, regardless of whether he or she reaches in any particular case to its completion.

The understanding that the criterion of validation of scientific explanations arises as a systematization of one normal operational manner of validation of our operation in daily life in the care of not confusing domains of experience, permits us scientists to become aware that the only thing special in us is our passion for explaining scientifically.

I consider that the greatest spiritual danger that a person faces in his or her life is to believe that he or she is the owner of a truth, or the legitimate defender of some principle, or the possessor of some transcendental knowledge, or the rightful owner of some entity, or the deserving meritor of some distinction, and so forth because he or she immediately becomes blind to his or her circumstance, and enters into the closed alley of fanaticism. I also consider that the second greatest spiritual danger that a person faces in his or her life is to believe in one way or another that he or she is not always totally responsible for his or her acts or for his or her desiring or not desiring the consequences of them. Finally, I consider also that the greatest gift that science offers to us is the possibility to learn, free from any fanaticism, and if we want it, to learn how to remain always responsible for our actions through recursive reflections on our circumstances.

Acknowledgement: This work was possible thanks to the generous support of the »Foundation for the Study of Human Cognition Inc.«.

Notes and References

1. See T.S. Kuhn, *The Structure of Scientific Revolutions*, Chicago, London: University of Chicago Press, 1962; E. Nagel, *The Structure of Science*, New York: Harcourt, Brace & World, Inc., 1961; and K.R. Popper, J.C. Eccles, *The Self and the Brain*, Berlin, etc.: Springer International, 1977.

2. See H.R. Maturana, *Biology of Cognition*, BCL Report No 9.0, Biological Computer Laboratory, Department of Electrical Engineering, University of Illinois, 93 p. 1970, Reprinted in H.R. Maturana, F.J. Varela, *Autopoiesis and Cognition*, Dordrecht: Reidel, 1980; »Biology of Language: Epistemology of Reality«, in G.A. Miller, E. Lenneberg (eds.), *Psychology and Biology of Language and Thought*, New York: Academic Press, 1978; »Reality: the Search of Objectivity, or the Quest for a Compelling Argument«, *The Irish Journal of Psychology* **9**, 1 (1988), 25-82.

3. Maturana, *op. cit.*, 1970, note 2; »Cognition«, in P.M. Hejl, W.K.Koch, G. Roth (eds.), *Wahrnehmung und Kommunikation*, Frankfurt, New York: Peter Lang, 1978, p. 29-49; *op. cit.*, 1988, note 2.

4. H.R. Maturana, »What is it to see?«, *Arch. Biol. Med. Exp.* **16** (1983), 255-269.

5. H.R. Maturana, *op. cit.*, 1970, note 2; *op. cit.*, 1978a, note 2, and *op. cit.*, 1988, note 2.

6. See H.R. Maturana, *op. cit.*, 1988, note 2.

7. H.R. Maturana, *op. cit.*, 1988, note 2.

8. H.R. Maturana, *op. cit.*, 1988, note 2.

9. See H.R. Maturana, *op. cit.*, 1970, note 2; and *op. cit.*, 1988, note 2.

10. See K.R. Popper, *The Logic of Scientific Discovery*, London: Hutchinson, 1959.

11. See H.R. Maturana, *op. cit.*, 1988, note 2.

12. See H.R. Maturana, *op. cit.*, 1978b, note 3; and *op. cit.*, 1988, note 2.

13. See H.R. Maturana, *op. cit.*, 1988, note 2.

14. See H.R. Maturana, *op. cit.*, 1988, note 2.

15. See H.R. Maturana, *op. cit.*, 1988, note 2.

16. See H.R. Maturana, *op. cit.*, 1988, note 2.

17. See H.R. Maturana, *op. cit.*, 1970, note 2; *op. cit.*, 1978a, note 2; *op. cit.*, 1978b, note 3; and *op. cit.*, 1988, note 2.

Self-Organization, Emergent Properties and the Unity of the World

GERHARD ROTH AND HELMUT SCHWEGLER
University of Bremen

1. Unity of the world and unity of science

Natural philosophers and scientists have always struggled for the unity of the world believing that there is only one reality as the ultimate framework for the infinity of phenomena that can be observed. However, to this date there is no accepted unified view of this world, and it is unlikely that there will be one in the near future. Instead of one all-explaining science or theory, there are many scientific disciplines, and most of them have little, if anything, in common. But even worse, within one and the same discipline, the logical and semantical connections between different areas are not always clear, for example, in physics between the microscopic and the macroscopic domains, in biosciences between molecular biology and organic biology, in social sciences between sociology and economics. Each of these disciplines and sub-disciplines seems to have phenomena, concepts and laws (if there are laws recognized at all) of its own irreducible to and unconnected with those of other disciplines and areas. At the end it seems that there is not one world, but as many worlds as disciplines and sub-disciplines of sciences exist, let alone all those worlds of non-scientific experience.

The reasons for this situation are unclear and may be of different kinds. It may well be that many gaps between domains of knowledge can be bridged by means of further development and refinement of theories and concepts and increased amount of empirical data. The history of science is filled with examples of bridging previously disconnected areas of knowledge (for example, optics and electrodynamics). But we cannot exclude that there are gaps so fundamental in their nature that they cannot be bridged regardless of how much empirical knowledge and theoretical insights we might accumulate. We will call these *ontological gaps* in the sense that they separate domains that contain completely independent phenomena and concepts. As a consequence, theories from ontologically different domains can neither be reduced one to another, nor can they be unified within the framework of a more comprehensive theory.

Recently, there have been renewed attempts to reach a unified view of the world on the basis of concepts such as »self-organization«[1] and »autopoiesis«[2]. These concepts, originating (partly) from biosciences, have quickly spread into other disciplines, especially into the social sciences and the humanities. Here, self-organization and autopoiesis became a sort of magic concept. Another recently much discussed concept dealing with the problem of diversity and unity in nature is emergentism. This concept has become extremely popular among scientists from all disciplines, particularly from biology.[3]

In this paper, we will discuss the questions whether a unified concept of the world is possible and whether ontological gaps exist in the world that would make such a view impossible. We will discuss the problems mainly with respect to biology, because

W. Krohn et al. (eds.), Selforganization. Portrait of a Scientific Revolution, 36–50.

biological systems are systems which, on the one hand, are strongly bound to physics and chemistry, but at the same time extend into the domains of psychology and social sciences, at least with respect to human beings. Thus, biology is central to any discussion about the unity of the world.

There are two eminent candidates for ontological gaps: the difference between the living and the nonliving world, and the difference between the material and the mental world. Accordingly, we will deal with two questions: (1) Is biology reducible to physics (and chemistry)? (2) Is mind reducible to brain processes? We will discuss the basic epistemological and methodological problems related to these questions, especially in the light of the concepts of self-organization and of emergent properties.

2. Can biology be reduced to physics?

There are two prevailing alternative answers to this question. One is physicalist reductionism claiming that biology can be reduced to physics in the sense biological phenomena can - at least in principle - be described in physical terms and explained by laws of physics and that essentially no biological laws exist.[4] Accordingly, biology has to be regarded either as a sub-discipline of physics like electrodynamics or solid-state physics, or treated, like chemistry, as a separate discipline for historical or practical reasons in which, however, all concepts and modes of descriptions can be expressed and explained in physical terms.

The other view is emergentism claiming that biology is autonomous with respect to physics in the sense that there is a large body of biological phenomena that cannot or not entirely be described in physical terms but need additional biological concepts, and that biological laws exist that are fundamentally different from physical laws. These properties and laws, mainly »higher-level« or organic ones, are called »emergent«, because they are »unexpected« and cannot be formulated and explained on the basis of even the most complete knowledge of the physical properties of the components of the system under consideration.[5]

The strongest argument in favor of the reductionist view is that with the great success of biochemistry, molecular and cellular biology there are major parts of biology where a description and explanation of biological processes in chemical and hence physical terms have been successfully achieved. This is true, for example, for metabolic processes inside the cell, for the processes of genome replication, transcription and translation, and for many processes at the organic level. Wherever, in this context, »gaps« exist, there is hope that these will be bridged as a function of progress in biochemistry and molecular and cellular biology.

There are »higher-level« fields of biology, for example developmental biology or neurobiology, where such a »reduction« is very far from being complete. How the fertilized egg (zygote), through cell division, differentiation and migration, develops into the adult organism, is not understood; how the incredible complexity of the brain is produced, how neural networks function in the context of perception, memory and control of behavior is not understood either. But whenever processes have been studied for a sufficient time and in sufficient depth within these fields, they turn out to be - at least in principle - explicable in chemical and physical terms. The establishment of the highly specific connectivity among nerve cells in the brain (»axon pathfinding»), once thought to be completely inaccessible to a physical approach, is close to its explanation as a highly complex interplay of chemical »guiding factors« (for example surface glycoprotein molecules) and response mechanisms inside the connecting neuron.

In certain fields of biology, for example evolutionary biology, however, a physicalist reduction seems to be completely out of reach. For the process of evolution, a description seems to need concepts and an explanation seems to need rules that cannot be borrowed from chemistry and physics. This seems to be especially true in the light of the paradox that the components of which organisms are built and which are accessible to a chemical and physical description, are highly uniform among even distantly related groups, but that the organisms at the same time show a high diversity of morphology, physiology, behavior and ecology. This diversity - so the argument - cannot result from the components but only from variation in composition which is governed by truly evolutionary, historical laws completely beyond physics and chemistry. The reductionist counter-argument states that if one accepts that evolution with all its phenomena results from the interplay between the organism (including all genetic and epigenetic processes) and the environment, both of which are processes and phenomena explicable in physical terms, then evolution should be explicable in those terms too. There is no evidence in evolution for something to come into play even in evolutionary biology that is »in principle« inaccessible to physics. The complexity of organic evolution could simply result from the complexity of the initial and boundary conditions of each single evolutionary process which, as a single process, is - at least in principle - describable in physical terms.

Before we discuss this reductionist view further, let us have a look at emergentism which today probably is the most popular scientific concept among biologists. According to Ernst Mayr, one of the leading proponents of emergentism, the characteristics of living organisms »cannot (not even in theory) be deduced from the most complete knowledge of the components, taken separately or in combination... When such systems are assembled from their components, new characteristics of the new whole emerge that could not have been predicted from a knowledge of the components«.[6] It is important to see that emergentists like Ernst Mayr do not want to be misunderstood as vitalists (in the sense that organisms contain some mystical »life-making« substance or principle). Mayr accepts that organisms consist of nothing but physico-chemical components. Yet, he sees that the organization of living systems is fundamentally different from that of non-living ones.

Emergent properties are not considered to be characteristic of biological systems only; they are believed to be quite universal and to occur in inanimate systems, too, but are considered especially important in biological systems.

In the reductionist as well as the emergentist view, the main problem consists in the question, to what degree can the properties of complex systems be explained on the basis of the properties of their components? Emergentism, as mentioned above, accepts that organisms consist of nothing but physicochemical components. However, for emergentism the knowledge of the properties of these components is not sufficient to deduce the properties of the system. These result from the organization of the system, including the existence of »hierarchical levels«.

Organization of a system is usually understood as the modes of interaction of its components. Accordingly, if a system has certain properties, these result from the modes of interaction of the components. If we accept such a definition, then the basic question is, how the properties of the components relate to their modes of interaction. If emergentism wants to be conclusive, it has to assume that there is no or no strict relationship between the two.

This view, however, neglects the simple fact that there is no way of distinguishing between »properties« and »modes of interaction«, because properties are defined through modes of interaction, and modes of interaction are defined through properties. In other words, no phenomenon or process has properties per se independent from its interactions with other phenomena and processes. We can describe phenomena and processes only in relation to other phenomena and processes. *So, every property results from modes of interaction, and every mode of interaction is a property.*

This becomes evident when we briefly discuss a standard example of emergent properties, the properties of sodium chloride, table salt, which consists of two chemically aggressive components, sodium and chlorine, but is harmless as a compound. A molecule (or atom) like sodium has a certain physical structure (nuclei and electron shells). This physical structure (for example the wave function of electrons) describes nothing but the modes of interaction of the molecule or atom which can be realized in certain experiments (i.e. interactions with certain media). So, a phenomenon which under all possible circumstances behaves (interacts with any kind of environment) like a sodium atom is a sodium atom. It makes no sense to claim that a sodium atom has certain properties *by itself* regardless of its modes of interaction.

Accordingly, properties of a phenomenon necessarily change whenever the modes of interaction change. These changes can be slight or profound; they may remain within the range of our expectation or may be surprising. If we could test a substance under all possible conditions of interaction with other substances, we would be able to predict any of those changes. However, this is not possible, and as a consequence we will often be confronted with »completely new« or »emergent« properties. The interaction of chemical substances can be such that the atomic structure of the two substances is completely altered (by fusion of the electron shells), as is the case with the sodium chloride molecule. Sodium chloride, in this sense, does not »consist« of sodium and chlorine atoms anymore. And it is, thus, not surprising that it has »completely new« properties. However, these new properties result from the properties of the components, viz. the capacity to drastically change their atomic structure in a given way. So, in this popular case of »emergent properties«, it is true that the properties of the sodium chloride molecule are by no means detectable at the level of the components; this is impossible because the molecule does not really consist of the elements sodium and chlorine, but of transformed states of these atoms. The same is true if we proceed to more complicated molecules, for example, biomolecules, and much more complicated systems such as cells and multicellular organisms.

Emergentism makes an unacceptable distinction between properties and modes of interaction. Accepting the fact that properties and modes of interaction are mutually defined, emergentism then becomes a truism because all properties are emergent properties.

Reductionism, in its common version (which is different from quantum chemical reductionism, see below), denies this fact, namely that due to a specific mode of interaction inside a given system, components can drastically change their properties so that they are not identifiable any more with the components prior to the composition of the system or with substances outside the system. This is especially true for living systems because they have two peculiar properties: they consist of unusual components, macromolecules of a very complex nature, and show unusual properties, those of self-production and self-maintenance.[7] The most important macromolecules of which organisms are built are amino acids, nucleotides, fatty acids and polysaccharides. Amino acids are the building blocks of proteins which, as structural proteins, constitute most of the structure of organisms and, as enzymes, regulate all physiological processes. Nucleotides, in the form of nucleic acids (DNA, RNA), are the components of the genome which serves as a »manual« for the enzymes to produce other enzymes and to control the physiological processes; as adenosine phosphates (for example ATP) they are the most important energy carriers and play an important role in the intracellular signalling system. Fatty acids (for example phospholipids) are the basic components of membranes inside and surrounding the cells and have also energy storage functions. Polysaccharides (»sugars») are the primary energy source for organisms.

All these are molecules that are not naturally found outside organisms. Their most important property is that they show the capability of self-organization or self-aggregation in the sense that the complex arrangement is reached »spontaneously« whenever the appropriate building blocks are present. This is not only true for the three-dimensional folding of enzymes, but also for more complicated structures such as enzyme complexes, ribosomes, membranes and viruses. Since many of these phenomena of self-organization and self-aggregation occur even in vitro, they must be caused by inherent properties of the macromolecules themselves. However, such first-order self-organizing processes - although being the necessary basis of life - have only a limited capacity for the generation of order. More complex structures and processes need the interaction of many self-organizing processes, for example, between enzymes and nucleic acids or between several proteins (second-order self-organization). Finally, when forming a closed circle of mutual production and maintenance, they form *autopoiesis* which, thus, can be understood as a circular concatenation of first- and second-order self-organizing processes.[8]

For a reductionist, autopoiesis, as the typical organization of life, is the direct result of the self-organizing properties of the components because only from them result the modes of interaction that lead to autopoiesis. In this sense, life and all its properties can be reduced to the properties of the building blocks. However, at present, it would be impossible to explain the properties of these biomolecules only on the basis of the properties of their components, for example predict the enzymatic activity of a protein from the amino acids of which it is built, if we had no substantial knowledge about how living systems are organized. So, reductionism in biology is always an ex-post reductionism: we can »deduce« the properties of the system from its components only after we have gained sufficient knowledge about the system as a whole.

The reductionism described so far holds that it can reduce the autopoietic organization of life to abiotic building blocks, the macromolecules, these to smaller molecules and these to atoms where the sequence of reduction stops. There is another and more sophisticated reductionism that does not rely on such a hierarchy but claims that everything has to be reduced to nuclei and electrons. We will call this quantum-chemical reductionism. This type of reductionism is fully aware of the above described situation in chemical reactions (for example, regarding our example of sodium chloride) and it does not reduce processes to a level of some objective components and their interactions, because quantum phenomena like electrons and nuclei adopt the appearance of reality only through interaction with (or measurement by) the observer. Therefore, reduction occurs by a construction of the properties of all (as the radical advocates of quantum-chemical reductionism believe) higherlevel objects and processes from quantum theory. However, even for the simplest molecules, this program can be carried out only by the use of drastic approximation and ad-hoc assumptions. The more complicated the molecules are, the more ad-hoc assumptions have to be applied that are guided by empirical (i.e. post-hoc) knowledge about the properties and the reactions of the molecules under consideration. Therefore, particularly for the large and complex biomolecules, a rigorous quantum-theoretical calculation »from first principles« or »ab initio« turns out to be a fiction, let alone a quantum chemistry of organic processes.

Thus, without any possibility to confirm or falsify this quantum-chemical fiction, there is no alternative to introducing the properties of those highly organized systems by their empirically determined modes of interactions which have to be studied under the conditions of life (or under conditions of which we know that they are very similar to those of life). Even the most detailed quantum-chemical knowledge about amino acids would not tell us what decisive role proteins (as sequences of amino acids) could possibly play when interacting with nucleic acids, phospholipids, polysaccharides and many more highly »improbable« chemical substances. So, we need at least partial knowledge about the organization of life in order to understand how life works. But this is not reductionism anymore because it says nothing but life originates from the interaction of specific building blocks under conditions that are typical for life.

So, to be able to understand, i.e. to explain, how life originates from its components, is not the same as to reduce life to its components. Rather, it is the opposite of reductionism in that we understand the properties of the components after we have understood their role inside organisms.

What we want to propose is a distinction between reductionism and physicalism. Physicalism is understood here as the attempt to describe all phenomena in terms of physics. The description of observable phenomena (even in atomic and nuclear physics and even more in biology) is possible on the basis of classical physics. In this sense, we can formulate a *non-reductionist physicalism* with respect to biology: all biological phenomena can be described in terms of classical physics. For explanations, low-level as well as high-level laws are allowed even if (what is usually the case) higher-level laws are not reduced to (explained by) lower ones. Also, »biological laws« are permitted, but, of course, no laws should contradict each other. This physicalism includes all kinds of »emergent« properties of complex systems, i.e. system properties that follow »higher-level laws« as compared to the properties of the components of the systems. What is excluded are »ontological« gaps between the different levels because the whole body of »reality« (see below) is described by the same language of classical physics.

3. Is mind reducible to the brain?

The relationship between mind and body, or in more modern terms, between mind and brain, has seen an almost sensational rebirth in recent years. This is not quite surprising, because the stormy progress of brain science and the equally stormy development of computer sciences and artificial-intelligence research has forced theoreticians of all kinds to thoroughly reconsider the mind-brain problem.[9]

It is not our task to present and discuss the numberless theories related to the mind-brain relationship. Many of them have become rather obsolete, viz., all those theories that see mind as being an independent entity with respect to the brain and the body. Empirical evidence shows that mind and mental capacities are strictly dependent on the presence of a brain and change with changes inside the brain. If we accept this, then the only question worth discussing is: how close is the relationship between mind and brain?

A now widely accepted view among philosophers and brain scientists is that mind and brain are identical and that they represent only two aspects of one and the same entity, an »external« (the material brain) and an »internal« one (conscious experience)[10]. Indeed, there seem to be good reasons to adhere to such a view. Modern experimental brain science and clinical neurology demonstrate a close correspondence between brain processes and mental processes. Since structure and functional organization of certain subsystems of the mammalian/human brain, for example the visual system or the motor system, are now well known, it is possible to correlate a lesion or dysfunction in a circumscribed brain area with a loss or impairment in visual perception, imagery, memory, or motor control. In some cases, this correspondence can be traced down to the level of single nerve cells. With the refinement of modern scanning techniques[11] it may soon be possible to »read« thoughts, i.e. to tell the actual mental activity of a subject, at least in some rough manner, from brain activity. Does this mean that mental activity, mind, can be indeed reduced to neural activity?

Our view is that even if modern brain science can demonstrate an arbitrarily close correspondence between neural states and mental states, this does not allow us to state an identity between both events. First and most trivially, an arbitrarily close correspondence between two phenomena A and B does not necessarily mean an identity of A and

B. B can simply be a consequence of A. Streets become wet when it rains, but no one would consider rain and wetness of the street to be identical phenomena. Likewise, we could say that mind is a function or consequence of brain events without implying that mind is identical with the brain.

Furthermore, not all brain processes are accompanied by mind (in the sense of conscious experience). As far as we know, this is true only for cortical processes, and not even for all of them. This means that neural events do not necessarily lead to mind, and that mind is not a consequence of the mere complexity of neural networks (as was often assumed). Cortical processes, in order to be mind-related, apparently need, in addition to sensory inputs, the highly dynamic interaction with several other subsystems, for example, centers in the brainstem reticular formation that control wakefulness and attention, and with the memory-access system in the limbic system. So, there are no single parts and centers of the brain with which we could identify mind, rather there is a very specific mode of interaction between parts of the brain that leads to or is accompanied by mind.

Finally, modern clinical scanning and EEG techniques have shown that the close correspondence between mind and brain may be realized differently in different individual brains, at least as regards »higher« cognitive functions, such as language and music recognition, planning and imagery; only in one and the same individual, is there a high degree of stereotyped correlation between these mental states and the activity of specific brain areas. Particularly the studies on the (asymmetric) distribution of cognitive functions in the cortex have revealed a high degree of inter-individual variability. While basic perceptive functions usually occur in the same brain areas in all human beings, due to the basic, experience-independent topography of the brain; with respect to »higher« cognitive functions the developing and self-organizing brain-mind system has a higher degree of freedom to specify which parts of the cortex will be involved.

This plasticity is possible only because neural activity is, as such, semantically neutral: a particular neural activity can achieve arbitrary meaning, depending on the context in the brain. This context is specified (i) by the topography of the brain area under consideration, i.e. the set of inputs and outputs of this area, and (ii) by the spatio-temporal processes within the larger networks in which the specific area is situated. Most non-cortical parts of the brain are highly complex in their cytoarchitecture and rigid in their processing capacities, and this means a low degree of variability in the relationship between structure and function. In contrast, the cortex is highly uniform architecturally, and this uniformity means a high degree of versatility with regard to the correlation between structure and function.[12] This constitutes the meaning of the cortical processes.

So, the strict correlation between complex cortical processes and their meaning is not pre-established, but is formed during development of the cognitive system through self-organization and self-specification. Initial neural networks develop a certain function that has certain consequences for other neural networks including motor areas that control behavior. These consequences are what we subjectively experience as meaning. They influence the further development of neural networks, and this again leads to modified functions which, through their behavioral or nonbehavioral (brain-internal) consequences, organize new networks etc. Thus, the relationship between brain and function/meaning within the cognitive system develops in an interactive and self-organizing manner during the development of the whole system.

A good example for the relationship between mind and brain is written language. There is no pre-established correlation between the word as a sequence of characters (and of the physical shape and nature of characters) and its meaning. Any word can, in principle, have any kind of meaning, and many words have indeed different meanings at the same time in different contexts or at different historical times. However, at a given time and in a given context, every word must have precisely one meaning, otherwise communication would be impossible. Thus, although word and meaning are by no means identical and can change in any direction, at a given time and in a given context we find a reliable correlation. As in the brain, this correlation between word and meaning is a product of self-organizing and self-specifying processes.

Mind is an emergent property of the brain as being a complex, self-organizing system. As such, mind cannot be reduced to properties of single nerve cells or networks, because what the activity of cells and networks means to the whole brain is asymptotically determined by the interaction of the whole brain, especially with respect to behavioral consequences. However, despite its non-reducibility, the emergence of mind depends on very specific properties and modes of interactions of the components, nerve cells and neural assemblies, with all their electrochemical and dynamical properties, and on a very specific interaction of certain parts of the brain, for example, cortex, brainstem, limbic system.

Thus, on the one hand, mental processes cannot be reduced to brain processes, because they are agents that - once arisen - are able to modify brain processes which in turn lead to the modification of mental processes. On the other hand, we have seen that all mental processes at a given time have exactly one neural equivalent. This results in the possibility that mind can be described in neural, i.e. physical terms, even if mind is not directly accessible to a physical description. So, if we really understood all the modes of interaction between the neural networks and all their initial and boundary conditions, we could, at least in principle, derive from the neural processes and their changes the activity of mind. In practice, however, this will never be possible, because by finding out the modes of interactions of neural networks and their initial and boundary conditions we would necessarily interfere with them and change them. What remains, however, is the physical description of the brain. With respect to the mind-brain relationship, as in the case of the biology-physics relationship, we end up with a non-reductionist physicalism.

4. *Reality* and *Actuality*

There is a fundamental reason why mind is not reducible to brain events, and this has to do with the fact that we try to describe and explain how mind results from the brain while being mind: the system that describes is identical with the system that is being described. Such a circular situation usually leads to a paradox. In order to cope with this problem, let us, for a moment, imagine what God sees while looking at the world. God sees a world, which we will call *reality*. In this world organisms exist that have to survive within their respective environments. Some of these organisms possess brains that generate and control behavior in such a way that the organisms are able to survive. At least some of these brains have conscious experience or mind that forms a world in itself, the phenomenal world or *actuality* (»Wirklichkeit»). An »Ego« exists that experiences itself as the subject of this phenomenal world. This Ego is part of the phenomenal world, *actuality*, which, in turn, is part of the objective world, *reality*. Or the other way around: *reality* produces *actuality* which produces the Ego. This Ego, by interaction with other Egos, produces the world of social events and processes.

Let us now turn to a subjectivistic scenario and consider the world in the way in which this Ego experiences its world, or, as we experience our world. There is the environment, filled with objects and processes of an infinity of qualities and quantities; there is our body and its various sensations: touch, position sense, pleasure, pain; and there is the domain of our emotions and thoughts, or »mental« states. These domains are experienced as being different from each other, although they are continuous in a certain sense: the environment and our body seem to belong to the same world, the physical world, which we encounter via our sense organs, although the sensation of our body is different from that of environmental physical objects and processes, and we are constantly aware of our body via a system called proprioception even if we do »nothing specific«. The body is felt as being »embedded« in the environment. Our emotions and thoughts seem very different from the phenomena of the environment, but they are intimately embedded in our body: our happiness, fear, anger and pleasure are able to affect our body completely; our thoughts can conquer our brain completely and bodily pain can deeply disturb our mental states.

A paradox arises when we try to combine these two views. Within the objectivistic scenario it seems clear that the phenomenal world, *actuality*, is part of, or produced by, the objective world, *reality*, via the nervous system and its functioning. In this sense, people say that our perception, cognition and feeling are produced by or arise »inside« our head or brain. But at the same time, it is said that it is the brain that constitutes all this world we experience! How can these two views be harmonized?

Let us suppose I am at this moment a patient undergoing brain surgery. I am lying in the operating theater with my skull open and my brain exposed. I can hear and see what is going on around me because brain surgery is usually carried out without anesthesia in order to test the site of important brain functions prior to removing parts of the brain (for example, those infected by a tumor). I can see and feel my body, and I can, with the help of a mirror or a video camera, see my own brain. This brain is, with no doubt, part of the environment, because I can look at it. At the same time, I am forced to assume that my brain is that system that produces everything I can see, hear, feel and even think. How, then, could this brain look at itself? How could it be outside itself? The whole operating theater, the persons surrounding me, my own body, even my brain should be inside my brain! Since the brain at which I am looking is part of the operating theater which is inside my brain, this would mean that my own brain were inside my own brain. This certainly is a deep paradox.

One could argue that this is a rather unlikely situation. However, it is simply a drastic illustration of the fundamental epistemological and ontological problems that arise whenever we try to develop a theory of cognition that is epistemologically and empirically well founded. This becomes evident, for example, in Maturana's theory as presented in his »Biology of Cognition«.[13] What Maturana was originally interested in was to answer the question: »How do we cognize and know?«. He realized that cognition could not be explained from itself, but only as a biological phenomenon; thus, in order to understand cognition he had to understand how living systems are organized. This led him to the question: how does the organization of the living beings condition cognition in general and self-cognition in particular? But in order to answer this question he had to analyze living systems, to be an observer. One of the central statements in the »Biology of Cognition« is that everything that is said is said by an observer. But this observer - as Maturana emphasizes - is himself a living system, and every explanation of cognition as a biological phenomenon has to contain an explanation of the observer and his role therein. In other words: the observer observes and explains cognition by observing and explaining living systems while being a living system and while exerting cognition.

Accordingly, in Maturana's theory we find a peculiar concatenation of three different types of theory: epistemology (observer), cognitive theory (cognition) and biological systems theory (autopoiesis). By linking these theories to an explanatory circle, Maturana hoped to come to a self-explanatory theory. He started to explain living systems as autopoietic systems that possess brains which, through self-interaction, produce cognition and, through self-description or language, develop the observer. This observer, then, is able to develop a theory of living systems as autopoietic systems that possess brains which, through self-interaction, produce cognition and, through self-description or language, develop the observer who, then, is able to develop a theory of living systems etc., etc. If this circle really worked it would yield the first complete theory in the history of philosophy and science, i.e. a theory that is able to explain its own logical, epistemological and empirical fundaments. And certainly, if it is true that both cognition and the observer are biological phenomena and that they are understood if and as soon as the autopoietic organization of living systems is fully understood, then we have indeed bridged the ontological gap between observer (mental states) and living system (physical states).

But how does one begin a description of this circular system of relations (in circular systems, beginning is always problematic)? Should we take the biological existence of the observer for granted and start with epistemological considerations about what the observer does, how he distinguishes objects from a background, identifies relations and properties, and finally specifies the organization of living beings, and himself as a living being? But then we would end up with the painful question about the ontological status of the statements about the organization of organisms. If they are observer-dependent, how can they answer the question about the »real« origin of the observer without running into a vicious circle? If they are observer-independent, how can an observer observe them at all, given the semantic closure of the cognitive system of the observer that is always emphasized by Maturana? Or should we, in contrast, take the epistemological conditions of the observer for granted and start with biological, even biophysical and biochemical investigations about how organisms are organized, how they interact with the environment, develop a brain as part of the autopoietic network that exerts cognition and, through self-description and language, generate the observer? But what epistemological and ontological status does, then, the description of the organization of living beings have? Are they statements about the »real« world? If this were the case, then the observer had access to this world, and this would again contradict the concept of semantic closure of the cognitive/observing system. But if they were not statements about the real world, how could we demonstrate the biological nature of the observer?

The true dilemma arises if we take both the biological and the epistemological status of the observer for granted and concentrate on the investigation of the nature of cognition. Do we, then, talk about cognition in an »objective« manner as resulting from the »real« organization of living organisms (as Maturana proposed) or in an observer-dependent manner? Is cognition a process belonging to the organism or to the observer? Does the observer result from the cognition of the organism? If yes, then the observer has to observe himself as emerging from the »real« organism. But if cognition results as an act of the observer, then we can say nothing about the emergence of cognition and of the observer from the »real« organism.

It becomes clear that by using the ingenuous circular relationship between organism, cognition and observer as proposed by Maturana, we run again and again into fundamental onto-epistemological problems. The idea of the circular relationship (or circular constitution) between organism, cognition and observer is a fatal idea because it is based on an ontological jump between *reality* and *actuality*, between the real biological system that has a brain and produces the observer, and the actual observer who is a cognitive system and describes biological systems in his phenomenal world. Those biological systems that are assumed to have brains that produce the observer do not exist within the ontological domain, *actuality*, in which the observer exists. They exist in the *reality* which is completely inaccessible to any observer (even though we can talk about this *reality* within *actuality*). In contrast, those organisms and brains which the observer can study and describe are not those organisms and brains that produce the observer. When I, as the above patient with an open skull, look at »my« brain, then this brain is, of course, not that brain that produces my phenomenal world, but it is a brain that is part of my phenomenal world. The »real« brain that is assumed to produce this phenomenal world including my own Ego does not appear in this phenomenal world.

The strongly seductive idea to establish a self-explanatory system by combining a theory of biological systems, a theory of cognition and a theory of the observer, led Maturana to become an objectivist despite all his attempts to demonstrate the semantic closure of the nervous system. This becomes evident in his writings when he specifies the »true«, i.e. observer-independent, nature of organisms (for example, the statement that organisms are mechanistic, purposeless, etc., systems). However, our phenomenal world is not transcendable; it is truly and completely closed, as Maturana rightly says.

Thus, we cannot reduce mind (observer) to matter/brain and explain how mind (observer) arises from matter/brain, because that brain and that matter that produces our mind (us as observers) is inaccessible to us, and that brain and that matter that is accessible is, as a part of *actuality*, not the producer of mind. The only thing we can do is to observe inside our phenomenal world the correlation between brain processes and behavioral reactions (including statements about mental states or self-experience in the case of a self-experiment). As mentioned above, this correlation seems to be very close. To what degree this holds for the relationship between the real brain and the mind that constitutes the phenomenal world in which we exist is a question that is unanswerable because the real world is truly and completely inaccessible. There are no words and no concepts by means of which we could describe a truly consciousness-independent world.

5. Conclusions

All we can say about the unity of the world and the unity of the description of this world refers to the phenomenal world. Thus, what we have proposed above concerns the unity of *actuality*. This world - the only world scientists can deal with - is neither reducible to a few basic phenomena and laws, as reductionism claims, nor does it consist of unbridgeable domains, as emergentism states. We propose a distinction between the possibility of a unified way of describing the phenomena of this world, and the possibility of a unified way of explaining it. There is the fact that at different levels of complexity systems show properties that as such are not reducible to the properties of their components and can be understood as being »emergent« properties. The reasons for such an emergence of properties result from the fact that properties and modes of interaction define each other mutually, and that whenever certain entities become components of a new system they will show new modes of interaction and, with this, new properties. To what degree these properties can be predicted ab initio from the properties of the entities prior to forming the system is a question of the complexity of the system and the amount of pre-knowledge. In many cases, especially with respect to biological systems, such an ab initio prediction is impossible, because the components, biomolecules, reveal their specific properties only or predominantly within the autopoietic network of the organisms. The only thing we could do (although even this is not possible at present) is to explain ex post how the different biomolecules interact in order to constitute life.

A similar situation is found with respect to the relationship between (actual) brain and mind. It is impossible to reduce mind to brain processes because brain processes as such are not mind, having no meaning as such. What meaning brain processes have depends on the functional organization of the brain, for example, its topography and the modes of interaction between neural networks. So, the whole system (or at least major subsystems of the brain) specifies the correlation between brain processes and meaning on the basis of the consequences of previous specification. What brain scientists can do is to analyze which permanent and actual properties of neural networks lead to specific mental states.

The language in which we describe the origin of life from biomolecules and the origin of mind from brain processes is the language of physics, i.e. a language that refers to processes of spatio-temporal interaction of tissues, cells and molecules. If we try to establish a unified description of the phenomenal world then there is no alternative to this language. This does not mean that we always have to use this language. It might be extremely cumbersome to describe behavior in terms of interaction of molecules. Many different level- and complexity-dependent languages are possible. It only means that all other languages used in biology have to be able to be translated in principle into the physical language. Accordingly, no concepts and terms are allowed that are not translatable into physical terms.

Also, this physicalism does not imply that everything can be reduced to physical laws. We believe that truly biological laws exist, for example, evolutionary laws, that can be expressed in physical terms without being reducible to physical laws. Our physicalism, thus, is a non-reductionist physicalism (we could also say, a physicalist emergentism, as opposed to a holistic emergentism), because the causes for law-like processes are not necessarily found at lower levels.

So, we believe in a unity of description of the phenomenal world that includes the acknowledgement of different levels of complexity of systems. The deep reason for the difficulties that many philosophers of science have with an understanding of the complexity of the world, which leads them either to reductionism or to a holistic emergentism, is the adherence to a substantialism that has dominated philosophy and science for more than two thousand years: the belief that objects have a true nature (substantia) and a true identity and possess properties (accidentia) that can change without a change in the substance of the objects. Any profound change of objects, then, either has to be denied (as does reductionism), or has to be considered inexplicable (as does emergentism). Our alternative view is that the properties of the objects and processes are constituted by interaction, which allows any kind of smooth or dramatic change of properties, and that there is no »true« nature of objects beyond these interactions.

Notes and References

1. E. Jantsch, »The Unifying Paradigm Behind Autopoiesis, Dissipative Structures, Hyper- and Ultra-cycles, in M. Zeleny (ed.), *Autopoiesis. Dissipative Structures, and Spontaneous Social Orders*, Boulder, Colorado: Westview Press, 1980, 81-87; *Die Selbstorganisation des Universums*, Munich: dtv, 1982.

2. H.R. Maturana, *Erkennen: Die Organisation und Verkörperung von Wirklichkeit*, Braunschweig: Vieweg, 1982.

3. E. Mayr, *The Growth of Biological Thought*, Cambridge, Massachusetts: Harvard University Press, 1982; *Die Entwicklung der biologischen Gedankenwelt*, Berlin, Heidelberg, New York, Tokyo: Springer-Verlag, 1984; »How Biology Differs from the Physical Sciences«, in D.J. Depew and B.H. Weber (eds.), *Evolution at a Crossroads: The New Biology and the New Philosophy of Science*, Cambridge, Massachusetts: MIT Press, 1985.

4. See F.J. Ayala, T. Dobzhansky (eds.), *Studies in the Philosophy of Biology*, Berkeley and Los Angeles: University of California Press, 1974.

5. Mayr, *op. cit.*, 1982, note 3; *op. cit.*, 1985, note 3.

6. Mayr, *op. cit.*, 1985, note 3.

7. See U. an der Heiden, G. Roth, and H. Schwegler, »Principles of Self-Generation and Self-Maintenance«, *Acta Biotheoretica* **34** (1985), 125-138; »Die Organisation der Organismen. Selbstherstellung und Selbsterhaltung«, *Funkt. Biol. Med.* **5** (1986), 330-346.

8. G. Roth, »Selbstorganisation-Selbsterhaltung- Selbstreferentialität: Prinzipien der Organisation der Lebewesen und ihre Folgen für die Beziehung zwischen Organismus und Umwelt«, in A. Dress, H. Hendrichs, and G. Küppers (eds.), *Selbstorganisation. Die Entstehung von Ordnung in Natur und Gesellschaft*, Munich, Zurich: Piper, 1986.

9. H. Gardner, *The Mind's New Science*, New York: Basic Books, 1985; P.S. Churchland, *Neurophilosophy*, Cambridge, Massachusetts: MIT Press, 1986.

10. H. Feigl, »The ›mental‹ and the ›physical‹«, in H. Feigl, M. Scriven, and G. Maxwell (eds.), *Minnesota Studies in the Philosophy of Science, Vol. II: Concepts, Theories, and the Mind-Body Problem*, Minneapolis: University of Minnesota Press, 1967, p. 370-497.

11. For example, a combination of PET-scan tomography and NMR technique, see M.E. Raichle, »Neuroimaging«, *Trends in Neurosciences* **9** (1986), 525-529.

12. O. Creutzfeldt, *Cortex Cerebri*, Berlin, Heidelberg, New York: Springer-Verlag, 1983; P. Rakic and W. Singer, *Neurobiology of Neocortex*, Chichester, New York, Brisbane, Toronto, Singapore: Wiley, 1988.

13. H.R. Maturana, *Biology of Cognition*, BCL, Urbana, Illinois: Department of Electrical Engineering, University of Illinois, 1970; *op. cit.*, 1982, note 2.

On a Fundamental Paradigm Shift in the Natural Sciences

BERND-OLAF KÜPPERS
Max Planck Institut für biophysikalische Chemie, Göttingen

1. There are in principle two ways of elucidating the conditions under which a branch of science is pursued. One way is to examine the structures to which this science has given rise; by »structures« I mean the products of scientific discovery which find their expression in hypotheses, models and theories about our world. The other way is to analyze the scientific process itself, the process represented in the activity of an individual scientist, in that of a group of scientists, and ultimately in the overall development of the branch of science in question.

The approach to science by structural analysis is necessarily incomplete, since it is based more or less on the momentary state of the science in question. The approach by analysis of process, on the other hand, is impeded by all the imponderabilia attached to historical reconstruction. Only a combination of the structure-analytical and process-analytical procedures, in which the two aspects complement one another, will provide us with a basis sufficient for the philosophical assessment and evaluation of science. So both procedures have their rightful place in the philosophy of science, and I shall appeal to both in the course of this lecture.

It is well known that science does not develop in the form of a steady accumulation of knowledge, but rather in leaps, as an irregular series of discoveries. Such bursts of development can take place within relatively short periods. Since they are most often associated with profound conceptual changes, such instances provide us with the possibility of examining directly the basic features of the development of science, and also with the opportunity to obtain fundamental insight into the assembly of structures in science. Such leaps or bursts of development, which are accompanied by a transformation of the inner constitution of a branch of science, are known to philosophers of science as »paradigm shifts«.

Today I wish to discuss a paradigm shift that is at present taking place in physics and in its neighboring disciplines such as physical chemistry and biophysics, and which may come to change our scientific world-view in a fundamental way. Perhaps one might ask whether a paradigm shift in the natural sciences is not fundamental by definition, implying that the title of this lecture conceals a philosophical pleonasm. In fact, I chose deliberately the adjective »fundamental«, because I wanted to make it clear that a paradigm shift of the kind in question has never before appeared in science and - as we shall see in detail - can never appear again. We are thus experiencing a singular phenomenon in the development of science, for which I can think of no better description than »fundamental«.

I must also confess that I am not sure whether the paradigm shift under discussion is confined to physics or to the natural sciences. There are indeed strong indications that the humanities are also affected by this paradigm shift. Had I fewer scruples, I might have entitled this lecture »On a paradigm shift in the sciences« - including the human sciences. In this case the word »fundamental« would presumably have been superfluous.

W. Krohn et al. (eds.), Selforganization. Portrait of a Scientific Revolution, 51–63.

In brief, I shall discuss a paradigm shift that, while having its roots in physics, manifests itself above all in interdisciplinary connections. It concerns the so-called »paradigm of self-organization« and its application in wide areas of the natural sciences, along with first indications of a possible application in the humanities.

Before discussing the paradigm of self-organization and its associated conceptional change in the natural sciences, I should like to make a brief comment on the term »paradigm«. It is well known that this term, as introduced into the philosophy of science by Thomas Kuhn, stands for a broad spectrum of characteristics that describe the collective behavior of scientists within a particular discipline of research.[1] For example, the decision of what field of problems shall be regarded as important or relevant is made within the context of a paradigm. Similarly, a paradigm is used to decide which methods of investigation and which solutions can be regarded as permissible. As a rule, a paradigm includes a generally-accepted theoretical foundation. This foundation is not necessarily a full-fledged theory, with a complete set of definitions and axioms; it can well have the character of a model or a phenomenological approach. Moreover, it is clear that an observation can never be absolutely independent of theory. Thus a paradigm is always one of the factors that determine the manner and the form in which a phenomenon is perceived and observed. In other words, the objects of any scientific investigation are to a high degree specified by a pre-existing paradigm.

In these circumstances there is little purpose in trying to define a paradigm shift by one of the criteria mentioned, especially as the various criteria of a paradigm clearly influence each other. Neither is it quite correct to speak of a »shift« of paradigm in the case under discussion. This is because we are not here abandoning an old and out-worn paradigm in favor of a new one; rather, we are removing a class of problems from an old paradigm and making it the subject of a new one, the paradigm of self-organization. This may turn out to be a meaningless distinction, since the two paradigms are fundamentally incommensurable and can therefore only be compared with one another to a very restricted extent. In addition to this, the paradigm of self-organization embodies the further difficulty that it does not yet contain a unified theoretical nucleus. We cannot yet speak of *the* theory of self-organization. The theoretical nucleus of the paradigm of self-organization is at present a conglomeration of widely differing theoretical approaches, and it remains to be seen whether and to what extent this nucleus will be capable of yielding a unified theory.

2. In the face of these rather blurred issues, I should like at first to concentrate upon the superficial appearance of the paradigm shift in question and then to pass on to its deeper structure. I would like to set out from the question of what the conceptual change in the natural sciences that has led to the paradigm of self-organization actually is.

It is a well-known fact that the basic disciplines of natural science, namely physics, chemistry and biology, have in recent years begun to overlap to an ever greater extent. Biologists are making increasing use of physical and chemical methods of investigation, while the phenomena of living nature have moved steadily towards the center of current debate in fundamental physics. The convergence of these disciplines is concomitant with a shift of concept within each of them. For example, biology is losing its original status as a descriptive and classificatory science, as molecular biology gradually becomes the basis of its explanatory models. Meanwhile, in physics, singular and irreversible phenomena are gaining in significance at the expense of reproducible and reversible phenomena. In a revolutionary change of attitude, physics no longer excludes the historical aspects of nature but now claims these as part of its realm of applicability.

The history of nature manifests itself in the irreversibility of spontaneous, natural processes. In open systems, that is, systems that exchange both energy and matter with their surroundings, such processes can lead to complex ordered states of matter. The highest states of order known to us are living systems and their collective superstructures, the ecosystems. Their irreversible, complexity-generating, natural processes are denoted by the general term »self-organization«. This gives us a first indication of which category of phenomena is embraced by the paradigm of self-organization: it is that of the complex and irreversible natural phenomena.

These developments in physics have led to substantive changes in the concept of Nature that has lain at the root of modern science. In its traditional form, physics has embraced only reversible phenomena. It is true that a universal empirical principle was formulated in the last century, in the form of the second law of thermodynamics, which expressed the irreversibility of natural events. However, repeated attempts to base this principle upon the laws of statistical mechanics have shown how deeply physical thinking is rooted in the concept of reversibility. The assertion that a process runs reversibly means that the system under observation is in a state of equilibrium at all times, so that the process is a passage through an infinite sequence of equilibrium states, each infinitesimally distant from the next. This in turn means that a process can only be reversible if the changes it necessitates proceed infinitely slowly. Reversibility is thus an idealized concept. In other words, reversible processes do not occur in the real world.

The conception of nature associated with the phenomenon of reversibility is that of material equilibrium. Its basis is the picture of a static and closed nature.[2] Or, expressed in a simple slogan, traditional physics is a »physics of being«. It is the physics of a reality symmetric with respect to time, from which all historical processes have been filtered out. In contrast to this, the paradigm of self-organization has established a picture of nature in which nature no longer appears as a huge equilibrium structure, a static assemblage, but instead as a non-equilibrium structure in all its dynamic variety. Real, that is to say irreversible and complexity-generating, processes are now becoming the center of attention. Again in a simple slogan, the present paradigm shift can be characterized as a transition from a »physics of being« to a »physics of becoming«.

At this point a brief observation from the history of science is in place. I have just outlined that the conceptional change which physics is now undergoing was set in motion by the opening-up of a new area of phenomena, above all the complex phenomena of living matter. A similar situation has existed once before in the history of science, at the turn of the nineteenth century.

The close of the eighteenth century was accompanied by two important scientific achievements: the displacement of the previously dominant phlogiston theory of chemistry by the modern theory of combustion and the discovery of galvanic electricity. In the wake of these discoveries, the fundamental scientific disciplines moved closer together. Typical processes of life, such as metabolism or nerve excitation and nerve conduction, now seemed as though they could be explained by the new paradigms of physics and chemistry.

At that time, the young Schelling, acquainted with these scientific discoveries, developed a novel natural philosophy, which conceptually is in many respects astonishingly similar to our present-day picture of nature that lies behind the paradigm of self-organization. Schelling, too, viewed nature as a dynamic, self-organizing structure.[3] He too underlined the importance of non-equilibrium processes for the explanation of natural processes. And, for Schelling as for the modern scientist, the unity of nature became apparent in the history of nature.

Independently of the fact that the real roots of Schelling's natural philosophy are found in the early phase of German idealism, one must still ask why it was that Schelling's ideas had no influence upon the conceptual foundations of physics and chemistry, especially as his ideas were distinctly topical and Schelling was widely read by natural scientists of his time.[4] There appear to be two decisive reasons for this. For the first part, the physics of the eighteenth century was still dominated by the mechanistic conception of Nature, the sweeping success of which was clearly visible in the growing industrialization of society. For the second, Schelling himself saw his natural philosophy not as a higher form of empirical science but rather as a »new organ for observing and understanding nature« that had no conceptual common ground with the empirical natural science of his day.[5]

I believe that the natural philosophy of Schelling gives us an impressive example of how the fundamental conceptual changes in science always come from within and not from outside influences. The biological problems that were in those days accessible to physical research methods found within equilibrium physics solutions that - judged by the norms of the time - were satisfactory; the question of conceptual re-orientation was therefore never raised. This is not the place for further discussion of this philosophically and historically interesting question, but its topicality in the present context is evident.

3. For a long time, the investigation of complex phenomena by physicists was characterized by the expectation that new, fundamental laws of physics might be found in this area.[6] However, these expectations have not been realized, and the prospect of the discovery of such laws is dwindling from day to day. If we therefore conclude the converse, that is, that even complex phenomena must be explicable on the sole basis of the known laws of physics, then we must seek new standpoints that reveal the complexity of a phenomenon.

These standpoints become clear in outline when we apply our attention to the logical structure of physical explanatory models. It is well known that physical theories possess a dual structure. They contain both general statements, in the form of laws with at least a local validity, and so-called initial or boundary conditions, which are not fixed by the theory but are specified in the area of applicability of the theory. The initial conditions are - mathematically speaking - the selection conditions of solutions actually present to differential equations that describe the physical laws.

I shall now attempt to justify the central assertion of this lecture which is that the complexity of natural phenomena lies in the complexity of their boundary conditions and not in the complexity of the natural laws to which the phenomena are subject. The term »boundary conditions« is understood in the physical sense just described. As we shall see, the boundary conditions constitute the conditional complex for a natural phenomenon. For the sake of clear terminology, I shall henceforth refer to conditional complexes in general as the »system« and to the natural processes thereby induced as the »system dynamics«. Thus, the important characteristics of such systems lie in their boundaries. If we can understand the boundaries and their development in the course of time then, at the same time, we understand complex systems and their dynamics.

To the best of my knowledge the first author to emphasize the significance of boundaries or boundary conditions for an understanding of complex phenomena was Michael Polanyi.[7] Although his conclusions - in particular with regard to biology - ultimately proved to be erroneous, it will serve our purpose well to follow his analysis for some of the way.

Consider first a complex system made by human hand, for example a sophisticated machine. We can use various criteria to characterize this system. First, we would state the purpose of the machine and explain how it functions. The purpose that the machine is intended to fulfil is dictated by human goals. Its mechanism, in contrast, is dictated by the obedience of matter to natural laws. We cannot construct a machine that functions in a manner contradictory to natural laws. There is, for example, no *perpetuum mobile* that can suspend the law of conservation of energy, or the second law of thermodynamics. A particular purpose, however, can often be achieved by several different mechanisms.

The function of a machine is laid down by the manner of its construction. This includes both the driving force, that is, the source from which the machine draws its energy, and the spatial structure of the machine. Let us ignore the energy source for a moment. The function of the machine is then determined unambiguously by the individual components, their form, and the way in which they are put together. We can say that the form of the components and the relationships between them represent a boundary condition under which natural laws become operational in the machine. The functional purpose that the machine serves is thus realized under specific boundary conditions. These must within certain limits remain stable if the functioning is to be sustained. Every machine is based upon a dualistic principle of this kind, that is, defined boundary conditions and the processes that are induced and constrained by them. The boundary conditions represent a selection condition which restricts the processes allowed by natural law that constitute the function of the machine.

I have chosen the example of the machine, above all, because it focusses attention upon the fundamental significance of the boundary conditions for our understanding of complex systems. This is of particular value here because the example of the machine can be transferred to the archetype of the complex system, the biological organism. This, too, behaves in many respects in a manner identical to that of a complex machine. It is not without significance that an entire field of research, cybernetics, has its origin in the machine theory of the organism. The organism is subservient to the manner in which it is constructed, whereby, as in the case of the machine, the functional order of all the parts of the organism in question is preordained. The principle of construction again represents a boundary condition under which the laws of physics and chemistry become operational in a system. In the case of the living system, the processes that occur under these boundary conditions are organized in such a way that the system is reproductively self-sustaining.

Clearly, it is the special structure of the boundary conditions that determines the overall construction of the system, both for the machine and for the living organism. We must therefore ask how, in turn, the structure of the boundary conditions is determined. In the case of the machine this is obvious. The boundary conditions are defined in the plans drawn up by the constructor of the machine. There also exists such a plan for living organisms. However, it resides in the DNA molecules of the organisms itself and - in contrast to the machine - it is part of the system. If a DNA molecule is placed in a suitable physico-chemical environment, such as an egg cell, then expression on the phenotypic level results; that is, the construction plans enciphered in the molecule are realized in matter, step by step. Thus, the structures of all the boundary conditions that appear in the complete, differentiated organism are laid down in the DNA molecule. The DNA molecule therefore represents a *primary* boundary condition which, in turn, determines all other boundary conditions in the living organism. A primary boundary condition of this kind we can also call an initial condition.

These arguments underline the importance of the concept of boundary conditions for the understanding of complex phenomena. However, up to now they have been based upon a plausibility argument. We must now crystallize the term »boundary condition« so as to bring it into line with the concept of boundary conditions in physics.

It will suit our purpose to start by considering the primary boundary conditions that lie behind the order present in living matter. As noted above, these reside in particular physical structures, the DNA molecules. These are long chain-like molecules built up from a single class of molecular building-blocks (nucleotides). All physical and chemical processes in the organism are ultimately based upon a structure of this kind. The structure of the DNA molecule is a genuine boundary condition in the physical sense. It is specified by stating the co-ordinates in space of all the atoms in the structure of the molecule, along with other physico-chemical environmental conditions such as the temperature, the ionic strength and so on. The structure represents a precise set of physical parameters which, in turn, determines all the physical and chemical processes that take place in the system.

However, these boundary conditions differ in one important point from the boundary conditions of simple physical systems. The dynamics of a complex system depends very sensitively upon its boundary conditions. Even the smallest changes in these can lead to the instability of the system. Thus, in living systems, even the exchange of a single nucleotide can lead to instability and collapse of the system. Simple physical systems behave differently in this respect. In a simple physical system, the nature of the physical processes induced by the boundary conditions will not change fundamentally if the boundary conditions are altered. For example, a body in free fall will always show similar physical behavior irrespective of the boundary conditions (e.g., various initial conditions). The boundary conditions of simple physical systems, such as the position and momentum of a body at a particular moment, are contingent quantities. »Contingent« means in this context that the boundary conditions can be selected arbitrarily; they are marginal quantities that may be chosen freely in an experiment and the values of which are restricted only by the range of validity of the natural laws in question.

The boundary conditions of complex systems, on the other hand, are not contingent quantities, since any significant alteration in these boundary conditions would also change significantly the dynamics of the system. In addition, the boundary conditions are usually unique in that they as a rule represent a restricted selection from a gigantic number of physically equivalent alternatives. For example, on the relevant structural level, the number of alternative structures available to the DNA molecule of a bacterial cell is no less than $10^{600\text{million}}$.[8]

So far, I have classified complex and simple systems according to the nature of their boundary conditions. Complex systems are associated with non-contingent boundary conditions and simple systems with contingent boundary conditions. The transition from one of these classes to the other is naturally a gradual one, in as much as the concept of contingency itself allows this. The concept of contingency is the logical antithesis of that of law, and therefore leaves open the question of how much a phenomenon is unrestrained by natural law. I wish now to go one step further and to use this classificatory characteristic as the basis for a definition. According to this definition, the complexity of a system will be given by the degree of non-contingency of its boundary conditions.[9]

4. After this exposition of the concept of boundary conditions, I shall now use it in an attempt to obtain further insight into the deeper structure of the paradigm shift in question. I will show that the concept of boundary conditions has given us a key to the understanding of the paradigm shift by way of theory dynamics.

The conquest of new phenomena by a branch of science results inevitably in changes in the theoretical nucleus of the branch of science in question. This is because a novel class of phenomena requires a new set of terms and concepts in order to describe it. However, terms and concepts only gain a precise meaning within a framework of wide knowledge and, ultimately, of a theory. The decisive step in a paradigm shift is always the expansion or the rebuilding of a basic theoretical structure.

The theoretical rearrangement associated with a paradigm shift can take place at two very different levels. It can, for example, be a change consisting in the modification and/or expansion of a basic theoretical frame. It might alternatively be that the basic theoretical frame requires no modification or expansion but merely an internal shift of emphasis. The former will always be the case when recurrent anomalies of explanation bring a branch of science into an internal crisis, one that can be resolved only by a radical rebuilding of its theoretical apparatus. It was doubtless such instances that Kuhn had in mind when he spoke of scientific revolutions occurring from time to time.[10] The latter case, that of internal shifts of emphasis, appears at first sight to be of less philosophical interest. However, we shall see that precisely this seemingly uninteresting case forms the basis for the paradigm shift under discussion here, and that its consequences will be in no way less revolutionary than those of Kuhnian scientific revolutions.

In order to see this, we must first express more precisely the idea of an internal shift of emphasis. I have already indicated that this shift of emphasis need not necessarily refer to the fundamental laws of physics. What, then, does this concept actually embrace? It clearly means more than the mere extension of interest from simple to complex phenomena, as already described. That would concern the range of applicability of the theory but not the theory itself. We must rather assume that the internal shift of emphasis concerns the inherent explanatory structure of the paradigm itself. We have therefore no choice but to draw the concept of explanation into our discussion.

In physics it is customary to say that there is a scientific explanation for a class of natural events if a general law can be formulated that allows the appearance of such an event to be deduced from its boundary conditions, for example, in an experiment in physics, nature is subjected to certain restrictions in order to observe its behavior under these restrictions. By variation or by narrowing or widening the experimental boundary conditions, the law-like behavior of matter can be detailed.

It will aid our discussion to formulate more exactly the concept of scientific explanation from a theoretical perspective. The most appropriate way to do this, especially in connection with explanations in physics, is the explanatory schema suggested by Carl Gustav Hempel and Paul Oppenheim within the framework of logical empiricism.[11] The Hempel-Oppenheim model of scientific explanation is based fundamentally upon two classes of statements:

(1) the statement $A_1,\ldots,A_m$, which describe the so-called antecedent conditions that precede the event to be explained or explanandum, and

(2) the statements $G_1,\ldots,G_n$, which describe the general laws that lie behind the explanandum.

According to the Hempel-Oppenheim schema, the actual process of explanation consists in the logical deduction of the explanandum from the premisses ($A_1,\ldots,A_m$, $G_1,\ldots,G_n$), which together make up the so-called explanans. For the deductive step from the explanans to the explanandum to be logically necessary, the antecedent conditions must be complete, that is, they must be not only necessary but also sufficient conditions for the occurrence of the event to be explained. If the conditional complex is incomplete, we refer to the explanandum as a random event. The deductive step from the explanans to the explanandum is then no longer logically necessary and the event occurs with a certain probability.

The antecedent conditions are therefore nothing other than the boundary conditions constituting the conditional complex for a natural phenomenon. In the traditional models of physical explanation the boundary conditions are irreducible premisses for formulated natural laws. They represent the physical quantities that express the contingency of the natural event. In contrast, in the context of the paradigm of self-organization, the boundary conditions of complex phenomena are regarded as non-contingent quantities; because of their origin under non-equilibrium conditions, these in turn are viewed as the result of an irreversible process of development, *i.e.*, of self-organization. In contrast to the traditional physical models of explanation, the boundary conditions must now themselves be explained so that they move into the center of the explanation.

This has brought us to the decisive characteristic feature of the shift of emphasis within theories, a shift that is in this case inseparably linked to the paradigm shift itself: the contingency of natural processes, previously firmly anchored in the contingency of the boundary conditions of a natural phenomenon now itself becomes the object of scientific explanation in the new paradigm. However, this does not mean that the contingent quantities can in this way be eliminated completely from our model of scientific explanation. For, when the boundary conditions of a system or process become the explanandum, this in turn presupposes the existence of other boundary conditions on a higher level of explanation. The resulting iteration either continues in an indefinite recursion or else breaks down when a level is reached on which the boundary conditions are no longer explicable and thus adopt the status of classical contingent initial conditions. In either case, this recursive explanatory procedure will considerably reduce the number of contingent phenomena. This is exemplified well by the theory of the self-organization of living systems which represents an attempt to explain the origin of non-contingent boundary conditions, *viz.* information-carrying DNA molecules, by appeal to contingent initial conditions, *viz.* random molecular structures.

The decisive step leading to the new paradigm of self-organization consists in the recursive shift of the explanandum to the boundary conditions. It is unique and thus represents a »fundamental« paradigm shift in physics. The physics of self-organization is still based upon the basic explanatory model as formalized by Hempel and Oppenheim. But now a whole series of explanatory steps are linked consecutively together so that one now can speak of the nested or hierarchically-structured Hempel-Oppenheim model. In this way, the hierarchically-organized structures and processes of living matter are directly reflected in the hierarchical explanatory structure of models of self-organization. This puts physics into a position to realize in a novel way its inner goal of demarcating the region of contingent phenomena and of tracing it back, by logical inference, to the law-like behavior of nature.

5. I should like finally to raise the question of what this paradigm shift implies in connection with the presumed unity of nature. I shall discuss this question first in the systematic context of the unity of the natural sciences and then in the context of a possible approach between the natural sciences and the humanities.

First of all we should clarify the question of what is to be understood by the systematic unity of science. The unity of the natural sciences is seen, for example, in the unity of their object: nature. The idea of unity can also be understood programmatically, that is - as Carl Friedrich von Weizsäcker has expressed it - as the hope that from the unity of the object a unity of the theory will emerge.[12]

According to Weizsäcker's ideas, for such a universal theory of natural phenomena - if it existed at all - to be complete and self-contained it must formulate the conditions of the possibility of experience, a conception in which physics is interpreted radically as a science of experience. But let us now leave aside the question of how the picture of a complete and self-contained natural science can be imagined. At any rate, it seems certain that a unity of science can only be attained by way of its theoretical ideas. One way of reaching this goal consists *inter alia* in tracing the multitude of empirical and semi-empirical laws back to a few universal principles. This has been achieved, for example, in the reduction of Kepler's laws to the Newtonian law of gravity, and the reduction of the empirical gas laws such as Boyle's law and Charles' Law to the more fundamental kinetic theory of gases.

The process of reduction of a special law to a general one possesses in itself the form of a scientific explanation. The special law, the explanandum, is deduced with the help of the more general law from certain premisses, which make up the antecedent condition.[13] Thus, as is well known, Kepler's laws follow from Newton's law of gravity, under the presupposition that the mass of the Sun is very large compared with that of the planets and that the planets do not approach each other too closely. These presuppositions, necessary as they are for the deduction of Kepler's laws, are in respect of physical reality contingent circumstances. They cannot be deduced from Kepler's laws themselves. In fact, they restrict the range of applicability of these laws. This shows that the term »contingency« always depends upon a reference system.

One might suppose contingency to occupy a central place in the relation between special and general laws. Seen from the perspective of the paradigm of self-organization, however, this question appears in a new light. This is because it cannot be excluded that the contingent subsidiary conditions in a statement of natural law may in some degree themselves be explicable by natural law. Indeed, first attempts have already been made to explain classical contingent quantities, such as the numerical value of the charge of the electron, from a theory of self-organization of matter.[14] Should such attempts prove successful, this would have the profoundest consequences for the problem of reduction of theories.

So much for the question of the contribution which the paradigm of self-organization to a unity of the natural sciences may one day be able to make. But how does the paradigm shift affect the relationship, already mentioned, between the natural sciences and the humanities? Will the new paradigm perhaps lead to a narrowing of the gap between them? These questions, which are the subject of increasing interest, confront us as usual with the task of first characterizing adequately these two currents of thinking and distinguishing them from one another.

A traditional position in relation to the demarcation between the natural sciences and the humanities, and one that seems of value for our discussion, was developed within the framework of neo-Kantianism. Wilhelm Windelband, in his famous rector's speech at Strasbourg in 1894 on »History and Natural Science«, introduced the pair of terms

»nomothetic« and »idiographic« in order to characterize the various thought patterns of the natural sciences and the humanities.[15] Using the example of psychology, he tried to demonstrate the differing goals of the two. This idea was later taken up by Heinrich Rickert and elaborated to a theory of the relationship between - as he called it - the natural and the cultural sciences which he then attempted to characterize by the terms »generalizing« and »individualizing«.[16]

Without entering more deeply into the complex problems of demarcation, we can exploit the pair of terms suggested by Rickert as a guide for our discussion. It has indeed hitherto been taken as a typical distinguishing feature that the natural sciences, as a generalizing branch of thought, are concerned exclusively with general, that is, reproducible phenomena, while the humanities attempt to understand individual, that is, unique and unreproduced phenomena. The observation that the new paradigm of self-organization now places the history of nature, and therewith unique and irreproducible phenomena in the center of scientific explanatory models, reveals the nature of the approach of the natural sciences and the humanities: the movement of interest within natural science from the general and the reproducible to the unique and the irreproducible. But in this connection, too, the question of greatest importance is whether indeed from the unity of the object a unity of the theory can be made.

Here we run up against a fundamental problem, for, as I have mentioned, the present state of the paradigm of self-organization is characterized above all by the great heterogeneity of the individual theoretical approaches. However, it does seem - if the ideas presented here are correct - that the various theoretical approaches will be capable of forming a unified theory that I should like to call the theory of boundary conditions. This theory, I believe, provides the approach to a new kind of structural science that occupies a middle position between the natural sciences and the humanities and which will be the means of the mutual approach of these two branches of thinking.

We are all acquainted with other examples of structural sciences that play a similar role, for example, mathematics, information theory, game theory and so forth. Such structural sciences belong neither to the natural sciences nor to the humanities.[17] The structural sciences are concerned with »structures *in abstracto*«, that is, structures independent of the question of which phenomena are associated with particular structures, or indeed of whether there exist at all phenomena with such structures. Weizsäcker has expressed the conjecture that progress within the structural sciences is perhaps the fastest and the most radical.[18] Let us ignore for the present the question of the rate of development of the structural sciences. If the paradigm of self-organization should really provide the means of approach to a new structural science then the progress thus made would indeed be radical. For a theory of boundary conditions would be the theory of irreversible and complexity-generating natural processes; as a structural science, it would be the most general of all theories of historical processes. As paradoxical as it may sound, this theory would describe the course of its own development. Not least in the anticipation of this thought lie the foundations of the establishment of models for the self-organization of science and research as attempted in recent years by sociologists of science.

Notes and References

* This paper was read in Bielefeld on May 31st, 1988, as a lecture in the series »Philosophy of Science«.

1. T.S. Kuhn, *The Structure of Scientific Revolutions*, second edition, Chicago: University of Chicago Press, 1970.

2. The term »static« refers here to the picture of Nature behind traditional physics. Quite obviously, a single movement, for example as described in classical mechanics, cannot be called static. However, the basic laws of mechanics are symmetric with respect to time, so that the historicity of Nature is not subsumed under the concepts of traditional physics.

3. Particularly revealing in this connection is F.W.J. Schelling, »Von der Weltseele, eine Hypothese der höheren Physik zur Erklärung des allgemeinen Organismus«, in K.F.A. Schelling (ed.), *Fr.W.J. v. Schellings sämmtliche Werke*, I. Abt., Bd. 2, Stuttgart: Cotta, 1857.

4. For the effect of the German natural philosophy upon the development of the natural sciences, see D. von Engelhardt, »Prinzipien und Ziele der Naturphilosophie Schellings - Situation um 1800 und spätere Wirkungsgeschichte«, in L. Hasler (ed.), *Schelling. Seine Bedeutung für eine Philosophie der Natur*, Stuttgart-Bad Cannstatt: Fromman Holzboog, 1981.

5. F.W.J. von Schelling, »Ideen zu einer Philosophie der Natur als Einleitung in das Studium dieser Wissenschaft«, in K.F.A. Schelling (ed.), *Fr.W.J. v. Schellings sämmtliche Werke*, I. Abt. Bd. 2, Stuttgart: Cotta, 1857.

6. Compare B.-O. Küppers (ed.), *Leben = Physik + Chemie? Das Lebendige aus der Sicht bedeutender Physiker*, Munich: Piper, 1987.

7. M. Polanyi, »Life's Irreducible Structure«, *Science* **160** (1968), 1308-1312.

8. A sequence consisting of n nucleotides, where there are ß classes of nucleotide, possesses $ß^n$ combinatorially possible sequences. For the sequence of a bacterial genome: ß = 4 and $n = 4 \times 10^6$.

9. It is worth noting that this definition would fit exactly for the so-called chaotic systems that are currently the object of intensive research effort in physics, and which we may regard as complex systems *par excellence*.

10. Kuhn, *op. cit.* 1970, note 1.

11. C.G. Hempel, P. Oppenheim, »Studies in the Logic of Explanation«, *Philosophy of Science* **15**, (1948), 135-175.

12. C.F. von Weizsäcker, *Die Einheit der Natur*, Munich: Hanser, 1971. English translation *The Unity of Nature*, New York: Farrar, Straus and Girous, 1980.

13. Compare E. Scheibe, »Gibt es eine Annäherung der Naturwissenschaften an die Geisteswissenschaften?«, *Universitas* **42** (1987), 5-18.

14. These considerations, which are highly speculative and have not yet been made available in published form, were initiated by H.-J. Queisser. Further details will be found in M. Eigen, »The Physics of Molecular Evolution«, *Chemica Scripta* **26B** (1986), 13-26.

15. W. Windelband, *Präludien*, Tübingen: Mohr, [9]1924.

16. H. Rickert, *Kulturwissenschaft und Naturwissenschaft*, Tübingen: Mohr, [2]1910.

17. Weizsäcker, *op. cit.* 1971, note 12.

18. Weizsäcker, *op. cit.* 1971, note 12.

The Cognitive Program of Constructivism and a Reality that Remains Unknown

NIKLAS LUHMANN
University of Bielefeld

I.

Interest in epistemological questions is not limited to philosophy today. Numerous empirical sciences have, in the normal course of their research, been forced to proceed from the immediate object of their research to questions involving cognition. Quantum physics is perhaps the best-known example, but it is no exception. In linguistics the question is raised today of what problems arise from the fact that research into language has to make use of language. Cognitive instruments have to be aquired via the object investigated by means of these very instruments and not, for example, through reflection of consciousness upon itself.[1] Brain research has shown that the brain is not able to maintain any contact with the outer world on the level of its own operations, but - from the perspective of information - operates closed in upon itself. This is obviously also true for the brains of those engaged in brain research. How does one come, then, from one brain to another? Or to take a further example: the sociology of knowledge had demonstrated at least the influence of social factors on all knowledge, if not their role as sole determinants. This is also true, then, for this statement itself since no justification for an exception can be found, in the sense, say, of Mannheim's »free-floating intelligence«. What conclusion is to be drawn from this? It was thought that one would have to found all knowledge on »convention«[2] or that knowledge was the result of a kind of »negotiation«.[3] But these attempts only wound up designating an ancient problem - that of the unity of knowledge and reality - by means of a new concept. Not without reason have these attempts been criticized for epistemological naiveté[4], since one either learns nothing about the relationship to reality or the connection is only made over theoretically unacceptable »both/and« concessions. There is little more to be gained by calling such »constructivism«, as has recently been done, »radical«[5] since what is identified here as »constructivism« hardly at first seems unfamiliar. It might be that the theory of knowledge - at least in some of its traditional variants - will be confirmed rather than caught unaware. Science is apparently reacting here to its own power of resolution. This can already be found in Plato who reduces everyday experience to mere opinion and raises the question of what reality lies behind it. As a result, these philosophic reflections were termed, at first, »idealism«. As we come to modern times the emergence of modern science led more and more to the conclusion that this »underlying« reality was knowledge itself. This altered the meaning of the concept of the subject, while it is only in our century that the name »idealism« has been replaced by »constructivism«. There was a shift in emphasis in the conflict between realism and idealism, but it is not easy to discover in this a new theory. There is an external world, which results from the fact that cognition, as a self-operated operation, can be carried out at all, but we have no direct contact with it. Without knowing, cognition could not

W. Krohn et al. (eds.), Selforganization. Portrait of a Scientific Revolution, 64–85.

reach the external world. In other words, knowing is only a self-referential process. Knowledge can only know itself, although it can - as if out of the corner of its eye - determine that this is only possible if there is more than only cognition. Cognition deals with an external world that remains unknown and has to, as a result, come to see that it cannot see what it cannot see.

So far there is nothing new here, unless it be in the definiteness and self-confidence with which all this is presented as knowledge. One has to look more closely at the theoretical distinctions with which this view of things is presented in order to discover something new. Insofar as constructivism maintains nothing more than the unapproachability of the external world »in itself« and the closure of knowing - without yielding, at any rate, to the old skeptical or »solipsistic« doubt that an external world exists at all - there is nothing new to be found in it. Nonetheless, the theoretical form in which this is expressed has innovative aspects - even such radical innovations - that it is possible to gain the impression that the theory of a self-referring cognition closed in upon itself has only now acquired a viable form. One can express this more precisely: it has only now acquired a form in which it can represent itself as knowledge. A problem arises here, however. With the word »constructivism« (taken over from mathematics) premature victories have been proclaimed, and one has to accept that there will be those who step aside, with a shake of the head, denying the validity of these claims. It is important, therefore, to investigate the question of what is new and convincing here - and this will lead the discussion far afield.

II.

For reasons that can only be clarified subsequently we begin our investigation with the question: by means of what distinction is the problem articulated? That is, we do not begin with the Kantian question: how is knowledge possible? We have avoided this form of the question because it might lead us to the premature response: in *this* way! At first the difference is of no great consequence. The one form of the question can be translated into the other (if one is not afraid to face problems of logical hierarchies as well as their failure). One can answer the question: »how is knowledge possible?«, with »by the introduction of a distinction«. In contrast with the tradition involving such concepts as »diapherein«[6] or »discernment«[7] here the concept of distinction is radicalized. For in order to recognize knowing it is necessary to distinguish it from what is not knowing. As a result, the question with regard to the foundation of knowledge is transformed into a question with regard to the distinction of distinguishing, that is, into an obviously self-implicative question.[8] The passage from the search for a founding - and therefore asymmetric - relationship with regard to some unity is transformed into a search for an operatively employed difference. It is, further, easy to recognize, that circularity and paradoxes can no longer be rejected but will come to play a role.

So, once again, the question is: by means of what distinction is the problem of knowledge articulated? (And, for the sake of clarity, let it be said once again: We are aware that with this question we have taken upon ourselves the difficulty of the distinction of distinguishing.)

In any case one will not be able to approach constructivism if one proceeds from the old controversy of whether the knowing system is a subject or an object. The subjectivist problem was to state and to show how it is possible by means of *introspection* - that is by passage to the self-reference of one's own consciousness - to form judgments about *the world of others*. That »intersubjectivity« is only a word which therefore does not solve the problem should be obvious. Objectivism, on the other hand, came up with the idea of describing knowledge as a condition or process in a particular object which was often called »organism«.[9]

The mistake here lies in the assumption that it is possible to describe an object completely (we won't go so far as to say »explain«) without making any reference to its relation to its environment (whether this relation be one of indifference, of selective relevance and capacity for stimulation, of disconnection, or of closure). In order to avoid these problems, which arise from the point of departure taken, both subjectivist and objectivist theories of knowledge have to be replaced by the system-environment distinction, which then makes the distinction subject-object irrelevant.

With this we have the distinction central to constructivism: it replaces the distinction transcendental/empirical by the distinction system/environment. The concept environment (as well as the corresponding one of system) was not available during Kant's day. What we call »environment« today had to be conceived of as the state of being contained and carried (periechon); and what we call »system« had to be thought of as order according to a principle. Both of these were already objects of knowledge. In order to answer the question of how knowledge is possible without falling into a self-referring circle the distinction transcendental/empirical was developed. Hardly anyone accepts this distinction today despite the labor that goes into the exegesis of historical texts. But if one drops this distinction how does one then avoid the circle of the self- founding of knowledge? Why must one avoid this circle? Can't one simply say: Knowledge is what knowledge takes to be knowledge?

The serving as medium foundation for dealing with these questions offers up the distinction system/environment and, in its context, a worked-out systems theory. This makes - virtually automatically - all the investigations and knowledge gained in systems theory of potential relevance for the theory of knowledge. In contrast to the procedure in transcendentalism, investigations bearing relevance for epistemological questions do not need to be carried out primarily with this end in mind. The relevance emerges as a side-effect of other investigations (e.g., of neurophysiological investigations or in the history of science) and one only has to take care that the transitions are smoothed over and now and then put in order, for example by adequate terminological recommendations. A good example of this is Humberto Maturana's use of the word »cognition« (»conocimiento«) for the extension of operations under the condition of interaction with the environment[10], however annoying this terminology might be for professional epistemologists afraid of a biological invasion of their domain.

It has been known for a quite some time already that the brain has absolutely no qualitative and only a very slight quantitative contact with the external world. All stimuli coming from without are coded purely quantitatively (principle of undifferentiated coding); furthermore, their quantity, as compared with purely internal processing events, plays but a marginal role.[11] Incoming stimuli are also erased in fractions of a second if they are not stored in internal storage areas with somewhat longer retention times (short-term memory) - an event which is more the exception than the rule. With this, even time is made to serve the internal economy of complex processes. Apparently it is fundamental for the functioning of the brain that selected information is enclosed and not that it is let through. As if it were already information (or data) before it motivates the brain to form a representation. Such knowledge as this was not made use of by theoretical epistemology and it is only a formulation in terms of systems theory that leads to an insight which must seem surprising to epistemologists[12]: only closed systems can know. The sociology of science has arrived at similar conclusions (which are still, for the most part, rejected as being too shocking).[13] Whoever still maintains that knowledge is the construction of a relation to the environment that fits things as they are, is welcome to his opinion, but he is forced to begin his theoretical reflections with a paradox: it is only non-knowing systems that can know; or, one can only see because one cannot see.

Philosophical epistemology has become marginal scientifically if not completely isolated; a situation that has often been lamented.[14] This *was* the case for the Neo-Kantians and *is* the case for the Neo-Wittgensteinians. Nonetheless, anyone familiar with both sides is aware of the numerous possibilities for contact. Systems theory or, more precisely, the distinction between system and environment, could play the role of mediator here.

The effect of the intervention of systems theory can be described as a *de-ontologization of reality*. This does not mean that reality is denied, for then there would be nothing that operated - nothing that observed, and nothing on which one would gain a purchase by means of distinctions. It is only the epistemological relevance of an ontological representation of reality that is being called into question. If a knowing system has no entry to its external world it can be denied that such an external world exists. But we can just as well - and more believably - claim that the external world is as it is. Neither claim can be proved; there is no way of deciding between them. This does not, however, call the external world into question but only the simple distinction being/non-being which ontology had applied to it. As a consequence, the question arises: why do we have to begin with precisely this distinction? Why do we wound the world first with this distinction and no other?

Systems theory suggests *instead* the distinction between system and environment.

III.

If one accepts this suggestion the answer to the question, how is knowledge possible?, is to begin with, as the operation of a system separated from its environment. If one, further, takes seriously that the system always has to be operationally closed then to the initial idea of separation assumptions are added regarding self-reference and recursivity. Operations of this kind are only possible within the context of a network of operations of the same system towards which they point and on which they are founded. There is no single operation that can emerge without this recursive network. At the same time the network itself is not an operation. »Multiplicity does not act as a relay«.[15] The whole cannot as a whole itself become active. Every operation reproduces the unity of the system as well as its limits. Every operation reproduces closure and containment. There is nothing without an operation - no cognition either. And every operation has to fulfil the condition of being one operation among many, as it cannot exist in any other form, cannot otherwise possibly be an operation.

As a result, for an observer the system is a paradox, a unity which is a unity only as a multiplicity, a *unitas multiplex*. Even when the system observes itself one has what is true for every observation. If a system wants to know what makes it possible that it can know, it encounters this paradox. All theory of knowledge has to begin with the resolution of a paradox.

A further consequence is: No system can perform operations outside its own limits. If new operations are integrated it means that the limits of the system have been extended. Consequently, the system cannot use its own operations to connect itself with its environment since this would require that the system operate half within and half without the system. The function of the boundaries is not to pave the way out of the system but to secure discontinuity. Whatever one wants to call cognition, if it is supposed to be an operation then the operation necessarily has to be one incapable of contact with the external world, one which, in this sense, acts blindly.

These ideas can be worked out further and the foreseeable extensions of a theory of closed, self-referring systems-in-an- environment will doubtless come to have over this route an influence on the theory of knowledge. But we will leave this question aside for the moment since we are now confronted with a fundamental question: is it possible, and is it acceptable, to call what here becomes perceptible »knowledge« at all?

In the search for an answer to this question it is advisable to introduce a second distinction between *operation* and *observation*. This distinction occupies the place that had been taken up to this point by the unity-seeking logic of reflection. (This means, therefore, a substitution of difference for unity).

An operation that uses distinctions in order to designate something we will call »observation«. We are caught once again, therefore, in a circle: the distinction between operation and observation appears itself as an element of observation. On the one hand, an observation is itself an operation; on the other hand, it is the employment of a distinction. An example would be that between operation and observation. A logic that would take its point of departure here could only be established as the unfolding of a circle, and it would have to make certain that the distinction can re-enter into what it has distinguished. Spencer Brown provides explicitly for this »re-entry« after deliberately ignoring it at the beginning with his instruction to an observer to »draw a distinction«. (Among other things this means that time is employed for the resolution of self-referring circles and paradoxes).

An observation leads to knowledge only insofar as it leads to re-usable results in the system. One can also say: Observation is cognition insofar as it uses and produces redundancies - whereby »redundancy« here means limitations of observation that are internal to the system. In consequence, particular observations are more or less probable.[16]

The passage to »constructivism« follows from the insight that *it is not only for negations that there are no correlates in the environment of the system but even for distinctions and designations (therefore for observations)*. This does not mean (to say it once again) that the reality of the external world is being called into doubt. It is also beyond doubt that an observer can observe that and how a system is influenced by its environment or deliberately and successfully acts upon its environment. Nonetheless, all distinctions and designations are purely internal recursive operations of a system (that is, operations that form or disturb redundancies). These are operations that are not able to go beyond the system and, as if at a distant remove, pull something into it. As a result, all achievements following from these operations, above all what is usually called »information«, are purely internal achievements. There is no information that moves from without to within the system. For even the difference and the horizon of possibilities on the basis of which the information can be selection (that is, information) doesn't exist in the external world, but is a construct - i.e. internal to the system. Does this mean, however - as is claimed in a direct line from Maturana - that the cognitive system operates »blindly«?

The metaphor of seeing and blindness can be retained as an abbreviated mode of speech, although it does not correspond to the current level of knowledge. One must also distinguish here: if every relation to the outer world is being denied in such a metaphor, too much is being called into question. On the other hand, it must be made clear that all observation (including the observing of observations) presupposes the operative deployment of a distinction which at the moment of its use must be employed »blindly« (in the sense of »non-observably«). If one wants to observe the distinction in its turn, one has to employ a different distinction for which the same is true.

There can be no doubt, therefore, that the external world exists or that true contact with it is possible as a necessary condition of the reality of the operations of the system itself. It is the differentiation of what exists that is contributed by the observer's imagination, since, with the support of the specification of distinctions an immensely rich structure of combinations can be obtained, which then serves the system for decisions about its own operations.

Expressed in other word, the unity of a distinction employed for observation is constituted within the system. It is only in the observing system that things distinguished are brought to the unity of being distinct. Cognition is neither the copying nor the mapping nor the representation of an external world in a system. Cognition is the realization of combinatorial gains on the basis of the differentiation of a system that is closed off from its environment (but nonetheless »contained« in that environment).[17] If a system is forced to cognize with the aid of distinctions and is unable to cognize in any other manner, it means further that everything that is for the system, and which therefore has reality, has to be constituted over distinctions. The »blind spot« of each observation, the distinction it employs at the moment, is at the same time its guarantee of a world. For example, social reality is what one, in observing a majority of observers, can observe to be uniform among them despite their differences.[18] Social reality exists only when an observer can distinguish a majority of observers (which may or may not include himself). By »world« is meant that which has to be assumed for every system to be the unity of the system/environment distinction (self-reference and external reference), when (and only when) this distinction is employed.

In conclusion we can say that knowing systems are real (empirical - that is, observable) systems in a real world. Without a world they could neither exist nor know. It is only cognitively that the world is unapproachable for them.

IV.

The contribution of the systems that makes cognition possible at all and which nothing in the environment corresponds to consists in the act of distinguishing. This recognition, which (as a distinction itself) implies its own limitation, has helped us as far as it goes. This would seem to answer the question usually raised in controversies about constructivism. But the interesting analyses are still to come: They involve not the question of a real agreement between knowledge and reality but questions of *time*. Cognitive systems (at least the brain, consciousness and the systems of communication called societies) operate on the basis of events that have only a momentary presence and that already begin to disappear at the moment of their emergence. Furthermore, these systems operate on the basis of events that cannot be repeated but which must be *replaced* by other events. Their structures must, therefore, provide for the passage from event to event - something for which there are also no equivalents in the environment. It is neither the case that the environment changes itself with the same tempo and rhythm (and this can only be spoken of on the basis of cognitive acts), nor can one find in the environment those autopoietic structures that suggest the one in the other. How then is the time relation between system and environment to be understood? The only answer can be: as *simultaneity*. The foundation for the reality of the system - whatever the contours of its own meaningful observations might be - is the simultaneity of its operation with the conditions of reality that sustain it. Whatever the system might contribute in the way of a non-present future and a non-present past - that is, of distinctions - the simultaneity of the environment and the eternally immediate present of the system is a condition that cannot be eliminated. Whatever is simultaneous cannot be influenced, cannot be integrated into the causal constellations of the system, cannot be synchronized, but is nonetheless the precondition for the application of distinctions in time. The system can place itself in relation to time between future and past, or as a moment in relation to duration or to eternity. Whatever might emerge from this, the system constructs time in relation to itself. What one does not have control over is the simultaneity that reemerges from moment to moment in all the operations of the system, the »common aging« in the sense of Alfred Schütz[19] or the splashing of the water on the bank of the Isle de

Saint Pierre, that »continuing noise that is, however, filled by intervals«, which, in convergence with internal movements, is sufficient »to make me sense my existence with pleasure, without my having to think«.[20] It is out of the unavoidable certainty of the simultaneity of the system and the environment that current time projections can arise. Examples of this can be found in the widespread »anticipatory reactions« in the plant and animal kingdoms, that is, in mere reactions to something assumed to be present on the basis of regularities that prove to be beneficial for the future, although they have not been perceived (i.e., have not been integrated into the processing of information).[21] Highly developed cognitive systems can, in addition, make prognoses, which does not mean that they can now perceive future present times. They are able to span this impossibility by means of constructions that organize their own information processing with the help of a distinction between what is past and what is to come that cannot appear in the external world as a *distinction*. Presumably, prognosis has to be understood as a product of our own imagination that can be evaluated by the memory[22], that is, as the creation of an excess of individual possibilities which is then offered up for selection according to self-constructed criteria of »suitability«. In other words, systems that make prognoses can prepare themselves for risks that they themselves have created and derive benefits from this.

Cognitive systems, therefore, have only a momentlike existence, as a result of the burden of simultaneity which keeps them on the ground. This existence must reproduce itself autopoietically in order to attain stability, even if it is only a dynamic one. They experience the world, therefore, with future and past - that is, as *duration* - only in the form of *non-presentness*. These systems can, therefore, consider their history to be finished insofar as they do not make present - as if in a dream - retrospective preferences. In the same way their future is full of enticing and threatening possibilities (although in reality there is no possibility at all, since everything is as it is). It is possible then to keep the non-present constant, which yields in turn the fascinating possibility for cognition of representing *changes* in the external world by terminological *constants* (instead of through changes in the system itself). Such systems need, as a result, records, which can, however, only be currently accessed; subsequently they help themselves with a kind of »vicarious learning«, with observing observations of others which have the same limitation. The vast unfolding of the world materially, temporally and socially is a construct[23] anchored in the simultaneity of the world which, in this regard, never changes but is nonetheless inseparable from every realization.

On the other hand, the freedom of cognition in its constructions is founded on a radical »de-simultaneity« of the world, on the reduction of the contemporaneous to an instant almost devoid of meaning. What is gained by this is a terrifying plethora of possibilities in which cognition has to find its way by its own guidance. This existential moment is doubtlessly only a moment for an observer who can see the limits of this presentness and can call it »existence«. Descartes was aware of this - and therefore made God responsible for continuity.

V.

The refined constructivist theory of knowledge that has been presented here not only dissolves the traditional rationalistic continuum of being and thinking - which presupposed the possibility of an agreement between both and had founded it upon such concepts as nature or creation. The theoretical transcendental position which had been first the reaction to the dissolution of this rationalist continuum is also renounced. Furthermore, the assumption is rejected of a subjective faculty of consciousness that can guarantee a priori the conditions of the possibilities of cognition. But then, it is not sufficient to replace this conception by the distinction between irritation (or perturbation) from without and self-determination from within, which simply gives the difference between inner and outer yet another formulation and weight.[24] What remains (and has to replace those assumptions) is the *recursivity* of observation and cognition.

A process is called »recursive« when it uses the results of its own operations as the basis for further operations - that is, what is undertaken is determined in part by what has occurred in earlier operations. In the language of systems theory (which is not quite suitable here) one often says that such a process uses its own outputs as inputs. In any case recursivity requires a continuous testing of consistency and it has been shown by investigations in perception and memory that this necessitates a binary schematization, even on the neurophysiological level, which holds in readiness the possibilities of acceptance and rejection.[25] The states of the system that have been produced by its own operation serve then as criteria for the acceptance or rejection of further operations; stimuli from the environment that effect the system can play a role here also. Decisive, however, is the continuous self-evaluation of the system - which always operates in a state of irritation or agitation by means of a code that permits acceptance and rejection with regard to the adoption of further operations. The brain functions in this way. And the same will be true for psychic and social systems. The codification true/false only gives this schematization its final finish and a form that is only used under very special circumstances.

One can, therefore, think of binarily schematized recursivity as a continuous calculation of operations on the basis of the current states of the system. The pleasure/pain mechanism also seems to function in this manner. With regard to observations, this structure makes possible the observation of observations. This can mean, first of all, that one repeats the same operation in order to see whether its results are confirmed or not confirmed. This leads then to a »condensation« of units of meaning whose verification can no longer be obtained by a single operation. More or less clear deviations can be built into such a replication. One observes the same thing at different times in different situations, under different aspects, which leads to a further enrichment of the condensed meaning and finally to the abstraction of denotation for what seems identical in the different observations. Thus, assumedly, the meaningful construction of the world comes about, gaining thereby a power no single operation can possibly dispose of. One speaks here, in the language of mathematics, of the »eigenvalues« of a system.[26] Again, no correspondence between system and environment is presupposed, but only the claim is made that it was possible to bring about these states.[27]

This theory provides a good explanation for the normal evolution of a knowledge that overcomes distance, so-called »distal knowledge«, as Donald Campbell, following Egon Brunswik, has called it.[28] If one takes into consideration the dependence of all observation on distinction, other possibilities of recursive observation emerge. The usual understanding of the observations of observation focusses above all on *what* an observer observes (distinguishing thereby between subject and object, but concentrating above all on the object). Constructivism describes an observation of observation that concentrates on *how* the observed observer observes. This constructivist turn makes possible a qualitative change, a radical transformation, in the style of recursive observation, since by this means one can also observe what and how an observed observer *is unable* to observe. In this case one is interested in his blind spot, i.e. the means by which things become visible or non-visible. One observes (distinguishes) the distinction used by the primary observer in his observing. Since this observer, in the midst of his observation, cannot distinguish this distinction, what is observed is something that remains unknown to him or incommunicable. In the terms of sociology one could also say that observation is directed now to the observed observer's *latent* structures and functions.

The kind of reality, the kind of »eigenvalue« produced by recursions of this type is still largely unknown, as the technique itself is no older than 200 years. It was probably first practiced in the novel, then in the Counter-enlightenment, and then in the critique of ideology that is always from a holier-than-thou perspective. The primary observer was placed into the domain of the harmless or the naive; or he was treated as someone who, without realizing it, had something to hide. This holier-than-thou perspective fed upon suspicion. And the generalization of the principle of suspicion made it possible for whole disciplines - from psychoanalysis to sociology - to establish themselves with additional credentials in a world in which everyone knows, or imagines he knows, the situation in which he acts and the reasons for his actions.

It does not seem a coincidence that this observation of latent structures developed parallel to transcendental theory - at first at the end of the 18th century and then with particular intensity a century later during the heyday of Neo-Kantianism. Apparently, something had been lacking in transcendental philosophy. All the same, a constructivist theory of knowledge goes beyond this state of affairs (again a hundred years later). Its concept of recursive observation includes the observation of latency, freeing it from the prejudice that latent structures give a false picture of the world, as it really is and as science sees it. The assumption - to be found above all in the classical sociology of knowledge - that latent structures, functions and interests lead to distortions of knowledge, if not to blatant errors, can and must be abandoned.[29] The impossibility of distinguishing the distinction that one distinguishes with is an unavoidable precondition of cognition. The question of whether a given choice of distinction suits one's latent interests only arises on the level of second order observation. The claim of ideological distortion can then be observed in the person making the claim (for which he has to be observed as observer, that is, in relation to what he is observing).[30]

The important question after all this is what »eigenvalues« a system is converging towards when it extends the recursivity of its observations in this direction - that is, when it continually turns its observations towards things other observers *cannot* observe. For the results of this method of observing we have, in the absence of anything better, a variety of different names: Gotthard Günther speaks of »polycontextuality«, others of »pluralism«, and still others of the postmodern arbitrariness in the emergence and passage of »discourses«. For constructivism this is, above all, an epistemological question and a kind of compensation for the limitations inherent in every act of cognition as a consequence of its dependence upon a distinction. One cannot draw the conclusion from the theory that now special »eigenvalues« of the social system will emerge that will be resistant to enlightenment for there is no guarantee that under all conditions such »eigenvalues« can be found and become stabilized. Still, the question can at least be raised and observation directed accordingly.

VI.

If one takes seriously the endeavor to set up a constructivist theory of knowledge an important question becomes shifted: that of the paradoxes. By a paradox is meant a permissible and meaningful statement that leads nonetheless to antinomies or undecidability (or, more strictly, a demonstrable proposition that has such consequences). Two possibilities for dealing with such a problem should be rejected. The first is used in the construction of formal systems and consists of an ad hoc procedure of exclusion. The paradoxes are eliminated by suitable precautionary measures. Structures that lead to paradoxes are forbidden - for example by the well-known but questionable theory of types. The epistemological questionability of such a procedure comes from its lack of justification; moreover, it usually has the consequence that it excludes more than just paradoxical possibilities for the construction of sentences.

As a result, philosophers have felt compelled to look for other means that would lead to a justifiable exclusion of paradoxes. MacKie, for example, suggests returning to a semantic theory of truth that would make it possible to say that the supposed objects designated by meaningful paradoxical propositions do not exist.[31] It is, however, not possible for a constructivist theory of knowledge to accept this way out, since what is claimed here as being non-existent, does not exist for constructivism anyway. Given that paradoxes re-emerge despite all the attempts to eliminate them, MacKie finally even calls for a »construction« of the paradoxical by adopting self-referring propositions into the construction and (at least implicit) quantification.[32] This suggestion is grist for the constructivist's mill: constructivism can view paradox as a problem in the machinery of the calculation of calculations, as a possible but nonetheless destructive construction. Should one look the Gorgon straight on - aware however that it is not the deadly Medusa one has before one, but her immortal sisters, Stheno (the Mighty) or Euryale (the Far Springer)?

We suggest instead a view from the side, the observing of observation. This enables one, when one includes observation of latency, to observe how other observers render invisible the paradoxes that get in their way, for example the paradox of each of our binary codes.[33] It is, therefore, not a psychoanalytical infection or a critical socio-ideological frivolity that brings us to include observation of the blind spot of the observer in the theory of knowledge. It is furthermore not simply an encouragement to propound values that are, in any case, irrational, as William James and Max Weber had thought. To see what others cannot see (and to accept that they cannot see what they cannot see) is, in a way, the systematic keystone of epistemology - taking the place of its a priori foundation.

It is, therefore, of importance that every observer involves himself in a paradox because he has to found his observing on a distinction. As a result, he is unable to observe either the beginning or the ending of his observing - unless it be by means of another distinction that he has already begun to make or by continuing with a new distinction after having ended.[34] This is why all projection, or the setting of a goal, every formation of episodes necessitates recursive observation and why, furthermore, recursive observation makes possible not so much the elimination of paradoxes as their temporal and social distribution onto different operations. A consensual integration of systems of communication is, given such conditions, something that should sooner be feared than sought for. For such integration can only result in the paradoxes becoming invisible to all and remaining that way for an indefinite future.

This remedy solves, as it were, the problem of the paradoxes by reference to a concrete theory: the theory of autopoietic systems, which by means of recursive operations produce and reproduce a network of such operations as the condition for the very possibility of this reproduction (a solution logicians will hardly find satisfying). In such systems (one of which is science) there is no operation without reference to other operations of the system. Even when one forms universal propositions that refer to all the operations of the system, and also when one exposes these universal propositions on the basis of the classic Cretan pattern to self-reference, one only produces an operation that is a point of departure for other operations. We simply claim: it *is* this way; and logicians who attempt to dispute this are, in consequence, punished by paradoxes.

VII.

Given all that has been said, what understanding of reality does constructivism have?

It may be useful here to review classical responses once again. As far as visual metaphors were used, two solutions were offered. Objectivists said that reality was manifold, which meant that there was no single observation point from which it could be seen in toto: what one sees conceals what one does not see. This deficiency can only be countered by changing the point of observation, that is by working sequentially or by a division of labor. Subjectivists could speak instead of a multiplicity of perspectives each of which makes possible a conditioned seeing, but which at the same time makes impossible or difficult the perception of the perspective one sees with.[35] More eyes - and therefore more emotions: that was Nietzsche's postulate in *The Genealogy of Morals*. Constructivism goes beyond these positions by radicalizing the relationship between cognition and reality. It is no longer a question of the difficulties that arise from a multiplicity of sides or perspectives, and the problem is no longer how one arrives, given this situation, at unity. This multiplicity, regardless of whether it is a multiplicity of sides or of perspectives, is itself a product of cognition, resulting from certain types of distinctions, which, as distinctions, are instruments of cognition. It is precisely by means of distinguishing that cognition separates itself from everything that is not cognition. Nonetheless, one is always dealing with concretely determined operations - even in the case of knowledge. Without water the jelly-fish goes limp. But in order to recognize that, distinctions are necessary: with/without water; not-limp/limp. These distinctions are codifications specific to cognition, which function independently of the environment (i.e. of stimuli), because there are not and cannot be any equivalents for them in the external world.

Cognitively all reality must be constructed by means of distinctions and, as a result, remains construction. The constructed reality is, therefore, not the reality referred to. This too can be recognized, but recognized only by means of precisely this distinction. For cognition, only what serves in a given case as a distinction is a guarantee of reality, an equivalent of reality. One could say more precisely: The source of distinction's guaranteeing reality lies in its own operative unity. It is, however, precisely as this unity that the distinction cannot be observed - except by means of another distinction which then assumes the function of a guarantor of reality. Another way of expressing this is to say that the operation emerges simultaneously with the world which as a result remains cognitively unapproachable to the operation.

The conclusion to be drawn from this is that the connection with the reality of the external world is established by the blind spot of the cognitive operation. Reality is what one does not perceive when one perceives it. In no way does this mean, however, that somewhere in the world there are states of affairs one cannot know, above all not in the old sense of the essence of nature's being secret. All that is meant is that the fruits of the concrete operation of cognition, which issue from the use of distinctions - that is, the proliferation of combinatorial possibilities - is due to an instrument requiring an operational closure specific for the given system. To attain this no »similarities« with the environment can be tolerated. If cognition demands meaning and meaning demands distinctions then the final reality must be thought of as devoid of meaning.

VIII.

If one compares this result with what has traditionally been called »idealism« one can recognize an important change. It affects the basic question to which an answer is sought and, therefore, the whole theoretical structure.

One had proceeded from the distinction between knowledge and object and, as a result, been forced to face the problem that could not be answered by means of this distinction: how does knowledge arrive at its object? In the final analysis, the problem lay then in the unity of the difference between knowledge and object. One answer was provided by the claim of a dialectical relationship. Dialectical theories proved to be the adequate form here and required hardly any further argument. If one accepts, however, the argument suggested above the distinction between knowledge and object is itself only a distinction, that is a construction used to wound, dissect, observe the world. The unity of this distinction is simply the blind spot used by someone who, by means of this distinction, produces observations and descriptions.

If one starts from the assumption, however - as constructivism does - that this is always a real process in a real environment, which is always subject to limitations coming from the environment, what might then be the problem?

The problem could reside in the question of how a system is able to transform such *limitations* into *conditions for increasing its own complexity*. The *non-arbitrariness* of knowledge would then be nothing other then the evolutionarily-controlled *selectivity* of this process of change. It assumes no operations of the system projecting into the environment, that is, no knowledge in the traditional sense. One has to postulate instead: Everything issuing from this process of a transformation of limitations into conditions for the increase of complexity is, for the system in question, knowledge.

In contrast with idealism, constructivist cognition neither seeks nor finds a ground. It reflects (when it reflects) the change in world-orientation from unity to difference. It begins and ends with distinctions, well aware of the fact that this is its own affair and not forced to this recognition by an unapproachable outer world. As the unity of the drawing of a distinction it can conceive of itself as a symbolic processing. The unity of the separated, the mutual suitability of the differentiated, is what serves as a symbol here. Francisco Varela has considered this, too, to be an operation or a value and called it »self-indication«.[36] We must leave the question open as to whether this leads to an effective calculus. On the other hand, it is easy to recognize that we are living in the world after the fall. We have eaten of the tree of the knowledge of good and evil. »Distinctions« can only be employed using »indications«. The symbol can only be employed diabolically; only what has been distinguished is integrable.

IX.

A few further thoughts on the matter will be given only cursorily. The concept of observing has been defined extremely formally as a distinguishing description. We reject, nonetheless, founding this formality »transcendentally«. With observing, distinguishing, designating we always mean an empirical operation that changes the system executing them - which means an operation which, in its own turn, is observable. No observer can avoid being observed, not even in its quality as »subject«.

On the other hand the formality of the concept leaves open which empirical operations are meant. Which organ - to speak in these terms - carries out the observation?

The abstraction of the concept is not meant to conduct one to a ground. Which results already from the fact that the operation of observing can lead to both true und false knowledge, as an observer can determine who observes observing by means of the distinction of true and false. The abstractness of the concept is not, therefore, intended to provide a grounding for knowledge, but only to keep open the possibility of observation operations' being carried out by very different empirical systems - living systems, systems of consciousness, systems of communication. The abstraction makes allowances for the very wide domain of the »cognitive sciences«, above all for the differentiation into disciplines in biology, psychology and sociology. Observation takes place when living systems (cells, immune systems, brain, etc.) discriminate and react to their own discrimination. Observation occurs when thoughts that have been processed through consciousness fix and distinguish something.[37] It occurs as well when a communicable integrable understanding of conveyed information - be it linguistic or non-linguistic - is attained (whatever psychic processes might occur in the minds of the participating individuals).

Given the state of research today one cannot get around taking into account the differences between these empirical realizations of distinguishing and designation (or should one perhaps for once say here: of discriminative focussing?). With this, the traditional attribution of cognition to »man« has been done away with. It is clear here, if anywhere, that »constructivism« is a completely new theory of knowledge, a post-humanistic one. This is not intended maliciously but only to make clear that the concept »man« (in the singular!), as a designation for the bearer and guarantor of the unity of knowledge, must be renounced. The reality of cognition is to be found in the current operations of the various autopoietic systems. The unity of a structure of cognition (or the »system« in the sense of transcendental theory) can only lie in the unity of an autopoietic system which reproduces itself with its boundaries, its structures and its elements.

By this means the significance of psychological epistemologies is considerably reduced, but relieved at the same time of the unreasonable expectation that they should provide more than individual-psychological knowledge. There is no such thing as »man«, no one has ever seen him and if one is interested in the system of observation that organizes its distinctions by means of this word or concept one discovers the communication- system called society. There are now approximately 5 billion psychological systems. It has to be asked which of these 5 billion is intended when a theory of knowledge employing a psychological reference system relates concepts such as observation and cognition to consciousness. If no answer is forthcoming, such a theory has to be characterized as practicing socio-communicative observation. And the suggestion would have to be made that it would be better if this practice were reflected upon.

Up to now, constructivism has profited mainly from research in biology, neurophysiology and psychology (Maturana, Varela, Piaget, von Glasersfeld), although it actually favors development of a sociological theory of knowledge. What we know as cognition is the product of the system of communication called society, where consciousness plays a permanent but always only fractional role. It is only in extreme exceptions that one has to know individual persons in order to know what is known - and these are typical instances (for example, statements by witnesses in court) in which direct perception plays a central role. Neither in its claim to validity nor in the evaluation of its possibilities for development is the fund of knowledge of modern society approachable through processes of consciousness. It is an artifact of communication - and what is amazing here is not so much that the world is as it is constructed by modern science, as that it is still possible to pursue communication under the conditions of this construction. It is obvious that this cannot be explained by some capacity of consciousness (which one?!) but by the possibilities of storage made available originally by printing and, more recently, by electronic data processing.[38]

This preference for society as a referential system (that is, as the choice of a system from the perspective of which something else is environment) becomes absolutely unavoidable when one takes into consideration the difference between everyday knowledge and scientific knowledge. Whatever this distinction might mean and whatever theory might offer it, it cannot be presented convincingly as a distinction between different psychic types of knowledge. The distinction is a consequence of the differentiation of the social system of society. And it is only from here that psychical systems can be influenced. No further argument is necessary when one recalls the well-known phenomena of exponential growth, increasing differentiation and specialization or the problems of the pace of change.

It is, finally, only in a sociological context that the ideas on recursive observation and second-order observation (i.e., the observation of observation) acquire their full significance. But why would an observer observe another observer as observer, as another psychical system? Why isn't the other system seen simply as a normal object in the external world, that is, why isn't it simply observed directly instead of as a pathway for the observing of its observing?

It is usually assumed that this is made possible by a sudden, intuitive analogy: the other is experienced as an alter ego, as operating like another I.[39] But we question, how does this occur? And further, is this phenomenon culturally invariant, independent of social structure? The usual answer describes only the result, is only another formulation of the problem and does not explain how it occurs.

Maturana avoids this problem by shifting to the mutually coordinating interaction of two organisms that interreact with each other in a sufficiently comparable area of interaction.[40] This makes it possible for him to explain the origin of language as a possibility of consensual coordination of the coordination of these interactions despite closure of the mode of operation of the participating systems. This still doesn't provide, however, a satisfying explanation of how the observation of observing emerges, that is, how observers construct the objects they have constructed as other observers.

A third theoretical suggestion (which draws on sociology, since psychology and biology have not sufficed) can begin with the assumption that the construction of the other observer is a necessary consequence of communication.[41] For communication is only possible when an observer is able, in his sphere of perception, to distinguish between the act of communication and information, that is, to understand communicative acts as the conveying of information (and not simply as behavior).[42] Out of this distinction - which only remains evolutionarily stable and only reproduces itself as a communication system when it is able to maintain itself - there emerges then a second one: that of subject and object. That communication can be continued requires no more than a kind of black-box concept for the subject and for the object, as far as the distinction operates. As a participant one can make use of one's own constructions, which can then be evaluated during the course of one's participation in the communication. One does not need to know what is going on »inside« the subject (and of course, could never know this) and also does not need to know the »essence« of things (which is of itself infinite): the filling necessary for the continuation of communication suffices. However, to the degree that systems of communication, in the course of their own evolution, become more sophisticated, differentiated and complex, more demanding concepts for subject and object are called for. It is in the course of this that one finally also learns to observe others as observers (even at times they are not communicating) and finally even to observe that others do not observe what they do not observe while observing. Society, finally, makes even latent observation of latent structures possible.

The question still has not been answered why communication together with its resulting achievements progresses. The answer can only be that the evolutionary force of a particular distinction - that between communication and information - has proven itself. This can, of course, be claimed of everything that exists, and is still not an explanation. Important, however, in the constructivist context outlined above is that this claim has been made for a *distinction*. With this another distinction has been added to those already used - system/environment and operation/observation: that of communication/information, which is of special importance for the analysis of social systems. The familiar distinction between ego and alter ego can be dealt with as derivative, and with it the whole theory of knowledge founded on the concept of intersubjectivity.

X.

The above has made it abundantly clear, we believe, that constructivism does not question the existence and reality of the world - but only constructs. But even after one has seen this, one can, as a sociologist, still ask why this happens, and why precisely today, after both ancient skepticism and idealism have been overcome, this constructivistic world-construction is of value. If a philosopher were to ask this question he would be faced with the difficult problem of a deeper analysis of Hegelian logic, which is the most profound scheme so far developed for the processing of distinctions of what is implied in them with regard to identical and contrary. For a sociologist the matter is simpler. He can take a theory of social evolution as his point of departure, a theory obviously, that itself is founded on a relevant distinction - for example, constructed on a Darwinian-scheme of variation and selection. It is possible then to understand constructivism as an epistemology suitable for a society with a highly differentiated system of the sciences.[43] In other words, in a society that can produce science in the modern sense, conceptual problems arise that can only be solved constructivistically - whatever one in this society might normally think about the world in which he lives and works, rides the bus and smokes cigarettes.

It shouldn't be very difficult to recognize that progress in science (whatever »progress« might mean here) is tied to even more sophisticated distinctions. This is, above all, the case for what Donald Campbell has called development in the direction of »distal knowledge« - that is, for the distinction between knowledge and the knower himself.[44] Divorce of the perspectives of comparison from the interests of the one doing the comparing also belongs here. One need, moreover, only think of the use of rigorously formal cognitive instruments - of logic, mathematics, quantification - as a form of representation of distinction in reality. This could still be dealt with under the concept of »idealism«, and it is in this context that Husserl makes his criticism of modern, »Galilean« science.[45] Today the »cognitive sciences« and the theory of self-referential systems add a new perspective which cannot be subsumed under »idealism« or criticized as »idealism«, that is, insight into the operative closure of self-referential systems. A theory of knowledge today that is to be compatible with the latest developments in science must be able to bear this new perspective. It is not surprising, therefore, that, after a period of open and rather irresolute epistemological pragmatism and a period in which highly formalized methodology was presented as epistemology - after James and Dewey, Baldwin, Rescher, Popper and others - epistemological constructivism is beginning to come into its own. Quantum physics, cytochemistry and neurophysiology, as well as historic-sociological relativism make this convergence necessary. If the task of epistemology is to analyze science as a social cognitive undertaking one will not be able simply to ignore scientific results. Constructivism is the form assumed in reflection on the system of science facing its own extravagances; it is the form in which an increasingly improbable distinguishing is finally recognized as the contribution of cognition. But it is also the form that can no longer mislead one to conclude it has nothing to do with reality.

A society that increasingly differentiates its most important sub-systems in relation to specific functions intensifies to a highly improbable degree its cognitive output in the area of science. If one then reflects on this situation, one arrives at theories that themselves seem improbable. For this reason epistemology cannot provide a foundation for the sciences. It cannot offer basic principles, arguments or even certainty. It can no longer be understood as a theory of the founding of knowledge. The opposite is true: it analyzes the uncertainty of knowledge and gives reasons for it. It therefore should come as no surprise that no theory of knowledge today can attain the degree of certainty to be found in quantum physics or biochemistry.

It is perhaps not the least important function of constructivist epistemology to make society irritatingly aware of the fact that it produces science.

Notes and References

1. As a typical solution the distinguishing of several levels of language or of cognition has been suggested, with the possibility of »autologic« relationships on the higher level. See L. Lofgren, »Towards System: From Computation to the Phenomenon of Language«, in M.E. Carvallo (ed.), *Nature, Cognition and System: Current Systems - Scientific Research on Natural and Cognitive Systems,* Dordrecht: Reidel, 1988,

p. 129-55. But this is a transparent stopgap solution, since a level derives its identity only from the fact that there are other levels that can be reached from it.

2. See D. Bloor, *Wittgenstein: A Social Theory of Knowledge,* London: Routledge & Kegan Paul, 1983, esp. p. 119ff. This »conventionalism« going back to Poincaré, has become a tradition, it meets with little opposition today as it has become almost reflex.

3. See D. Bloor, *Wittgenstein: A Social Theory of Knowledge,* London: Allen & Unwin, 1979, passim., for example, p. 95.

4. This can be found in the recent publication by A. Chalmers, »The Sociology of Knowledge and the Epistemological Status of Science«, *Thesis Eleven* **21** (1988), 81-102: The argument presented, however, shows no progress.

5. See P. Watzlawick (ed.) *Die erfundene Wirklichkeit,* Munich: Piper, 1981; H. Gumin/A. Mohler (eds.), *Einführung in den Konstruktivismus,* Munich: Oldenbourg, 1985; S.J. Schmidt (ed.), *Der Diskurs des radikalen Konstruktivismus,* Frankfurt am Main: Suhrkamp, 1987; E. v. Glasersfeld, *Wissen, Sprache und Wirklichkeit: Arbeiten zum radikalen Konstruktivismus*, Braunschweig: Vieweg, 1987.

6. See Plato, Theaetetus, 208 C.

7. See, for example, C. Buffier, *Cours de sciences sur des principes nouveaux et simples*, Paris 1732, reprint Genf: Slatkine 1971, p. 800ff. where this concept dealt with at length in the »Traite des vérités de conséquences« (not in the »Traité des première vérités«).

8. See in this regard, as well as for the limiting cases of universal distinction (nothing is excluded) and elementary distinction (nothing is included), R. Glanville/F.J. Varela, »Your Inside Is Out and Your Outside Is In« (Beatles 1968), in G.E. Lasker (ed.), *Applied Systems and Cybernetics, Vol. II*, New York: Pergamon Press 1981, p. 638-641. Following S. Brown, the authors distinguish inclusion and exclusion by using the concept of form in order to distinguish the act of distinguishing. Their argument corresponds, moreover, exactly to the idea Nicolas Cusanus used to found the coincidentia oppositorum and upon it his concept of God beyond all distinctions.

9. See, for example, the »object-psychological« epistemology of A. Naess, *Erkenntnis und wissenschaftliches Verhalten*, Oslo: Dybverd, 1936, in particular p. 103ff., where the author demands that all epistemology should be limited to description »of the process in the internal functional space« of an organism. See also p. 105: »The common distinction between ›situation‹ and ›behavior‹ is, viewed psychologically, a distinction between two kinds of ›behavior‹«.

10. See H.R. Maturana, *Erkennen: Die Organisation und Verkörperung von Wirklichkeit*, Braunschweig: Vieweg, 1982, esp. p. 32ff. (German translation from the English); H. R. Maturana/F.J. Varela, *Der Baum der Erkenntnis: die biologischen Wurzeln des menschlichen Erkennens*, Bern: Scherz, 1987, esp. p. 31ff. See the critical discussion precisely this connection between biological systems theory and epistemology, in G. Roth, »Autopoiese und Kognition: Die Theorie H.R. Maturana und die Notwendigkeit ihrer Weiterentwicklung«, in Schmidt, *op. cit.*, 1987, note 5, p. 256-286.

11. See G. Roth, »Die Entwicklung kognitiver Selbstreferentialität im menschlichen Gehirn«, in D. Baecker, et.al. (eds.), *Theorie als Passion*, Frankfurt am Main: Suhrkamp, 1987, p. 394-422 (419f).

12. See the contributions of H. v. Foerster, *Sicht und Einsicht: Ausgewählte Arbeiten zu einer operativen Erkenntnistheorie*, Braunschweig: Vieweg, 1985. See also F.J. Varela, »Living Ways of Sense-Making: A Middle Path for Neuroscience«, in P. Livingston (ed.), *Disorder and Order: Proceedings of the Stanford International Symposium* (Sept. 14-16, 1981), Saratoga, California: Anma Libri 1984, p. 208-223; G. Roth, »Selbstorganisation-Selbsterhaltung-Selbstreferentialität: Prinzipien der Organisation der Lebewesen und ihre Folgen für die Beziehungen zwischen Organismus und Umwelt«, in A. Dress, et al. (eds.), *Selbstorganisation: Die Entstehung von Ordnung in Natur und Gesellschaft*, Munich: Piper, 1986, p. 149-180, esp.

168ff.; G. Roth, »Erkenntnis und Realität: Das reale Gehirn und seine Wirklichkeit«, in Schmidt, *op. cit.*, 1987, note 5, p. 229-255.

13. »The natural world has a small or non-existent role in the construction of scientific knowledge«, is claimed, e.g., by H. Collins, »Stages in the Empirical Programme of Relativism«, *Social Studies of Science*, **11** (1981), 3-10 (3). See also, H. Collins, *Changing Order*, London: Sage Publications, 1985. There would be far less controversy if one read the latter work in conjunction with works on brain research and not as an alternative to them. The question isn't whether brains *or* language construct the world; the claim is that if it's brains, then it must be language, and *vice versa*.

14. See, for example, D.T. Campell, »Descriptive Epistemology, Psychological, Sociological and Evolutionary«, William James Lectures at Harvard University, 1977; noted from an unpublished manuscript.

15. M. Serres, »Dream«, in Livingston (ed.), *op. cit.*, 1984, note 12, p. 225-239 (238).

16. See with regard to this systems-theoretic use of the concept of redundancy: H. Atlan, *Entre le cristal et la fumée*, Paris: Editions du Seuil, 1979; or, H. Atlan, »Disorder, Complexity and Meaning«, in Livingston, *op.cit.*, 1984, note 5, p. 109-128.

17. On closure as »enclosure« see H. v. Foerster, »Entdecken oder Erfinden. Wie läßt sich Verstehen verstehen?«, in Gumin, Mohler, *op. cit.*, 1985, note 5, p. 27-68.

18. Subtle analyses of this question are to be found in F.H. Tenbruck, *Geschichte und Gesellschaft*, Berlin: Duncker & Humblot 1986, resp. p. 175ff.

19. A. Schütz, *Der sinnhafte Aufbau der sozialen Welt: Eine Einleitung in die verstehende Soziologie*, Vienna: J. Springer, 1932, p. 111f.

20. J.J. Rousseau, *Les rêveries du promeneur solitaire, Cinquième promenade, quoted from Oeuvres complètes* (Éd. de la Pléiade), Vol. 1, Paris: Gallimard, 1959, p. 1040ff (1045).

21. Of course, these are always the statements of an observation that itself sees more time from the observed system does. An extensive analysis of these questions can be found in R. Rosen, *Anticipatory Systems. Philosophical, Mathematical and Methodological Foundations*, Oxford: Pergamon Press, 1985.

22. The same is true for language and this similarity points to a close evolutionary and even neurophysiological association. See on this question H.J. Jerrison, *Evolution of the Brain and Intelligence*, New York: Verlag?, esp. p. 426f.

23. For greater detail see N. Luhmann, *Soziale Systeme: Grundriß einer allgemeinen Theorie*, Frankfurt am Main: Suhrkamp, 111f.

24. This is the formulation of Glasersfeld, *op. cit.*, 1987, note 5, in his presentation of »radial constructivism«. Maturana also uses such formulations to explain his constructivist position.

25. See H. v. Foerster, »What is Memory that it May Have Hindsight and Foresight as Well?«, in S. Bogoch (ed.), *The Future of the Brain Sciences*, New York: Plenum Press, 1969, p. 19-64.

26. See v. Foerster, *op. cit.*, 1985, note 12, esp. p. 205ff.

27. In Maturana's theory the corresponding concept is »Conservation of Adaption«. See Maturana/Varela, *op. cit.*, 1987, note 10, p. 113f. or, in greater detail, H.R. Maturana, *Evolution: Phylogenetic Drift through the Conservation of Adaptation*, Ms. 1986. It is crucial that adaptation can only be preserved, not improved. A system is adapted for the processing of its autopoiesis in its environment or it isn't, and is destroyed. There is no more or less in this regard, just as the operations of the system either can take

place or can't take place. Every other judgement is the affair of an observer and can only be observed in an observer.

28. S. Donald, T. Campbell, »Natural Selection as an Epistemological Model«, in R. Navoll/R. Cohen (eds.), *A Handbook of Method in Cultural Anthropology*, Garden City, New York: The National History Press, 1970, p. 51-85.

29. And has been given up today. See, for example, B. Barnes, *Scientific Knowledge and Sociological Theory*, London: Routledge & Kegan Paul, 1974; D. Bloor, *Knowledge and Social Imagary*, London: Routlege & Kegan Paul, 1976.

30. Quite consistently, Marxists learn about the critique of political economy from Marx; they don't turn to political economy for this. But the result is that the common views of the political economy of Marx' day are discussed with reference to Marx works, that Marx himself seems like a political economist (not completed without this being his own fault) and that changes in the critique that have occurred over the best 150 years are not sufficiently taken into account.

31. See J.L. MacKie, *Truth, Probability and Paradox: Studies in Philosophical Logic*, Oxford: Clarendon Press, 1973, ch. 6 in association with ch. 2.

32. J.L. MacKie, *op. cit.*, 1973, note 31, p. 273.

33. As case studies in this question of N. Luhmann, »The Third Question: The Creative Use of the Paradoxes in Law and Legal Theory«, *Law and Society Review* **15** (1988), 153-165.

34. See R. Glanville, »Distinguished and Exact Lies«, in R. Trappl (ed.), *Cybernetics and Systems Research* 2, Amsterdam: North-Holland Publications, 1984, p. 655-662.

35. N. Rescher expresses this perspective on perspectivity: »Perspectives are diaphanous, and one tends not to see them as such«, in his *The Strife of Systems: An Essay on the Grounds and Implications of Philosophical Diversity*, Pittsburgh: University of Pittsburgh Press, 1985, p. 187.

36. See F.J. Varela, »A Calculus for Self-Reference«, *International Journal of General Systems* **2** (1975), 5-24.

37. See in this regard N. Luhmann, »Die Autopoiesis des Bewußtseins«, in A. Hahn/V. Kapp (eds.), *Selbstthematisierung und Selbstzeugnis: Bekenntnis und Geständnis*, Frankfurt am Main: Suhrkamp, 1987, p. 25-94.

38. Furthermore, what is offered as epistemology in the context of the cognitive sciences is inconceivable without computers - both from the theoretical aspect. See F.J. Varela, *The Sciences and Technology of Cognition: Emerging Trends*, Ms., Paris, 1986. This is also true of logical theories and their truth procedures.

39. This is also held by strict constructivists. See, for example, E. v. Glasersfeld, »Konstruktion der Wirklichkeit und des Begriffs der Objektivität«, in Gumin, Mohler, *op. cit.*, 1985, note 5, p. 1-26 (20f).

40. See Maturana, *op. cit.*, 1982, note 10, p. 52ff. The literature that follows uses »parallelion« or »parallelizing«, instead of »analogy«. See, for example, P.M. Hejl, »Konstruktion der sozialen Konstruktion: Grundlinien einer konstruktivistischen Sozialtheorie«, in Schmidt, *op. cit.*, 1987, note 5, p. 303-339.

41. Related to this, but nonetheless to be distinguished from it, are the well-known attempts of ... to use the process of learning a language for the elucidation of epistemological questions. See, for example, W. v. O. Quine, *Word and Object*, New York: Wiley, 1960. See also the less well known ideas of D.T. Campbell, »Ostensive Instances and Entita.... in Language Learning«, in W. Gray/N.D. Rizzo (eds.), *Unity Through Diversity, A Festschrift for Ludwig von Bertelanffy*, New York: Gordon and Breach, 1973, Vol. II, p. 1043-1057. The question dealt with here is the claim that the learning of language is not

possible without reference to things of the external world, which means that language can never completely construct reality out of itself.

42. See Luhmann, *op. cit.*, 1984, note 23, p. 191ff.

43. See here also N. Luhmann, »Intersubjektivität oder Kommunikation: Unterschiedliche Ausgangspunkte soziologischer Theoriebildung«, *Archivio di Filosofia* **54** (1986), p. 41-60.

44. See, for example, Campbell, *op. cit.*, 1970, note 28, p. 51-85.

45. See E. Husserl, »Die Krisis der Europäischen Wissenschaften und die transzendentale Phänomenologie«, *Husserliana*, Vol. VI, Den Haag: Nighoff, 1954. See also A.N. Whitehead, *Science and the Modern World* (Lowell Lectures 1925), quoted from the edition New York: Free Press, 1954.

II. Selfreference and Selfregulation in Social Systems

How the Law Thinks: Toward a Constructivist Epistemology of Law*

GUNTHER TEUBNER
University of Bremen/European University Institute Firenze

I. Jabberwocky

> »Twas bryllig, and the slythy toves did gyre and gymble in the wabe: all mimsy were the borogoves; and the mome raths outgrabe.«
>
> American law professor commenting on Niklas Luhmann, »The Unity of the Legal System«.

European and American scholars of law and society apparently have problems in communicating with each other. To invoke Lewis Carroll's authority on a piece of legal theory indicates how serious the problems are. After all, traced to its true origins, »Jabberwocky«, the famous »Stanza of Anglo-Saxon Poetry«[1] means »weeks of woe« in its original German version.[2] And inextricably involved in the interpretation of the poetry is a certain Hermann von Schwindel...

This lack of mutual understanding is only a recent phenomenon. Communication was still easy when Merton's regime of middle-range theories was governing law and society. There was a consensus that from the patient observation of the real law in the real world, a body of non-speculative, non-metaphysical theories would evolve. And this consensus was reflected in a common, sober, professional, comprehensible language. However, with the return of Grand Theory, with the invasion of post-structuralism, critical theory, self-referentiality and autopoiesis in the socio-legal world, the unified discourse of law and society is falling apart again into different cultural provinces. The deplorable result is a fragmentation of theory languages, the »Jabberwocky« of socio-legal theory.

Obscurity of language, then, is the most common critical comment on those recent European theory-fashions be they of Parisian, Frankfurtian or Bielefeldian origin. The language is said to be overly complex, often incomprehensible, and to conceal usually trivialities behind a smoke-screen of trendy words like legal discourse, communicative rationality or legal autopoiesis.

Of course, bad translations play an unfortunate role in this exchange of ideas. And national cultural contexts are still today so diverse that the transplantation of a theory from one context to the other leads to a degree of incomprehensibility which can only be gradually reduced by careful explanation. And one should also concede that sometimes personal idiosyncracies of theorists render their texts needlessly difficult to understand. However, the core of the problem lies elsewhere. It is a question of whether the language is complex enough to match the complexity of the subject matter. The new theories on law claim to construct socio-legal realities that cannot be adequately

W. Krohn et al. (eds.), Selforganization. Portrait of a Scientific Revolution, 87–113.

expressed by ordinary language. For them, to give in to the demands of easy comprehensibility would be to compromise on the content of their message.

Let us take a concrete example. In the context of legal autopoiesis, several authors are working on a new theory of the legal person.[3] In their language, »... the social reality of a legal person is to be found in the collectivity: the socially binding self-description of an organized action system as a cyclical linkage of identity and action«. ... What? More Jabberwocky? Do organizations think? How can they have the capacity to describe themselves? Linkage of identity and action? All this sounds like those infamous mystifications of collectivities. Obviously, collectivities do not act, but only individuals, and it is nothing but individual actions that are aggregated into collective action. So why not go back to Max Weber's more sober and comprehensible formulation of the same subject matter: »These concepts of collective entitities... have a meaning in the minds of individual persons, partly as of something actually existing, partly as something with normative authority. This is true not only of judges and officials, but of ordinary private people as well. Actors thus, in part, orient their action to them, and in this role such ideas have a powerful, often a decisive, causal influence on the course of action of real individuals«.[4]

But is it still the same? Certainly, one can now easily understand the words. The message, however, is lost. The novelty of the construction lies in the following issues which depart point by point from the world views invoked by ordinary language: 1. Organizations do not consist of human individuals as members, but of communications, more precisely of decisions as their self-constituted elements; 2. organizations do »think«; it is via internal communication that they construct social realities of their own, quite apart from the reality constructions of their individual members; in short, organizations are epistemic subjects; 3. organizations are not *per se* capable of collective action; they transform themselves into collective actors by communicatively constituting their identity; 4. the capacity for collective action emerges when organizations in their collective identity produce actions and vice versa organizational action produces their collective identity. Obviously, these four issues suggest a social reality of the legal person that lies far beyond the well-known territories of fiction, group or entity theories of corporate personality.[5]

This example should have made clear that the above mentioned communication problem is not due to obscurity in language but to the limited capacity of our language to express the construction of newly perceived social realities. This, at least, is what the following new theories on legal epistemology have in common. It is true, Michel Foucault, Jürgen Habermas and Niklas Luhmann »gyre and gymble in the wabe«, but they do so because they imagine social realities the reconstruction of which clearly goes beyond the limits of ordinary language. What makes them seemingly incomprehensible is their radical departure from epistemological premises that are deeply embedded in contemporary thinking on law and society, particularly from what Pizzorno polemically calls the reification of a »*metafisica quotidiana*« - epistemological realism and methodological individualism.[6] Although poststructuralism (Foucault), critical theory

(Habermas) and the theory of autopoiesis (Luhmann) develop quite different visions of modern law, they converge in their anti-realism and their anti-individualism.[7]

One should hasten to add that anti-realism does not mean epistemological idealism, and anti-individualism does not mean methodological holism/collectivism. We are not confronted with a revival of the old dichotomies realism/idealism and individualism-/collectivism that dominated the legal theory debates in the first half of the 20th century. It is neither Kelsen nor Duguit who are on the agenda of legal theory today. Rather, in the return of grand theory, epistemological realism is replaced by a new epistemological constructivism and the agents of methodological individualism are replaced by constructs such as discourse, social self-reflection and self-organization. What does this radical re-orientation of social theory mean for law?

For law, the crucial point is the combination of both the change in epistemology and the new perception of individuality: epistemological constructivism rules out the naive reality assumption that human actors through their intentional actions make up the basic elements of society. From this combination follow the main theses of this article:

1. Under a constructivist social epistemology, the reality perceptions of law cannot be matched to a somehow corresponding social reality »out there«. Rather, it is law as an autonomous epistemic subject that constructs a social reality of its own.

2. It is not human individuals by their intentional actions that produce law as a cultural artifact. On the contrary, it is law as a communicative process that by its legal operations produces human actors as semantic artifacts.

3. Since modern society is characterized on the one side by a fragmentation into different epistèmes, on the other side by their mutual interference, legal discourse is caught in an »epistemic trap«. The simultaneous dependence on and independence from other social discourses is the reason why modern law is permanently oscillating between positions of epistemic autonomy and heteronomy.

»Social construction of reality« apparently has become, after Berger and Luckmann, received wisdom in sociology.[8] However, our three theses show that there is a more profound version of social epistemology than the usual understanding of how social institutions, scientific communities and laboratory cultures influence individual perception. There is more to social epistemology than the »interests« of social agents that are responsible for the manipulation of knowledge. The three new theories under consideration here - poststructuralism, critical theory and theory of autopoiesis - have radicalized the notion of the »social« in social epistemology which is worthwhile being examined in our context of legal cognition. What is the precise meaning of the somewhat ambiguous statement that law constitutes an autonomous reality? Similarly, what is meant by saying that the individual is a mere construct of society and law? And, above all, how does the law »think«?

II. Discourse and Autopoiesis

»It is comforting, however, and a source of profound relief to think that man is only a recent invention, a figure not yet two centuries old, a new wrinkle in our knowledge, and that he will disappear again as soon as that knowledge has discovered a new form.«

It is not only law and economics that are irritated by Michel Foucault's anti-realist and anti-individualist provocation[9], but most strands of social theory that are influencing modern legal thought feel uncomfortable with poststructuralism's decentering of the subject. Under the enormous influence of the »founding fathers« of methodological individualism, Hayek[10] and Popper[11], the quasi-natural reality of individual human actors is assumed by contemporary economic and social theories, such as theories of micro-foundations[12] and rational actor theories[13] which demand that any collective phenomenon be reduced to intentional actions of human individuals. In an analogous fashion, the reduction of social macro-phenomena to characteristics of individuals is quasi-axiomatic for sociological behaviorism.[14] But also for theories on law in the tradition of Max Weber's interpretive sociology, the reality of the acting individual is a fundamental assumption. »After all, the actions of individuals form society«.[15] And even social theorists pursuing structuralist and systemic approaches feel compelled to correct them with an infusion of individualism.[16]

And it is indicative of the epidemic character of the individual-as-reality-syndrome that even critical legal authors who are deeply influenced by Foucault's ideas and enthusiastically take over his political messages plainly refuse to draw the epistemological consequences. Duncan Kennedy, in his recent analysis of legal indeterminacy[17] reveals a highly individualist bias for the reflective legal subject and a benign neglect of law's communicative aspects. Thus, concentrating on the individual judge's reflections and strategic considerations, he is as far away from a discourse analysis as are his »liberal« adversaries. And Robert Gordon[18] explicitly rejects the anti-individualist tendencies in structuralism and post-structuralism as undermining the humanistic intentions of critical legal thought.[19]

What makes this combination of realism and individualism in contemporary legal thought so viable is not so much its inherent virtues but the lack of credible alternatives. The traditional alternatives, epistemological idealism and methodological collectivism, are seen as unattractive - and rightly so. But is it true that the only available alternatives are those that read »... as if these impersonal structures had a life of their own and human beings were enslaved to the needs of that life-cycle, building or demolishing as the World-Spirit might dictate«?[20]

As I will discuss in the following pages, there are alternatives to the prevailing realist and individualist modes of legal thought. From the diffuse contemporary movement toward »social construction of reality« and »decentering of the subject«, I would like to single out three theorists who have contributed to a more profound understanding of socio-legal epistemology and who represent at the same time important intellectual strands in Western Europe: Michel Foucault (post-structuralism), Jürgen Habermas (critical theory) and Niklas Luhmann (theory of autopoiesis). What they have in common is to replace the autonomous individual, not with supra-individual entities, but with communicative processes. They differ, however, in their identification of the new cognizing unit. In Habermas' version of critical theory, correspondence theories of truth are overturned by consensus theories and »intersubjectivity« takes the place of the epistemic subject. Foucault and Luhmann are more radical in their disenchantment of the human individual. For Foucault, the human individual is nothing but an ephemeral construction of an historically contingent power/discourse constellation which dictates the »epistème« of a historical epoch. Luhmann completely separates psychical processes from social ones and perceives the human individual in society as a communicative artifact, as a product of self-observation of social autopoiesis. The new epistemic subjects are autopoietic social systems.

III. Jürgen Habermas: Intersubjectivity and Consensus

To arrive at a legal epistemology that really deserves its name, three important changes in our perception of law and society have to be made: first, from realism to constructivism; second, from individual to social construction of reality; and third, from law as a rule system to law as an epistemic subject. While the first one leads to a certain modification of Kantian positions, the other two changes break new ground in social and legal theory. The second change reveals the social foundations of cognition in a more radical way than traditional sociology of knowledge ever has done, and the third one attributes to the discursive practices of law the production of an autonomous social reality.

In this reorientation of social and legal epistemology, Habermas's theory of communicative rationality[21] plays a prominent role. Habermas's key concept of »rational discourse« highlights the crucial role of procedure in empirical and normative cognition and at the same time his »universal pragmatics« takes account of the social dimension in moral and legal cognition, as against a predominantly individualist epistemology.

Habermas rejects traditional correspondence theories of truth (from Aristotle to Tarski) according to which statements are true if they correspond to an external reality. Instead, he follows a consensus theory of truth which declares as criterion of truth the »potential« consensus of all discourse participants.[22] This move, of course, creates the need to identify an independent criterion in order to distinguish true from false consensus. Going through a sequence of different criteria, Habermas finally finds it in the presupposition of an »ideal speech situation« which in itself is defined by certain formal and procedural characteristics.[23]

It is this proceduralization of the truth criterion which has rendered Habermas's discourse theory so important for law.[24] It makes the theoretical-empirical discourse of the sciences directly comparable to the practical-normative discourse in politics, morals and law: their validity claims depend on the correctness of procedure.[25] And it opens the way to a rethinking of the modernity of law in which Max Weber's thesis of the materialization of formal law is replaced by concepts of proceduralization of law.[26]

Habermas' other main contribution to an epistemology of law is to take account of the social element in empirical and normative cognition. His philosophy attributes »epistemic authority« no longer to the autonomous subject, but to the communicative community.[27] While traditional epistemology situates cognition exclusively in the consciousness of the (empirical or transcendental) subject, Habermas recognizes that cognition is basically a communicative process. »Intersubjectivity« takes the place of the Kantian epistemic subject. It is the authentic consensus of the communicative community and not the consciousness of the autonomous individual that determines truth in cognitive and normative issues. Thus, Kant's famous question: »What are the conditions for the possibility of cognition?« is redirected from the conditions of consciousness to those of communication. And even transcendentalism becomes socialized. The new *a priori* is represented by the »ideal speech situation,« the presupposition of which is a condition of the possibility of communication.[28]

However, the »*a priori* of the communicative community«[29] is at the same time one of the great problems of this theory. With the apriorization of certain features of communication, Habermas attempts to escape from the »paradoxes of self-reference«[30] that necessarily emerge from his hierarchy of discursive justification. The core of Habermas's theory is in the self-application of discursive practices: the procedures of discourse can be justified only by discourse whose procedures in turn have to be justified by discourse.[31] And in order to avoid infinite regression or circularity, Habermas resorts to communicative transcendentalism.

Closely related to the transcendentalist foundation of rational discourse are the ambiguities of »intersubjectivity« that represent the other principal unresolved problem in Habermas's epistemological account. What is meant: elements or relation? Consciousness or communication? Psychical or social processes? Habermas's epistemic subject oscillates between these two positions without ever finding its identity in one world or the other.[32] It seems as if Habermas again attempts to avoid the paradoxes of self-reference in discourse, this time by changing the system reference. If discourse can be founded on discourse only recursively, need it not then be founded on human consciousness?

IV. Michel Foucault: Discourse and Epistème

Foucault's ideas on discourse and power can be read as a radicalization of Habermas's epistemological position. Indeed, Foucault directly attacks what we have just described as the main unresolved problems in Habermas's account: the foundation of discourse in a communicative *a priori* and the ambiguous role of individual consciousness in intersubjectivity. Foucault's main contribution to a social epistemology is to liberate the core concept of »discourse« from any transcendental or any psychical foundation. Of course, this does not save him from the traps of self-referentiality. Foucault's escape is at the same time the most famous and the weakest point of his theory - the ubiquity of power.

Foucault's starting point is constructivist: reality is not something external to cognition but is constituted, »constructed« by cognition itself. However, in sharp contrast to the classical tradition, it is not the individual consciousness of the subject that constitutes reality. Nor is intersubjectivity, as in Habermas' theory, the communicative result of interaction between human actors. Rather it is »discourse« - an anonymous, impersonal, intention-free chain of linguistic events.[33] One should hasten to add that this is not a structuralist position.[34] Discourse in Foucault's account is much richer than the abstract orders of signs in structuralism. It is social practice, not social structure; it is *parole*, not *langue*. The basic elements of the discourse are not signs, but »énoncés«, i.e. social usage of language that constructs reality. The task of discourse analysis does not consist, in Foucault's words, of »treating discourses as groups of signs (signifying elements referring to contents or representations) but as practices that systematically form the object of which they speak«.[35] Discourse is both event and structure, »a stream of linguistic events in space and time as well as a highly selective organization of linguistic events«.[36] And it is this historically contingent social practice of discourse that dictates the »epistème« of a certain historical epoch, that defines the conditions for the possibility of cognition, not in an atemporal universal manner, but temporally, concretely, locally.[37]

Such a radical social epistemology has no place for individual consciousness and the intentional actions of human subjects and no need for an *a priori* foundation. The human subject is no longer the author of the discourse. Just the opposite: the discourse produces the human subject as a semantic artifact.[38] At the same time, discourse formations are historically contingent, lacking any *a priori* foundation. Every society has its own order of truth, its own politics of truth.

Now, it would be a consequence of this way of thinking that discourse formations, those highly autonomous social practices, would themselves produce the criteria for their own transformation. Dreyfus and Rabinow, for example, clearly see this necessary self-referentiality as a condition for structural change of discourses, since »... he is committed to the view that discursive practices are autonomous and determine their own context ... he must locate the productive power revealed by discursive practices in the regularities of these same practices. The result is the strange notion of regularities which regulate themselves«.[39] Foucault, however, stops short of those paradoxes of self-reference. He withdraws from the consequences of his own construct and introduces the concept of power in order to externalize self-referential relationships. In his later thinking he gives up the idea of the autonomous discourse as the new epistemic subject and resorts to the ubiquity of power as a quasi-transcendental foundation of discursive practices.[40]

V. Niklas Luhmann: Constructivism and Autopoiesis

The paradoxes of self-reference seem to be the principal obstacle to the development of an authentically social epistemology. Habermas and Foucault have made important contributions, but the radical consequences of their ideas seem to be blocked by self-referential structures (circularity, tautology, infinite regression, paradox) in their specific versions of discourse theory. How can rational discourse be justified, if not by rational discourse itself (Habermas)? How can those discourse formations that govern the epistème of a whole historical epoch be transformed if not by those discourse formations themselves (Foucault)? Both authors are well aware that these questions necessarily lead to paradox, but their solution is to avoid the paradox at any cost. Of course, in the end, the paradoxes of self-reference cannot be avoided, they simply reappear at the termination of their escape route. When Habermas finds the transcendental foundation of communication in the distinction between the ideal speech situation and real speech situations, is this distinction, then, in itself empirical or is it transcendental? Alternatively, when he reintroduces the subject to the discourse, the classical paradoxes of the self-reflecting subject are obviously bound to reappear. When Foucault identifies the foundations of discourse in ubiquitous power-constellations he does so at the price of the self-referential paradoxes of power.

The theory of autopoiesis[41] deals with these paradoxes of self-reference in a different way: do not avoid paradoxes, but make productive use of them! If social discourses are autopoietic systems, i.e. systems that recursively produce their own elements from the network of their elements, then they are founded on that very self-referentiality that Habermas and Foucault are desperately trying to avoid.[42] As autopoietic systems, discourses cannot but find justification in their own circularity and cannot but produce regularities that regulate themselves and that govern the transformation of their own regularities. The paradox of self-reference then is not a flaw in our intellectual reconstruction of discourse which we have to avoid at all costs, but is its very reality which we cannot avoid at all. And the recursive application of operations to the results of these very operations does not necessarily lead to paradoxical blockage paradoxes or to sheer arbitrariness, but, under certain conditions, to the emergence of »eigenvalues«.[43] From continual recursive »computation of computation«, social discourses »blindly« learn those modes of operation that are valid in coping with their environment to which they have no direct access.[44] The epistemological consequence is a radical constructivism.[45] Any cognition - be it psychical or social, be it scientific, political, moral or legal cognition - is a purely internal construction of the outside world; cognition has no access whatsoever to reality »out there«. Any cognitive activity - be it theory or empirical research - is nothing but an internal construction by the cognizing unit; and every testing procedure that pretends to examine the validity of internal constructions against outside reality is only an internal comparison of different world constructions.

In this radicalized version of the »social construction of reality«, there again is no place for individual action and thought.[46] Social autopoiesis is exclusively based on communication - defined as the synthesis of utterance, information and understanding - that recursively reproduces communication.[47] Social construction of reality is sharply separated from psychical construction of reality. Here lies the important difference from Habermas, who in the ambiguous concept of intersubjectivity blends communication and consciousness, and also from Foucault, for whom the subject is nothing but a historically contingent construct of shifting discourse/power constellations. For the theory of autopoiesis, psychical processes form a closed reproductive network of their own -

psychical autopoiesis - accessible only to themselves and inaccessible to any communication. Communication in turn forms a closed autoreproductive network of its own - social autopoiesis - accessible only to communication and inaccessible to any psychical processes. Certainly, human individuals reappear in this world of communication, but only as communicative constructions, as semantic artifacts that have no correspondence to consciousness, to the autopoietic processes in the psychical world.[48] Psychical and social processes do co-exist. They are »coupled« by synchronization and co-evolution, but there is no overlap in their operations. There is nothing but a symmetry of reality constructions: psychical processes produce mental constructs of society, and social processes produce communicative constructs of the psyche.

In these two aspects - radicalization of constructivism and de-individualization of discourse - Luhmann is expanding on what Habermas and Foucault have developed in their versions of social epistemology. However, there is a third aspect in Luhmann's theory of autopoiesis that clearly goes beyond discourse analysis in its Parisian or Frankfurtian version - this is the view of modernity as an irreconcilable conflict of different »epistèmes«.[49] While Foucault sees in history the ruptures of discourse formations that dictate one paradigmatic society-wide »epistème« for a certain historical epoch and interprets the modern epoch as the governance of one pervasive »subjectivist« epistème following the Kantian revolution[50], Luhmann views modernity as the fragmentation of society into a plurality of autonomous discourses, as the multiplication of »epistèmes« in society. The crucial feature of modern society is the loss of a unifying mode of cognition. Society is seen as fragmented into a multiplicity of closed communicative networks. Each communicative network constructs a reality of its own that is, in principle, incompatible with the reality constructions of other networks. At the same time, there is a multiplication and fragmentation of individualities which corresponds to the multiplication and fragmentation of social discourses. On the basis of its specific code and programs, each specialized communicative network produces »persons« - semantic artifacts of individual actors - to which actions are attributed.[51] The »Multiple Self«[52] is the product of the fragmentation of social discourses in modernity.

This fragmentation of society into different epistèmes is one of the strongest points in Luhmann's theory - and at the same time its »blind spot«. The emphasis on fragmentation, differentiation, separation, closure and self-reference of social epistèmes creates problems, to say the least, as to how their interconnection, interference, openness and hetero-reference can be theoretically reconstructed.[53] Unlike Habermas and Foucault who, at any cost, try to avoid the traps of self-reference, Luhmann courageously faces self-referential realities in law and society. He even declares law to be founded on the paradoxes of self-reference.[54] But a theory that deals extensively with self-reference, may ultimately be caught in the self-created closure of self-referential constructions. And the obvious problem that autopoiesis theory has to face is how to deal with the interrelations of different autonomous epistèmes, their conflicts, their incompatibilities, their interferences.[55] The open questions for a theory of fragmented epistèmes are: is there something like an epistemic minimum in modern society that serves as a common base for the autonomization of social discourses? Does one find co-variation or even co-evolutionary trends among autonomous social epistèmes? Or is the only way to connect them through the reconstruction of an epistème within the framework of another epistème? These questions will reappear when we examine in detail, on the basis of the foregoing discussion, how a constructivist epistemology of law reconstructs legal cognition in its conflicts with other modes of cognition in society.[56]

VI. Law - an Epistemic Subject?

How does the law think? Mary Douglas, in a recent book, has again raised the old question: »How Institutions Think«.[57] After an exciting flirtation with Emile Durkheim's *»collective consciousness«* and Ludwig Fleck's *»Denkkollektiv«*, she finally finds her way back to good old individualism: of course, it is the individual member of the institution that thinks. However, his/her thinking is influenced by an institutional context. In this version of social epistemology, the social element is represented by socialization of the individual mind. That's it. Collectivism is banned and individualism happily survives after a healthy dose of socialization.

From our selective reconstruction of Habermas, Foucault and Luhmann on social epistemology, the picture changes dramatically. It is true that individual cognition is shaped by social institutions such as law, via socialization (and here constructivism would add that since there is no access from communication to consciousness, socialization can only be self-socialization). But this is only half the story. The other half is that institutions such as law do »think« independently from their members' minds. The law autonomously processes information, creates worlds of meaning, sets goals and purposes, produces reality constructions and defines normative expectations - and all this quite apart from the world constructions in lawyers' minds. Such a constructivist legal epistemology is at the same time non-individualist and non-collectivist. It needs no recourse to individual actors and intentions; at the same time, it does not presuppose the existence of a supra-individual collective entity, *Denkkollektiv*, *conscience collective*, World III, legal consciousness, *Weltgeist*...

Law is communication and nothing but communication. By this very conceptualization it is possible to avoid the traps of methodological individualism that would define law as a set of rules constraining individual action and that, apart from the catch-all phrase of unintended consequences, has no tool with which to analyze of the autonomy of the social, not to speak of the »legal proprium«.[58] At the same time it avoids the traps of collectivism that views law as a supra-individual subject and that cannot explain who is, in fact, acting in the name of the *Weltgeist*.

The precise construction is as follows[59]: law is defined as an autopoietic social system, i.e. a network of elementary operations that recursively reproduces elementary operations. The basic elements of this system are communications, not rules; law is not, like analytical-normativist legal theories have it, a system of rules. On the other hand, the sociological-realist definitions of law as a system of legal professionals and organizations are problematic as well because they see human actors as the basic elements of law and other social institutions. The self-reproductive character of law as a social process becomes intelligible only if one chooses communications as the law's basic elements. Law as an autopoietic social system is made up neither of rules nor of legal decision-makers, but of legal communications, defined as the synthesis of three meaning selections: utterance, information and understanding. These communications are interrelated to each other in a network of communications that produces nothing but communications. This is what is basically meant by autopoiesis: the self-reproduction of a network of communicative operations by the recursive application of communications to the results of former communications. Law as a communicative network produces nothing but legal communications.

Legal communications are the cognitive instruments by which the law as social discourse is able to »see« the world. Legal communications cannot reach out into the real outside world, neither into nature nor into society. They can only communicate about nature and society. Any metaphor about their access to the real world is misplaced. They do not receive information from the outside world which they would filter and convert according to the needs of the legal process. There is no in-struction of the law by the outside world; there is only con-struction of the outside world by the law. This is not to say that the law arbitrarily »invents« social reality. A constructivist perspective should not be confused with »methodological solipsism«[60]; it rather looks for a »middle path« between representationalism and solipsism.[61] Legal constructivism, then, presupposes the »existence« of an environment for the law. The point is not a monadological isolation of the law but the autonomous construction of legal models of reality under the impression of environmental perturbations. Legal order from social noise!

What about the world perceptions of lawyers and lay people? Is it not their aggregation that forms the collective world view of the law?[62] Of course, the communicative process of law needs lawyers and lay people; it would not work without their intentions, strategies and actions. But their (subjective, internal, psychical) intentions never enter the (objective, external, social) communication of law. They only make up part of the psychical processes, accompanying the social process of law and co-evolving with it. Law as a communicative process is not accessible to any of those accompanying psychical processes of lawyers and lay people, and, vice versa, it has no access to them. They work only as »perturbations«, as *»chocs exogènes«*[63] under the pressure of which the communicative process of law builds up its own autonomous order and creates the world of legal meaning.[64]

But does the law as a social process not constantly deal with real people? Is the law not driven by actual motives, strategies, actions of clients, professionals, judges and legislators? Does the law not constantly refer to mental states of real people, to their intentions, goals, consent, dissent, errors, negligence, *mens rea*? - Obviously, the law does so. But the »persons« the law as a social process deals with are not real flesh-and-blood people, are not human beings with brains and minds, are not the above mentioned autopoietic psychical systems. They are mere constructs, semantic artifacts produced by the legal discourse itself. Mental states are »in reality (sic!) constructs of practical discourses, necessary for the formation of communicative circles, of discursive communities«.[65]

As social constructs, they are indispensable to legal communication because law as a social process needs to attribute communication to actors (individual or collective ones) in order to continue its self-reproduction. But these »actors« are only role-bundles, character-masks, internal products of legal communication.[66] The densely populated world of legal persons (the plaintiffs and defendants, the judges and legislators, the parties to a contract, the corporations and the state) is an internal invention of the legal process. Not only the corporation, but any legal person - be it collective or individual - is nothing but that famous »artificial being, invisible, intangible, existing only in contemplation of law«, rediscovered by Chief Justice Marshall in the celebrated case of Dartmouth College v. Woodward, 4 Wheaton 518, 627 (1819).

So human actors have a »double identity« in the world of autopoiesis. While in their social existence they are pale constructs of autopoietic social systems, among them the law; in their psychical existence, they are themselves vibrant autopoietic systems. It is plainly wrong to argue, as some critics do, that autopoiesis de-humanizes society[67], has no place for actors and intentions[68], does not account for the individual as epistemic subject.[69] The point is not the individual subject withering away but the multiplication of centers of cognition. Social discourses are the new epistemic subjects that compete with the consciousness of the individual. Insofar as autopoiesis insists on the epistemic autonomy of a multiplicity of social discourses, it takes part in »decentering the subject«, i.e. moving the subject away from its privileged position as the sole and ultimate center of cognition. To repeat, if we talk about human actors in the law we have to carefully distinguish between the autopoietic reproduction of human consciousness, i.e. the operative reality of psychical processes, and the autopoietic reproduction of the social life of law in which human actors are not elements but constructed social realities.

VII. The Epistemic Trap

While discourse analysis in the tradition of Foucault sees the modern epoch in the grip of one pervasive »epistème«[70] and views law like other disciplines only as a particular expression of the power/knowledge complex[71], autopoiesis theory characterizes modern society as fragmented into multiple autonomous »epistèmes«.[72] Autopoiesis, thus, throws modern legal discourse into an irreconcilable conflict between epistemic autonomy and heteronomy.[73] The dynamics of social differentiation force legal discourse to produce reality constructions of its own, but the very same dynamics make law dependent upon a multiplicity of competing autonomous »epistèmes«.

The epistemic autonomy of law results from the fragmentation of modern society that drives the law into second order autopoiesis.[74] In the dynamics of social evolution, self-referential relations are multiplying within the legal process, culminating in a hypercyclical linkage of the law's components. The law becomes autonomous from general social communication. It develops into a closed communicative network that produces not only legal acts as its elements and legal rules as its structures, but legal constructions of reality as well. The autonomy of modern law refers primarily to its normative operations which become independent of moral and political normativity[75]; and secondarily, autonomy refers to the law's cognitive operations that - under the pressure of normative operations - construct idiosyncratic images of reality and move them away from the world constructions of everyday life and from those of scientific discourse.[76]

In this context, Baudrillard[77] speaks of »hyper-reality« as a movement from reality-dependent theory to theory-dependent reality. In an autopoietic reformulation, one would describe this process as an autonomization of specialized social discourses in which reality constructions of general social communication are increasingly replaced by reality constructions of the specialized discourses. The legal discourse invents and deals with a juridical »hyper-reality« that has lost contact with the realities of everyday life and at the same time superimposes new realities on everyday life. It is an »efficacité quasi magique«, as Bourdieu calls it, which law possesses in its practices of »world making«.[78] »Institutional facts« such as corporate personality, contract, and the will are only the tip of an iceberg of legal reality constructs drifting in an ocean of »brute facts« of diffuse social communication. Legal discourse increasingly modifies the meaning of everyday world constructions and in case of conflict replaces them by legal constructs.

From a constructivist perspective, there is no way to challenge the epistemic authority of law, either by social realities themselves, or by common sense, or by scientifically controlled observation. A social epistemology on a constructivist basis can explain why law appears to be an »essentially self-validating discourse« which one should expect to be »largely impervious to serious challenge from other knowledge fields«.[79] It is simply naive to invoke social »reality« itself against legal conceptualism, against the »heaven of legal concepts«[80] or against the law's »transcendental nonsense«.[81] There is no direct cognitive access to reality. There are only competing discourses with different constructions of reality, and all that Jhering and Cohen have to offer is their own »transcendental nonsense« in a different »heaven of legal concepts«. Is there any reason to believe that *Freirecht*, sociological jurisprudence or legal realism have made the legal discourse more realistic? Not at all. They have not moved legal concepts closer to social reality »out there«. They have just replaced one conceptual jurisprudence with another conceptual jurisprudence. »Social interests«, the atoms of realistic jurisprudence, are unreal fictions, artificial semantic products, just as much as the »legal subjects«, the atoms of classical jurisprudence.

»Law and society«, »law and economics« are not doing any better if they pretend to invoke the authority of controlled scientific observation against the lawyers' »mystifications« of the social world.[82] If epistemological constructivism does anything, it is to deconstruct the claims of modern science to having privileged access to reality.[83] Science does not discover any outside facts, it produces facts. »Science is in a literal sense constructive of new facts«.[84] Radical constructivism maintains that »...science produces a construction of the world which is validated by its distinctions and not by the world as such. Thus, science cannot claim the authority to discover the only and the correct access to the real world and to communicate this to others«.[85] If we can believe constructivist reconstructions of the scientific process, then the celebrated controlled experiment is not what it pretends to be, a test of an internal theory against external reality, but is a mere internal coherence test comparing two constructs that are produced according to different procedural requirements: the logic of theoretical reasoning and the logic of the laboratory.

Social science theory, say, on the relation between organization and collective action, is not in any way superior to legal doctrine on the relation between the corporation and legal personality; both are discursive artifacts the construction of which is not arbitrary, but rationally guided by specific codes and programs. Similarly, empirical facts, say, about dysfunctions in organizational life, hard facts that result from scientifically controlled inquiry, are in no sense more »true« than legal facts about the violation of corporate duties which are produced under the firm guidance of the rules of law of evidence. In both cases, rational procedures and conventions of factual inquiry lead to empirical statements about reality. They serve as »hard« evidence confirming or refuting »soft« claims based on theoretical speculations or on legal reasoning. And if these empirical facts conflict with each other - which is not so rare - then there is no superiority of scientific constructs over legal constructs, as some sociologists would like to have it.[86] Epistemic authority is claimed by both scientific discourse and legal discourse - and rightly so. What a naive realism would call the observation of »facts« is in both cases the production of artifacts the truth of which is guaranteed by formalized procedures of factual inquiry, procedures that differ considerably in law and science. These procedures in turn are conventions, not arbitrary ones but structural selections which reflect choices made in the history of scientific and legal discourse.[87]

The epistemic authority of legal discourse is an undeniable fact of modernity, and we have found ways and means to cope with the fact of multiple truths - scientific truth, legal truth, political truth ... . *Res judicata* is the classical example of an institutionalized conflict between legal facts and scientific facts. Even if it can be proven with scientific evidence that a factual statement in a legal procedure was blatantly wrong, the factual statement of the court and, what is worse, its legal, economic and social consequences will not be reversed (apart from very few, narrowly defined exceptions), if the procedural requirements are fulfilled and the appeal procedures exhausted. Obviously, scientific facts collide with legal facts, but we are used to living with this collision, rationalizing it by invoking higher values like legal certainty or appealing to the relativism of our cultural provinces.

However, things are not quite so easy. Windscheid's notorious »lawyer as such«[88] who is entitled by the law of social differentiation not to be »concerned with ethical, political or economic considerations« is forced by the same law to give up the entitlement and to incorporate those non-legal considerations into his/her autonomous reasoning. This is what I would call the »epistemic trap« of modern law. Law is forced to produce an autonomous legal reality and cannot at the same time immunize itself against conflicting realities produced by other discourses in society.

The underlying reason for this confusion is »interference«, that is, the mutual diffusion of law and other social discourses.[89] This is one of the most challenging problems for autopoietics if this theory intends to avoid the fallacies of solipsism and monadism.[90] Although the legal discourse is closed in its self-reproduction and produces its own constructions of reality, it remains always social communication and uses the general social constructions of reality and influences general social communication by its specific world constructions. Any legal act is at the same time - *uno actu* - an event of general social communication. One and the same communicative event, then, is linked with two social discourses, the specialized institutionalized discourse of law and the diffuse and general social communication. Interference of law and other social discourses does not mean that they merge into a multidimensional super-discourse nor does it imply that information is »exchanged« among them. Rather, information is constituted anew in each discourse and interference adds nothing but the simultaneity of two communicative events.[91] Thus, juridical constructs are exposed to the constructs of other discourses in society, particularly to the constructs of science. They are exposed to a test of »social coherence« that replaces the old fiction of a test of correspondence with outside reality.

In the world of non-legal communication legal constructs inevitably lose in this epistemic competition. Here, science has the advantage of having specialized in procedures for purely cognitive operations, while law uses cognitive operations only secondarily and has, thus, shaped the procedures of cognition in a different institutional context. But what about the world of legal communication in court rooms, law offices and legislative chambers? Here, the legal discourse claims to be entitled to »enslave« cognitive operations according to normative context and institutional purpose. The »empirical« models of legal communication are in the firm grip of »strategic« and »operative« models.[92] However, it is the institutional context of the legal process itself that produces an internal contradiction. While it requires idiosyncratic reality constructions through legal communication, it forces legal communication to re-construct the scientific constructs of reality and to expose - even within the law's domain - juridical constructs to the »higher« authority of science in cognitive questions. The conflictual character of legal procedures - litigation as well as legislation and scholarly disputes - forces legal discourse to examine any piece of new knowledge produced outside the legal world if only it is »relevant« to the law. Any practicing lawyer who does not challenge legal evidence in the light of a new scientific research method would act against his/her interests and violate his/her professional duties. In the legislative process, political opponents on, say, health-legislation will challenge legal measures once there is credible scientific evidence that the presupposed nexus between a disease and certain causal factors does not exist. And scholars in law and economics reap their highest reputational profits when they inform courts about their naive pre-scientific models of human behavior and propose scientifically proven alternatives.

The epistemic trap of modern law, therefore, produces a challenge of the first order to legal doctrine, legal theory and legal sociology. Relentlessly, legal doctrine - through the mouths of judges and law professors - comes up with positive proposals as to how to escape from the trap. Reflexively, legal theory helps to broaden the escape routes, generalizing particular solutions and importing supportive knowledge from other disciplines. And positive legal sociology zealously studies the correlations between those legal semantics and the broader socio-cultural context, while it remains the privilege of critical socio-legal studies to »trash« those attempts, to demonstrate to lawyers in a merciless deconstructive analysis that they are still in the old trap.

VIII. Escape Routes

To renounce epistemic authority, at least partially, would be the easiest way for legal discourse to escape from these troubles. Indeed, Luhmann who probably underestimates the possibility of conflict in authority among social *epistèmes*, seems to favor this escape route when he discharges the law from re-examining everyday interpretations and scientific constructs, like »woman«, »cylinder capacity«, »inhabitant«, »thallium«. »Should questions such as whether women, etc., really exist arise, they can be turned aside or referred to philosophy«.[93] Unfortunately, such a clean separation of social spaces does not exist. Moreover, with such a division of labor among social discourses one would not exploit the richness of the autopoiesis concept and would have to face empirical counter-evidence. In the day-to-day practice of legal decision making, law is constantly forced to decide autonomously on cognitive questions that are supposedly within the competence of scientific inquiry or of common sense. If the normative context of law requires cognitive statements on specific matters, then it is true that the law may start its operations with common sense understanding and with reference to science. But whenever in the legal process these cognitive statements become controversial - and this is usually the case for the politically and legally »hot« issues - then law can no longer turn them aside or refer them to philosophy. Then, *hic et nunc*, the legal process must provide for procedures to settle these divergences and must make a decision that is based on a legal determination of those questions, even if they are controversial or actually non-determinable in the sciences. More particularly, political and juridical conflicts in the environmental law area requiring much extra-legal scientific and technical expertise, show the great degree to which legal decisions have to be based on a specifically juridical assessment of scientific controversies or have to be made without any guidance from scientific results.[94]

The other main escape route from the law's epistemic trap is the integration of law and social sciences. Instead of clearly separating the realms of juridical cognition from those of scientific cognition, the legal discourse is supposed to incorporate social knowledge into its world constructions and permanently revise legal models of social reality according to the accumulation of knowledge in the social sciences. From the times of Jhering, Gény and Pound to the most recent variations of the »law and ...« movements, this has been the most challenging intellectual adventure of modern legal thought.

What can legal epistemology learn from almost a hundred years of experimentation with »law and social sciences«? Although social science thinking has been remarkably successful in influencing legal practice[95], the great expectations of legal enlightenment raised in academia have been dashed in the courtrooms. Psychiatry, sociology, policy analysis, economic analysis have successfully entered the legal sphere, but the result is not a greater degree of isomorphy of law and social reality that would result in more rational legal policies. Rather, the social science enlightenment of law has resulted in unanticipated consequences - the production of hybrid artifacts with ambiguous epistemic status and unknown social consequences.

»Interest analysis«, for example, is a surprising success of the efforts of »sociological jurisprudence« to replace formalist, conceptually derivative legal reasoning.[96] Today, interest analysis practically dominates legal decision-making in the courts: the courts analyze legal conflicts in terms of underlying conflicting social interests and »balance« them against each other according to standards that they infer from legislative goals expressed in a comparable context. But what is sociological about this type of sociological jurisprudence? No sociologist whatsoever would dare to follow lawyers in their attempts to conceptualize, operationalize and empirically identify those phenomena called »social interests« that figure prominently in legal decisions[97], not to speak of the juridical methods of »balancing« them. There are just too many explicit and implicit normative assumptions based on a complex network of legal-doctrinal considerations that enter into legal interest analysis. Juridical interest analysis cannot be legitimated from the standpoint of sociological theories or methods. In practice, interest analysis is a new conceptual jurisprudence that originally was subsidized by social science constructs but has for a long time gained an autonomy of its own. It may very well be that »interest analysis« contains elements of a new legal rationality[98], but they are surely different from the original goals of sociological jurisprudence and they evolve by institutional experimentation, not by the incorporation of sociological knowledge.

»Policy analysis« tells a similar story. Basically, it is a method of decision-making inspired by the instrumental use of social science knowledge.[99] Define the goals consented upon in the political process, determine the factual conditions of the regulatory situation, choose among the regulatory instruments according to nomological knowledge about means-ends relations, take into account side-effects, and, if you can, learn from practice about unanticipated consequences and perverse effects! But what has legal practice made of this »rational jurisprudence«? The lawyers have simply shifted their scholastic methods of doctrinal reasoning from the level of rules to the level of »policies«, purposes, goals and principles supplanting social science analysis by the obscure hermeneutics of »teleological« interpretation. Legal consequentialism has in practice become a caricature of a scientifically controlled, causal analysis supported by empirical evidence.[100] What counts as a relevant consequence of a legal rule or decision derived from legal doctrine is in a circular fashion defined by legal doctrine itself. Thus, doctrine which originally was supposed to be controlled by its social consequences now controls its social consequences. Moreover, the rational calculation of probable consequences of decision-making in practice turns out to be nothing but the common sense projection of judges. And consequentialism is taken seriously only on the level of rules and not on the level of individual decisions that are in practice never reversed if the calculation of consequences turns out to be wrong. Again, we are faced not with social science in law but with a new type of legal doctrine constituting »policies« as the new legal artifacts that replace old-fashioned rights and duties.

One could continue with the »poverty of psychiatry«. Is it conceivable, from the point of view of a positivist science, that a psychiatric expert give an opinion of how to distinguish, abstractly and/or concretely, between guilt and causality?[101] Although from a scientific standpoint, any notion of individual guilt is nothing but a »trans-scientific issue« - that is questions of fact that can be stated in the language of science but are unanswerable by science[102] - forensic psychiatrists routinely give such opinions because they allow the law to »enslave« the basic concepts of their discipline.

»Economic analysis of law« is a more recent battlefield for epistemic competition. It has still to be seen if economic imperialism will prevail or if, vice versa, juridical dogma will colonize economic thought. Especially in the hands of economizing lawyers, analytical concepts of economics undergo a subtle (and often not so subtle) change into normative constructs that serve as cornerstones for legal-doctrinal edifices. If, for example, one examines the new legal economics literature on the firm as a network of contracts[103], what is left from the methodological principles of economics, formulated by Williamson:[104] theoretical openness, readiness to learn from other fields of experience, refutability of implications and exposure to empirical falsification? Judge Easterbrook's piece on »Corporations as Contracts«[105] in any case is a prototype of ideological orthodoxy, doctrinal rigidity, and conceptual immunization against contradicting experience.

These polemical remarks should not be misunderstood. They are not meant to defend the purity of scholarly conceptualization against strategic misuse by lawyers with ulterior motives. On the contrary, they are meant to demonstrate that social science constructs are not only transformed or distorted but constituted anew, if they are incorporated into legal discourse.[106] They are not imported into the law bearing the label »made in science«, but are re-constructed within the closed operational network of legal communications that gives them a meaning quite different from that of the social sciences. It is not a question of the same thing being looked at from different angles, appropriately to different disciplinary interests, methods, etc.[107] This would be to presuppose an underlying reality that is capable of unifying the diverse aspects stressed in different disciplines and of deciding between conflicting descriptions. Rather, the differences are to be found in the realities themselves that are produced by different discourses and that can be neither unified nor reconciled.

Thus, the incorporation of social scientific knowledge is not really an escape from what we called the epistemic trap of modern law. It does not solve the conflict between juridical and scientific realities, but adds a new reality that is neither a purely juridical construction nor a purely scientific construction. The constructs of sociological jurisprudence, legal economics, »legal politology«, etc. are hybrid creatures, produced in the legal process with borrowed authority from the social sciences. However, epistemic authority and responsibility are no longer with the social sciences but with the law. And their »truth«, their social adequacy, their viability will be decided no longer in the process of scientific inquiry but in the process of legal communication. For instance, certain psychoanalytic constructs, as well as fully deterministic models in psychology will never be viable constructs in a juridical world that is based upon assumptions of individual guilt and responsibility. Or to take another field, the relative success of legal economics compared to sociological jurisprudence has probably nothing to do with the intrinsic »scientific« values of the models involved, but with their structural affinity to traditional legal doctrine. If courts considering questions of, say, negligence, public policy, fairness or properties of the »reasonable man«, resort to »social norms«, a sociological conceptualization would require time, energy and money for extended empirical research while an economic conceptualization in terms of transaction costs requires an armchair ...

It would be wrong, however, to view the incorporation of social knowledge as »irrational«. Given the inherent tension between scientific and juridical realities and the authority of modern sciences, it seems quite rational for the law to attempt to make the legal reality constructs at least compatible with recent developments in the sciences. In this respect, law resembles religion[108] For legal dogma and theological dogma alike, it is advisable to keep the world of faith compatible with the world of scientific truth. However, there is more to the integration of law and social sciences than merely making contradictory world constructions compatible. The »law and ...« movement, it should be admitted, has benign effects for the decision-making quality of modern law in terms of justice and utility. The most recent results of the social sciences and the permanent challenge which they represent can serve as a »variety pool« for legal innovation. It is a tremendously rich source for an ongoing re-construction of the legal world, comparable only to the richness in what people find litigable and which creates legal conflicts. However, what happens to those constructs once they enter the legal scene is no longer in the hands of the social sciences. Selection and retention of these variations is the job of legal evolution.

There are indications today that this legal reconstruction of scientific knowledge, if carried too far, becomes risky in itself. In the environmental law area, for example, Gerd Winter[109] felt a growth in the »judges' anxiety« about technical risk assessment and other legal incorporation of scientific findings together with a tendency to reduce the scope of legally relevant issues. This looks like a return to the first mentioned escape route in the permanent oscillation between epistemic autonomy and heteronomy. However, there are other attempts to cope with this situation, experimentation with a third solution, a kind of middle path between the two main escape routes. These more promising attempts can be summarized in the following formula. Law cannot take over full epistemic authority and responsibility for the reality constructions involved, but at the same time it does not totally delegate epistemic authority to other social discourses. Rather, as a precondition for the incorporation of social knowledge, the legal system defines certain fundamental requirements relating to procedure and methods of cognition.

A case in point is the decision of the German Supreme Court on co-determination in economic organizations.[110] For years, constitutional lawyers had judged the constitutionality of labor participation on the basis of its economic effects - in terms of efficiency of the firm, performance of the West German economy, and its position in international competition. In this way, the collective actors involved, i.e. firms, employer associations, labor unions, government and parliament, had prepared legal reality constructions in their briefs with detailed scenarios about the socio-economic consequences of co-determination, either with catastrophic or beneficial consequences whichever was appropriate to their position.[111] In addition, economic and sociological experts had been mobilized on both sides. The court refused to take a substantive position on these scenarios about possible consequences and resorted to a »procedural« solution. Instead of confirming or rejecting reality constructions, the court allocated risks of information and risks of prediction among the collective actors involved, including the court itself, and created a new legal duty for the legislature: to reverse its decisions if the predictions on which they were based should turn out to be wrong.[112] In several more recent decisions this tendency has been strengthened: to abstain from a material construction of reality and to proceduralize the legal solution; to delegate epistemic authority to different collective actors, i.e. regulatory agencies, private firms, labor unions, research institutions, interest associations, governmental organizations, parliament, courts; to allocate risks of information and prediction; to define procedures and methods; to decide which collective actor must bear the »burden of proof« for reality constructions; and to define responsibilities for failures in information and prediction.[113]

To a certain degree, a constructivist perspective would favor such attempts to »proceduralize« the conflict between epistemic autonomy and heteronomy in modern law. Indeed, when correspondence theories of truth have to be replaced by consensus theories and coherence theories, when the authority of science is based only on its internal procedures of validation, when institutional contexts like the law are condemned to epistemic autonomy and cannot resort to external authorities, then practical and theoretical attention must focus on the procedures that dictate the premises, content and consequences of institutional constructions of social reality.

Notes and References

* This is a revised version of an article which appeared in Law and Society Review 1989. For helpful criticism I would like to thank Michael Donnelly, Reiner Grundmann, Christian Joerges, Wolfgang Krohn, Giandomenico Majone, David Nelken, Helga Nowotny, Alessandro Pizzorno, Joyce Reese, Sean Smith.

1. L. Carroll, *Stanza of Anglo-Saxon Poetry*, Misch-Masch, 1855; *Through the Looking Glass and What Alice Found There*, London: Macmillian, 1871, (cited after the edition of 1960, Bramhall House, New York, p. 191).

2. R. Scott (alias Chatterton), »The Jabberwock Traced to its True Source«, Macmillan's Magazine, February 1872.

3. Collective actors, corporate personality; see N. Luhmann, *Soziale Systeme. Grundriß einer allgemeinen Theorie*, Frankfurt am Main: Suhrkamp, 1984, p. 270ff; G. Teubner, »Enterprise Corporatism: New Industrial Policy and the ›Essence‹ of the Legal Person«, *The American Journal of Comparative Law* **36** (1988), 130-155, 130ff; D. zu Knyphausen, *Unternehmungen als evolutionsfähige Systeme: Überlegungen zu einem evolutionären Konzept für die Organisationstheorie*, Herrsching: Kirsch, 1988, p. 120ff; G. Vardaro, »Before and Beyond the Legal Personality: Group Enterprises and Industrial Relations«, in D. Sugarman, G. Teubner (ed.), *Regulating Corporate Groups in Europe*, Berlin: de Gruyter, 1989; M. Hutter, *Die Produktion von Recht: Eine selbstreferentielle Theorie und der Fall des Arzneimittelpatentrechts*, Tübingen: Mohr & Siebeck, 1989, ch. 4; K.H. Ladeur, *Zu einer Grundrechtstheorie der Selbstorganisation des Unternehmens: Festschrift für Helmut Ridder*, Neuwied: Luchterhand, 1989.

4. M. Weber, *Economy and Society*, Berkeley: University of California Press, 1978, p. 4.

5. For the ongoing discussion in terms of those classical theories, see M. Horwitz, »Santa Clara Revisited«, *West Virginia Law Review* **88** (1985), 173; M. Dan-Cohen, *Rights, Persons, and Organizations. A Legal Theory for Bureaucratic Society*, Berkeley: University of California Press, 1986; S.A. Schane, »The Corporation is a Person: The Language of a Legal Fiction«, *Tulane Law Review* **61** (1987), 563; C.M. Roos, *Corporate Personality and Contractual Structure: Legal Aspects on the Firm as a Naxus of Treaties*, Uppsala: Ms., 1988.

6. A. Pizzorno, »Individualismo metodologico: Prediche e ragionamenti«, in L. Sciolla, L. Ricolfi (eds.), *Il soggetto dell'azione: Paradigmi sociologichi e immagini dell'attore sociale*, Milano: Angeli, 1989.

7. Given the humanistic orientation of critical theory, it might sound strange to characterize this theory as anti-individualistic. However, we are not talking about moral-political options, but theory constructions. In a threefold sense, this theory is anti-individualistic: (1) in its critique of methodological individualism in economic and rational actor theories, (2) in its replacement of monological theories of norm formation by dialogical ones, (3) in locating the discourse in the center of cognition, and not the classical epistemological subject (see »communicative versus subject-centered reason« in J. Habermas, *The Philosophical Discourse of Modernity*, Boston: Beacon Press, 1987, ch. 11; and J. Habermas, *The Theory of Communicative Action, Vol. 1: Reason and the Rationalization of Society*, Boston: Beacon Press, 1984. ch. 3 in general).

8. P.L. Berger, T. Luckmann, *The Social Construction of Reality: A Treatise in the Sociology of Knowledge*, New York: Doubleday, 1966. See, for example, D. Bloor, *Knowledge and Social Imagery*, London: Routledge & Kegan Paul, 1976; B. Latour, S. Woolgar, *Laboratory Life: The Construction of Scientific Facts*, Princeton: Princeton University Press, 1979; K. Knorr-Cetina, *Die Fabrikation von Erkenntnis: Zur Anthropologie der Naturwissenschaften*, Frankfurt am Main: Suhrkamp, 1984; N. Gilbert, M. Mulkay, *Opening Pandora's Box*, Cambridge: Cambridge University Press, 1984; H. Collins, *Changing Order: Replication and Induction in Scientific Practice*, Beverly Hills: Sage, 1985; S. Fuller, *Social Epistemology*, Bloomington: Indiana University Press, 1988.

9. M Foucault, *The Order of Things: An Archaeology of the Human Science*, London: Travistock, 1974, xxiii.

10. F.A. Hayek, *Individualism and Economic Order*, London: Routledge & Kegan Paul, 1948; *Law, Legislation and Liberty. Vol.1 Rules and Order*, London: Routledge & Kegan Paul, 1973.

11. K. Popper, *The Poverty of Historicism*, London: Routledge & Kegan Paul, 1953.

12. E.R. Weintraub, *Microfoundations. The Compatibility of Microeconomics and Macroeconomics*, Cambridge University Press, 1979; A. Nelson, »Some Issues Surrounding the Reduction of Macroeconomics to Microeconomics«, *Philosophy of Science* **51** (1984), 573.

13. J. Elster, *Explaining Technical Change*, Cambridge University Press, 1983; *Making Sense of Marx*, Cambridge: Cambridge University Press, 1985.

14. G.C. Homans, *Social Behavior: Its Elementary Forms*, London: Routledge & Kegan Paul, 1961.

15. V. Aubert, *In Search of Law. Sociological Approaches to Law*, Oxford: Robertson, 1983, p. 119.

16. For example, M. Crozier, E. Friedberg, *L'acteur et le system. Les contraintes de l'action collective*, Paris: Seuil, 1977; A. Giddens, *Social Theory and Modern Sociology*, 1987, p. 98ff; for the legal system, A. Febbrajo, »The Rules of the Game in the Welfare State«, in G. Teubner (ed.), *Dilemmas of Law in the Welfare State*, Berlin, New York: de Gruyter, 1985, p. 136; M. van de Kerchove , F. Ost, *Le systeme juridique entre ordre et desordre*, Paris: Presses Universitaires de France, 1988, 157ff; F. Ost, »Between Order and Disorder: The Game of Law«, in G. Teubner, *Autopoietic Law. A New Approach to Law and Society*, Berlin, New York: de Gruyter, 1988, p. 87.

17. Duncan Kennedy, »Freedom and Constraint in Adjudication: A Critical Phenomenology«, *Journal of Legal Education* **36** (1986), 518.

18. R. Gordon, »Critical Legal Histories«, *Stanford Law Review* **36** (1984), 57, 117f.

19. It is true, there are important exceptions among the critical scholars, who develop serious alternatives to the prevailing individualism, above all Thomas Heller, »Structuralism and Critique«, *Stanford Law Review* **36** (1984), 127, and »Accounting for Law«, in Teubner (ed.), *op. cit.*, 1988, note 16; and David Kennedy, »Critical Theory, Structuralism, and Contemporary Scholarship«, *New England Law Review* **21** (1985), 209. But these exceptions confirm our rule: it is their language, even in their own intellectual circles, that has to struggle with the Jabberwocky syndrome.

20. Gordon, *op. cit.*, 1984, note 18, p. 117.

21. J. Habermas, *Knowledge and Human Interest*, Boston: Beacon Press, 1971; »Vorbereitende Bemerkungen zu einer Theorie der kommunikativen Kompetenz«, in J. Habermas, N.Luhmann, *Theorie der Gesellschaft oder Sozialtechnologie - Was leistet die Systemforschung*, Frankfurt am Main: Suhrkamp, 1971; *Communication and the Evolution of Society*, Boston: Beacon Press, 1974; *Legitimation Crisis*, Boston: Beacon Press, 1975; *op. cit.*, 1984, note 7; *op. cit.*, 1987, note 7; *The Theory of Communicative Action. Vol.2*, Boston: Beacon Press, 1987; *Nachmetaphysisches Denken: Philosophische Aufsätze*, Frankfurt am Main: Suhrkamp, 1988.

22. Habermas, *op. cit.*, 1971, note 21, p. 123; »Wahrheitstheorien«, in H. Fahrenbach (ed.), *Wirklichkeit und Reflexion*, Pfullingen, 1973.

23. Habermas, *op. cit.*, 1984, note 6, ch. 3.

24. See, for example, R. Alexy, *Theorie der juristischen Argumentation. Die Theorie des rationalen Diskurses als Theorie der juristischen Begründung*, Frankfurt am Main: Suhrkamp, 1978; K. Günther, *Der Sinn für Angemessenheit: Anwendungsdiskurse in Moral und Recht*, Frankfurt am Main: Suhrkamp, 1988.

25. Habermas, *op. cit.*, 1984, note 6, ch. 3.

26. J. Habermas, »Law as Medium and Law as Institution«, in Teubner (ed.), *op. cit.*, 1985, note 16, p. 215f; »Wie ist Legitimation durch Legalität möglich?«, 20 *Kritische Justiz*, 1987c, p. 1; R. Wiethölter, »Materialisation and Proceduralization in Modern Law«, in Teubner (ed.), *op. cit.*, 1985, note 16; »Social Science Models in Economic Law«, in T. Daintith, G. Teubner (ed.), *Contract and Organization. Economic Analysis in the Light of Economic and Social Theory*, Berlin: de Gruyter, 1986; Günther, *op. cit.*, 1988, note 24; Ch. Joerges, »Politische Rechtstheorie and Critical Legal Studies«, in Ch. Joerges, D. Trubek (ed.), *Critical Legal Thought: An American-German Debate*, Baden-Baden: Nomos, 1989; K.H. Ladeur, »The Law of Uncertainty«, in Joerges, Trubek (eds.), *op. cit.*, 1989, note 26; U.K. Preuss, »Rationality Potentials of Law: Allocative, Distributive and Communicative Rationality«, in Joerges, Trubek (ed.), *op. cit.*, 1989, note 26.

27. J. Habermas, *Moralbewußtsein und kommunikatives Handeln*, Frankfurt am Main: Suhrkamp, 1983, p. 26; *op. cit.*, 1988, note 21, p. 63ff, 80.

28. Habermas, *op. cit.*, 1971, note 21, p. 136; *op. cit.*, 1983, note 27, p. 53; *op. cit.*, 1984, note 6, ch. 3.

29. K.O. Apel, »Das Apriori der Kommunikationsgemeinschaft und die Grundlagen der Ethik«, in K.O. Apel, *Transformation der Philosophy*, Frankfurt am Main: Suhrkamp, 1973; D. Boehler, *Rekonstruktive Pragmatik: Von der Bewußtseinsphilosophie zur Kommunikationsreflexion: Neubegründung der praktischen Wissenschaft und Philosophy*, Frankfurt am Main: Suhrkamp, 1985.

30. C.P. Wormell, »On the Paradoxes of Self-Reference«, *Mind* **67** (1958), 267; W.V. Quine, *The Ways of Paradox*, Cambridge, Massachusetts: Harvard University Press, 1976; K. Krippendorff, »Paradox and Information«, in B. Dervin, M. Voigt (eds.), *Progress in Communication*, 1984; J. Barwise, J. Etchemendy, *The Liar. An Essay in Truth and Circularity*, New York: Oxford University Press, 1987.

31. The problem of infinite regression/circularity in Habermas' theory of discursive justification is perhaps most clearly expressed in Habermas, *op. cit.*, 1971, note 21, p. 123ff, and *op. cit.*, 1973, note 22, p. 255ff. Any one of those independent criteria that are supposed to distinguish true from false consensus, to judge the competence of the speakers, or to decide on the authenticity of their utterances, has to be subjected to consensus again. And even the cognitive schemes that guide the universalization of needs within the discourse have to be examined in a metadiscourse that in its turn is guided by cognitive schemes...

32. For the controversy on intersubjective versus communication, see Habermas, *op. cit.*, 1987, note 6, ch. 12; *op. cit.*, 1988, note 21, 95ff; N. Luhmann, »Intersubjektivität oder Kommunikation: Unterschiedliche Ausgangspunkte soziologischer Theoriebildung«, *Archivio di Filosofia* **41** (1986), 41, 41ff.

33. M. Foucault, *The Archaeology of Knowledge*, New York: Harper Colophon, 1972, ch. 2.

34. See H.L. Dreyfus, P. Rabinow, *Michel Foucault: Beyond Structuralism and Hermeneutics*, Chicago: Chicago University Press, 1982, p. 44ff.

35. Foucault, *op. cit.*, 1972, note 33, p. 49.

36. A. Honneth, *Kritik der Macht: Reflexionsstufen einer kritischen Gesellschaftstheorie*, Frankfurt am Main: Suhrkamp, 1985, p. 164.

37. Foucault, *op. cit.*, 1974, note 9, ch. 2,3,7, for the sciences; *Discipline and Punish*, New York: Vintage/Random House, 1979, for law.

38. Foucault, *op. cit.*, 1974, note 9, ch. 9.

39. H.L. Dreyfus, P. Rabinow, 1982, note 34, p. 84.

40. For a critique, see Honneth, *op. cit.*, 1985, note 36, p. 168ff.; Habermas, *op. cit.*, 1987, note 26, ch. 10.

41. H.R. Maturana, F.J. Varela, *Autopoiesis and Cognition*, Boston: Reidel, 1980; H. von Foerster, *Observing Systems*, Seaside, California: Intersystem Publication, 1981; Luhmann, *op. cit.*, 1984, note 3.

42. N. Luhmann, »The Autopoiesis of Social systems«, in F. Geyer, J. van der Zouwen (eds.), *Sociocybernetic Paradoxes: Observation, Control and Evolution of Self-Steering Systems*, London, Berverly Hills: Sage, 1986, p. 172; »The Theory of Social Systems and its Epistemology: Reply to Danilo Zolo's Critical Comments«, *Philosophy of the Social Sciences* **16** (1986), 129; »The Third Question: The Creative Use of Paradoxes in Law and Legal History«, *Journal of Law and Society* **15** (1988), 153.

43. von Foerster, *op. cit.*, 1981, note 41, p. 274; »Entdecken oder erfinden? Wie läßt sich verstehen verstehen?«, in H.Gumin, A. Mohler (eds.), *Einführung in den Konstruktivismus*, Munich: Oldenbourg, 1985, p. 36.

44. For an elaboration of these somewhat jabberwocky remarks, see G. Teubner, »And God Laughed...: Indeterminacy, Self-Reference and Paradox in Law«, in Joerges, Trubek (eds.), *op. cit.*, 1989, note 26.

45. J. Piaget, *The Construction of Reality in the Child*, New York: Ballantine, 1971; E. Glasersfeld, »Radical Constructivism and Piaget's Concept of Knowledge«, in F.B. Murray (ed.), *Input of Piagetian Theory*, Baltimore: University Pask Press, 1975; »An Epistemology for Cognitive Systems«, in G. Roth, H. Schwegler (eds.), *Self-Organizing Systems. An Interdisciplinary Approach*, Frankfurt am Main: Campus, 1981; »Konstruktion der Wirklichkeit und des Begriffs der Objektivität«, in Gumin, Mohler (eds.), *op. cit.*, 1985, note 43; Maturana, Varela, *op. cit.*, 1980, note 41; von Foerster, *op. cit.*, 1981, note 41; Luhmann, *op. cit.*, 1984, note 3, p. 647ff; M.A. Arbib, M.B. Hesse, *The Construction of Reality*, Cambridge: Cambridge University Press, 1986; S. Schmidt (ed.), *Der Diskurs des Radikalen Konstruktivismus*, Frankfurt am Main: Suhrkamp, 1987.

46. For the relation of individual and social observation, see N. Luhmann, »Individuum und Gesellschaft«, *Universitas* **39** (1983), 1; »Die Autopoiese des Bewußtseins«, in A. Hahn, V. Kapp (eds.), *Selbstthematisierung und Selbstzeugnis: Bekenntnis und Geständnis*, Frankfurt am Main: Suhrkamp, 1986, p. 25; »The Individuality of the Individual: Historical Meaning and Contemporary Problems«, in T.C. Heller, M. Sosna, D.E. Wellbery (ed.), *Reconstruction Individualism: Autonomy, Individuality, and the Self in Western Thought*, Palo Alto: Stanford University Press, 1986, p. 313.

47. Luhmann, *op. cit.*, 1984, note 3, p. 193ff; *op. cit.*, 1986, note 42, p. 172ff.

48. Luhmann, *op. cit.*, 1984, note 3, p. 158ff, *op. cit.*, 1986, note 46, p. 313ff.

49. N. Luhmann, »Closure and Openness: On Reality in the World of Law«, in Teubner (ed.), *op. cit.*, 1988, note 16, p. 335ff.

50. Foucault, *op. cit.*, 1974, note 9, ch. 9f; *op. cit.*, 1979, note 37, ch. 4.

51. Luhmann, *op. cit.*, 1984, note 3, p. 155ff.

52. J. Elster (ed.), *The Multiple Self*, Cambridge: Cambridge University Press, 1986; A. Etzioni, *I & We*, New York: Free Press, 1989, 11ff.

53. For a more detailed critique, see G. Teubner, »Social Order from Legislative Noise? Autopoietic Closure as a Problem for Legal Regulation«, in G. Teubner (ed.), *State, Law, Economy as Autopoietic Systems*, Milano: Giuffrè, 1990.

54. Luhmann, *op. cit.*, 1988, note 42.

55. For first steps in this direction, see Luhmann, *op. cit.*, 1988, note 49; »The Coding of the Legal System«, in Teubner, *op. cit.*, 1989, note 53.

56. See below VII. and VIII.

57. M. Douglas, *How Institutions Think*, Syracuse: Syracuse University Press, 1986.

58. P. Selznick, »Law: The Sociology of Law«, *International Encyclopedia of the Social Sciences* **9** (1968), 50.

59. For an elaborate discussion of the characteristics of autopoietic law, see G. Teubner, »Evolution of Autopoietic Law«, in Teubner (ed.), *op. cit.*, 1988, note 16; »Hypercycle in Law and Organization: The Relationship between Self-Observation, Self-Constitution and Autopoiesis«, *European Yearbook in the Sociology of Law* **43** (1988); »Social Order from Legislative Noise? Autopoietic Closure as a Problem for Legal Regulation«, in Teubner (ed.), *op. cit.*, 1989, note 53.

60. J.A. Fodor, »Methodological Solipsism Considered as a Research Strategy in Cognitive Psychology«, *Behavioral Brain Science* **3** (1980), 63.

61. F.J. Varela, »Living Ways of Sense-Making: A Middle Path for Neuro-Science«, in P. Livingstone (ed.), *Disorder and Order: Proceedings of the Stanford International Symposium*, Saratoga, California: Anma Libri, 1984, p. 217.

62. See the actor-based objections against an autopoietic law by Febbrajo, *op. cit.*, 1985, note 16, p. 134ff; Kerchove and Ost, *op. cit.*, 1988, note 16, p. 157ff and Ost, *op. cit.*, 1988, note 16, p. 87ff.

63. Kerchove and Ost, *op. cit.*, 1988, note 16, p. 159.

64. See von Foerster, *op. cit.*, 1981, note 41; Teubner, *op. cit.*, 1989, note 53.

65. Pizzorno, *op. cit.*, 1989, note 6.

66. For an elaboration on collective actors, see Teubner, *op. cit.*, 1988, note 3, p. 133ff.

67. H. Grünberger, »Dehumanisierung der Gesellschaft und Verabschiedung staatlicher Souveränität: das Politische System in der Gesellschaftstheorie Niklas Luhmanns«, in I. Fetscher, H. Münkler (ed.), - *Pipers Handbuch der politischen Ideen*, Munich: Piper, 1987.

68. U. Schimank, »Der mangelnde Akteurbezug systemtheoretischer Erklärungen gesellschaftlicher Differenzierung - Ein Diskussionsvorschlag«, *Zeitschrift für Soziologie* **14** (1985), 421; R. Mayntz, »Steuerung, Steuerungsakteure und Steuerungsinstrumente: Zur Präzisierung des Problems«, *HiMon-DB* **70**, Siegen: Universität, 1986; Ost, *op. cit.*, 1988, note 16, p. 87ff; H. Rottleuthner, »Biological Metaphors in Legal Thought«, in Teubner (ed.), *op. cit.*, 1988, note 16, p. 122.

69. K. Podak, »Ohne Subjekt, ohne Vernunft. Bei der Lektüre von Niklas Luhmanns Hauptwerk ›Soziale Systeme‹«, *Merkur* **7** (1984), 734; G. Frankenberg, »Der Ernst im Recht«, *Kritische Justiz* **20** (1987), 296.

70. Foucault, *op. cit.*, 1972, note 33, ch. 2; *op. cit.*, 1974, note 9, ch. 9.

71. Foucault, *op. cit.*, 1979, note 37, ch. 4.

72. Luhmann, *op. cit.*, 1988, note 49, p. 335ff; »The Coding of the Legal System«, in Teubner, *op. cit.*, 1989, note 53.

73. For two types of epistemic conflict between social systems, see Teubner, *op. cit.*, 1989, note 44, *op. cit.*, 1989, note 53.

74. For elaboration, see H.G. Deggau, »The Communicative Autonomy of the Legal System«, in Teubner (ed.), *op. cit.*, 1988, note 16 p. 128; Heller, *op. cit.*, 1988, note 19, p. 283; K.H. Ladeur, »Perspectives on a Post-Modern Theory of Law«, in Teubner (ed.), *op. cit.*, 1988, note 16, p. 242; Teubner, *op. cit.*, 1988, note 53, p. 217; *op. cit.*, 1988, note 59, p. 60.

75. See L. Mengoni, »La questione del ›diritto giusto‹ nella societa' post-liberale«, *Fenomenologia e Societa'* **11** (1988), 15.

76. For an elaboration on the ›facts of law‹, see P. Nerhot, »The Fact of Law«, in Teubner (ed.), *op. cit.*, 1988, note 16.

77. J. Baudrillard, *L'echange symbolique et la mort*, Paris: Gallimard, 1976.

78. P. Bourdieu, »La force du droit: Elements pour une sociologie du champ juridique«, *Actes de la recherche en sciences sociales* (1986), 13.

79. R. Cotterell, »Law and Sociology: Notes on the Constitution and Confrontation of Disciplines«, *Journal of Law and Society* **13** (1986), 15.

80. R. von Jhering, *Scherz und Ernst in der Jurisprudenz*, Leipzig: Breitkopf & Haertel, 1884.

81. F.S. Cohen, »Transcendental Nonsense and the Functional Approach«, *Columbia Law Review* **35** (1935), 809- 849.

82. See, for example, Aubert, *op. cit.*, 1983, note 15, p. 98ff; H. Rottleuthner, »Diskussionsvotum zum vorstehenden Beitrag«, in E. Blankenburg, E. Klausa, H. Rottleuthner (ed.), *Alternative Rechtsformen und Alternativen zum Recht*, Opladen: Westdeutscher Verlag, 1980, p. 137ff, for sociology; M. Adams, *Ökonomische Analyse der Gefährdungs- und Verschuldungshaftung*, Heidelberg: Decker & Schenk, 1985, for economics.

83. Bloor, *op. cit.*, 1976, note 8; B. Barnes, *Scientific Knowledge and Sociological Theory*, London: Routledge & Kegan Paul, 1974.

84. Arbib, Hesse, *op. cit.*, 1986, note 45, p. 10.

85. N. Luhmann, *Wissenschaft*, Bielefeld: Ms., 1988, p. 2,9.

86. K.D. Opp, *Soziologie im Recht*, Reinbek: Rowohlt, 1973.

87. Thus, the resulting relativism of different social discourses is not »anything goes« relativism. It is a relativism that invites to »raise the status of the other ›mythologies‹ by a more careful investigation of their methodological and cognitive credentials« and to examine »the various kinds of criteria of acceptability that apply to different kinds of constructed models and myths«, Arbib, Hesse, *op. cit.*, 1986, note 45, p. 10.

88. B. Windscheid, »Die Aufgaben der Rechtswissenschaft«, in B. Winscheid, *Gesammelte Reden und Abhandlungen*, Leipzig: Duncker, 1904.

89. See Mengoni, *op. cit.*, 1988, note 75, p. 23.

90. See above V.

91. For details, see Teubner, *op. cit.*, 1989, note 44, *op. cit.*, 1989, note 53.

92. For an elaboration of the mutual constraints exerted among different internal models of the outside world, see G. Teubner, »Generalklauseln als sozio-normative Modelle«, in H. Stachowiak (ed.), *Bedürfnisse, Werte und Normen im Wandel Bd.1.*, Munich: Fink & Schöningh, 1982, p. 96ff.

93. Luhmann, *op. cit.*, 1988, note 49, p. 340.

94. See for the German situation, H. Kitschelt, *Der ökologische Diskurs. Eine Analyse von Gesellschaftskonzeptionen in der Energiedebatte*, Frankfurt am Main: Campus, 1984; R. Wolf, *Der Stand der Technik*, Opladen: Westdeutscher Verlag, 1986; G. Winter, »Die Angst des Richters bei der Technikbewertung«, *Zeitschrift für Rechtspolitik* **20** (1987), 425.

95. See, for example, R. Cotterell, *The Sociology of Law: An Introduction*, London: Butterworths, 1984, p. 253ff.

96. For a recent analysis of the German and French practice in administrative law and its sophisticated interpretation, see K.H. Ladeur, *»Abwägung« - Ein neues Paradigma des Verwaltungsrechts. Von der Einheit der Rechtsordnung zum Rechtspluralismus*, Frankfurt am Main: Campus, 1984, p. 11ff, p. 57ff.

97. For example, the legal concerns of creditors, debtors, neighbors, corporations, regions, states.

98. In terms of flexibility, openness and learning capacity, see Ladeur, *op. cit.*, 1984, note 96, p. 216ff.

99. For a recent statement, see H. Albert, »Law as an Instrument of Rational Practice«, in Daintith, Teubner (ed.), *op. cit.*, 1986, note 26, p. 34ff.

100. In Germany, see the lively debate on »Folgenkontrolle«, N. Luhmann, *Rechtssystem und Rechtsdogmatik*, Stuttgart: Kohlhammer, 1974; H. Rottleuthner, »Zur Methode einer folgenorientierten Rechtsanwendung«, *Archiv für Rechts- und Sozialphilosophie*, Beiheft **13** (1979); G. Lübbe-Wolf, *Rechtsfolgen und Realfolgen*, Freiburg: Alber, 1981.

101. For example H. Prins, *Offenders, Deviants or Patients? An Introduction to the Study of Socio-Forensic Problems*, 1980, ch. 2.

102. See A.M. Weinberg, »Science and Trans-science«, *Minerva* **10** (1972); G. Majone, »Process and Outcome in Regulatory Decision-Making«, in C.H. Weiss, A.Barton (eds.), *Making Bureaucracies Work*, Beverly Hills: Sage, 1979.

103. A.A. Alchian, H. Demsetz, »Production, Information Costs, and Economic Organization«, *American Economic Review* **62** (1972); E. Fama, M. Jensen, »Agency Problems and Residual Claims«, *Journal of Law and Economics* **88** (1983); R.C. Clark, »Agency Costs Versus Fiduciary Duties«, in J.W. Pratt, R.J. Zeckhauser (eds.), *Principals and Agents: The Structure of Business*, Boston: Harvard Business School Press, 1985; E. Schanze, »Potential and Limits of Economic Analysis: The Constitution of the Firm«, in T.Daintith, G.Teubner (eds.), *Contract and Organization: Legal Analysis in the Light of Economic and Social Theory*, Berlin: de Gruyter, 1986; »Contract, Agency, and the Delegation of Decision Making«, in G. Bamberg, K. Spreman (eds.), *Agency Theory, Information, and Incentives*, Berlin: Springer, 1987; Roos, *op. cit.*, 1988, note 5.

104. O.E. Williamson, *The Contractual Logic of Internal Organization*, Firenze: EUI Conference Paper, 1987.

105. F. Easterbrook, *Corporations as Contracts*, Stanford: Conference paper, 1988.

106. »Much depends on noticing that law's autonomy lies not in its freedom from being influenced by external causes and influences but in the way in which it incorporates and responds to them«, D. Nelken, »-Criminal Law and Criminal Justice: Some Notes on Their Irrelation«, in I. Dennis (ed.), *Criminal Law and Justice*, London: Sweet & Maxwell, 1987, p. 139.

107. V. Aubert, »On the Relationship between Legal and Sociological Concepts«, in E. Blankenburg, Klausa, Rottleuthner, *op. cit.*, 1980, note 82; op. cit., 1983, note 15, p. 98ff; Rottleuthner, *op. cit.*, 1980, note 82, p. 137ff.

108. For a constructivist account of the conflict between science and religion, see Arbib, Hesse, *op. cit.*, 1986, note 45, p. 16ff, 197ff.

109. Winter, *op. cit.*, 1987, note 94.

110. Bundesverfassungsgericht, *BVerfGE* 50, 290.

111. See P. Badura et. al., *Mitbestimmungsgesetz 1976 und Grundgesetz: Gemeinschaftsgutachten*, Munich: Beck, 1977, p. 137ff, p. 246ff; F. Kübler et. al., *Mitbestimmung als gesetzgebungspolitische Aufgabe: Zur Verfassungsmäßigkeit des Mitbestimmungsgesetzes*, Baden-Baden: Nomos, 1978, p. 35ff, p. 99ff, p. 145ff, p. 197ff.

112. For an in-depth analysis of such a »proceduralization« of institutional cognition, see Wiethölter, *op. cit.*, 1985, note 26; *op. cit.*, 1986, note 26; »Proceduralization of the Category of Law«, in Joerges, Trubek (eds.), *op. cit.*, 1989, note 26; Majone, *op. cit.*, 1979, note 102; *Evidence, Argument and Persuasion in the Policy Process*, New Haven: Yale University Press, 1989.

113. See for environmental law, *BVerfGE* 49, 89; for corporation law, *BVerfGE* 72, 155; for the law or property *BVerfGE* 74, 264.

Self-Regulation in Social Systems[1]

PETER M. HEJL
University of Siegen

1. Introduction

It was E. Durkheim who put the problem of social self-regulation on the agenda of sociology when he wrote his *Division of Social Labor.* Though mostly not read from this perspective, Durkheim's model of the relation between social differentiation and individualization can be seen as a breakthrough toward a better understanding of self-organization as a process that has self-regulating effects.[2] Whereas in Durkheim's writing the problem of self-regulation is not clearly distinguished from his notion of social solidarity, a reformulation of the problem within the conceptual frame of self-organization[3] seems to provide a new conceptual solution.

2. Self-Regulation of Social Systems

Social Systems: The Conceptual Frame

Speaking of social systems presupposes a clear understanding of which kinds of phenomena may be called a »social system«. For our purpose[4] it may suffice to define a social system as a group of individuals who 1. have, through social interactions, generated a common set of reality constructs[5] together with a set of actions and behaviors deemed adequate to handle the so defined realities, and who 2. interact with respect to these realities by means of socially defined actions and behavior.

Components are, according to the definition, individuals as they fulfill both conditions stated above. If we ask what makes an individual a member of a social system it is evident that it is exclusively the display of those capacities of interaction that define a given individual as a member of a particular social system.[6] In a generalized form, components are therefore defined exclusively by those interactive capacities (taken as the sum of all observable interactions) they exhibit in the system.

Any treatment of components raises the question of how their activities are coordinated or regulated in such a way that the composite unity they constitute forms in fact a unity or a whole. The answer has been given by systems theory: *organization.* Fundamentally, every organization can be characterized by its selectivity. Apparently selectivity is a necessary condition for any specific coordination of the interactions of the components. In every social system there exist (due to a formal organization, to interaction routines, traditions, etc.) such selective patterns of interactions between its members that are stable for a certain period of time. Hence this allows us to describe the *organization* of a social system as the pattern of interaction that is observed as being stable over a certain period of time in a system.

W. Krohn et al. (eds.), Selforganization. Portrait of a Scientific Revolution, 114–127.

If we now look at the dynamics by which a system produces its behavior (leaving aside the interactions with other social systems), the specific function of the organization becomes visible. As N. Luhmann[7] has argued, an organization operates and gains an importance »of its own« due to the very fact that it allows the individual member to act without deciding every minute, from scratch, what to do with whom.

Now, if an individual member or a small number of the participants, e.g., in a research group, modify their ideas and their behavior (e.g., convictions of what constitutes a legitimate research problem and adequate methods for the group), the interactive capacities (with all the concomitant and interrelated reality constructs) of other group members will not change both simultaneously and in an adjusted manner. More probably they will go on just as before. Due to its organization, the system is in principle independent of or autonomized with respect to particular actors, although it consists only of individual actors. As there is in most cases a well-established net of interactions, a member who changes his behavior for whatever reason continues to be treated in the »normal way«, i.e. the way defined by the reality construct and the behavioral programs of the system. Put in a more formal manner, if a component changes its behavior, it still receives to a large extent the same inputs as before, at least for a certain period of time.[8] These inputs trigger spontaneous activities that in turn are determined by the dynamics of the components. Seen from this perspective, the organization provides inputs that stimulate the component to produce the output it is able to produce at the given point in time. All the same, this output depends entirely on the component's eigenbehavior.

Self-Regulation as Organizational Change[9]

If we take »regulation of a system« as the process by which its generation of behavior is modified, »self-regulation« then designates that set of regulating influences on a system that is generated by the system itself. From a certain level on, modifications of the components result in a modification of the organization. For example, if a sufficient number of members of a research group or of an invisible college[10] modify the definition of what constitutes their subject matter and how to investigate it, the general pattern of their interactions will change as well.[11] This analysis can be pushed a step further. Cognitive - and hence interactive - developments of the members of a social system do not engender only organizational alterations. Due to the selective triggering of component behavior through the organization, organizational variations in turn precede modifications of components. I propose to call the complete cycle of modifications »self-organizing«. »Self-organization« thus means an interaction between the components of a system and its organization that leads to modifications on both sides. Consequently, a system is self-organized to the extent that its behavior results from these interactions.[12]

If we look at self-organization from the point of view of regulation, it is evident that every variation of behavior that springs from self-organization is at the same time an example of self-regulation. The path of self-organizational changes of a system can be seen from this perspective as a series of self-regulations. But these self-regulations have an important trait: they are linked to organizational evolution. This makes them rather stable but slow to bring about and slow to alter. The situation is different with the second type of self-regulation. It is not linked to organizational change but to a certain type of organization.

Flexible Self-Regulation

Whereas in the case of self-organization, regulation is merely a consequence of the self-organizing processes, the same does not hold for the second type of self-regulation. We must therefore look briefly at the regulation problem itself rather than examine the traditional way to regulate systems. This enables us to underscore what the problem of regulation is and why the traditional hierarchical way to solve it is not compatible with the requirements of self-regulation.

One of the factors that distinguishes social systems from other types of system is the fact that in social systems *every* component is part of the boundary (notwithstanding the fact that there might be specific components that fulfill specific functions - for example, in the interactions with other social systems). Moreover, and peculiar to social systems, is that components belong (as cognitive processes and the corresponding behavior) to individuals who participate at the same time in other social systems. Both features influence the dynamics of the components and have to be taken into account. They contribute, together with the proper dynamics of the components, to the problem of regulation in social systems.

According to the definition of social systems, the behavior of the system is the behavior of its components as it is determined by the proper dynamics of the components at any given point in time and triggered by the inputs they receive due to the pattern of selective connections that constitute the system's organization. If all or most input events are responded to by behavior that is seen as »normal« in the context of the system, and hence accepted as »right« or »adequate« or »well understood«, there is of course no need for regulation, simply because there are no conflicts that have to be solved and no decisions that have to be either agreed upon or imposed. In this case, all members of the system will »spontaneously select« the same behavior, so there is no need to take special actions.

Now, if the dynamics of components differ - due, for example, to differential experiences of the members of a social system - the same event might constitute different inputs. It is possible as well that different events, seen from an external observer, are taken as identical by component individuals. Even more, the same might happen at the output side. With respect to an observer, the same inputs can consequently lead to different outputs or different inputs to the same outputs. If such divergences between components happen in a more or less regular way, they probably stimulate self-organizing processes and lead to internal differentiations that can be qualified as »functional«.

The situation is different, however, if divergences between components are produced in an unsteady or unpatterned way, so that the possibilities of self-regulation via self-organization do not suffice any longer or are inadequate because they are too slow in occurring or too coercive to be accepted by the members of the system[13] (although self-organization may take place as well). In fact it can be argued with respect to Durkheim's theory of the division of social labor that socially produced individualization plays a major role in the processes we are looking at. If it increases above a certain level, or, equally, if for reasons of internal differentiation, the system constructs an environment so complex that there are events to cope with that demand more flexibility than the functional differentiation of self-organization allows for, self-regulation through self-organization becomes insufficient. To put the problem in a more general form, a system faced with a need to decide continuously between different and changing possibilities of action generated as a result of the dynamics of its components is confronted with a regulation problem that cannot be solved through self-organization. Such a system must develop possibilities to cope with its regulation problem or it will disintegrate. In other words every system that does not completely control the behavior of its components must have regulating devices to cope with the resulting decision problems.

The traditional answer to the regulation problem was hierarchy formation in various forms. But hierarchies have specific problems that prevent their strict realization in social systems. At this stage and as individualization becomes more prevalent, self-regulation becomes more important. To gain a clearer understanding of what it might mean, we have to understand hierarchies.

One of the most common or at least most visible ways to solve decision problems is the formation of hierarchies, both cognitive[14] and organizational. The structure of all hierarchies is linear. The classical examples are value systems, military command chains, or traditional bureaucracies. This linearity is by no way contradicted by the frequent existence of tree-like structures (e.g., several subordinates per superior) or of feedback loops (decisions are made on top, but on the basis of reports from the bottom of the organization). The core idea of hierarchy is that of the existence of a decision center endowed with ultimate power and knowledge. Components that are »higher,« or that come »earlier« in the hierarchy, are thought to be nearer to that decision center. Seen from a structural point of view, the idea of hierarchy is linked to that of a linear concatenation of processes. But it is not just ordinary linearity, because it is thought to have its origin in a fixed point or center of knowledge and power. Ultimately, this center is derived from the idea of an almighty god or, in its secularized version, an absolute monarch or complete knowledge.

Of course we know that even in strongly hierarchical social systems there is never a completely hierarchical organization. Nobody is endowed with absolute knowledge, and information is chosen on lower levels to trigger decisions on top of the hierarchy. This being widely accepted as the fundamental problem of hierarchical organizations in the social domain, we might think the problem of hierarchy to be less important. But seen from a logical (and hence theoretical) point of view, there is a general incompatibility between hierarchical organization and self-regulation. It stems from the fact that a hierarchical regulator is defined as the component of the system that regulates the behavior of other components without being itself regulated by other components of the same system. This is of course the paradox of the barber who shaves everyone who does not shave himself. The question: »Who shaves the barber?« becomes in the case of regulation the question: »Who regulates the regulator?«.

Although these logical and practical problems exist, hierarchies are important, and their potential should not be dismissed too hastily. It seems to me that there is at least one strong pro-hierarchy argument: hierarchies are needed for decision making. As N. Luhmann[15] has argued, hierarchies can be seen as a priori decisions on how decisions shall be made. In this sense, they in fact reduce divergences and allow quicker decision making.

This argument appears in many forms and must be taken seriously. When analyzing the hierarchy problem, we encounter once more the system-theoretical pair component/organization.

From the early 1970s on, many researchers engaged in the debate on political planning were interested in W.R. Ashby's »law of requisite variety.« Its basic idea was derived from game theoretical considerations, namely, from the fact that a player can only win if he can counter any possible move of his adversary. This implies that he has to have at least as many different moves (in other words as many degrees of freedom or the same complexity of thinking and behavior) as his adversary (plus the advantage - normally not mentioned - of the last move). In its popular version, the »law of requisite variety« states that »only variety can destroy variety«.[16] The idea of requisite variety describes a relation between a system and its environment (or parts of it). It stipulates that the system's success in manipulating or controlling this environment depends on the degree to which the required correspondence is achieved. Consequently, the concept was used to advocate planning and advisory institutions that were, to a certain extent, to model the domain for which they were planning or advising. Although Ashby's law holds for the situation in games and game-like situations, it is, to say the least, not applicable to the problem of self-regulation that is fundamentally different from the competitive situation in games. This is demonstrated by a logical analysis of one of its implications, the need for complete (self-)knowledge. Exploring this implication leads directly to one of the two core problems of self-regulation we are interested in here.

The idea of requisite variety leads logically to the requirement of modeling the whole system in itself. To be precise, it has to be modeled as properties of the regulating component which is of course part of the system. This consequence was clearly seen by Ashby himself, as one of the last papers he published demonstrates. Coauthored by R. C. Conant and published in 1970, its title asserts: »Every good regulator of a system must be a model of that system«.[17] Now, this requirement leads in fact to an infinite regress of models in models in models, and so forth. The interesting point, however, is that this prerequisite is tantamount to the claim that any self-regulation that uses the idea of requisite variety poses as its condition of success (»Every good regulator ...«) that the regulating component should possess a complete knowledge of the system to which it belongs. This necessarily includes knowledge about itself and its role in the system as well as knowledge about the dynamics of the behavior of the other components and hence about the surrounding world. We encounter, then, disguised in the technical claim of requisite variety, the fundamental idea of hierarchy, which is the idea of a center of absolute knowledge (including absolute self-consciousness) and power as the superior regulator. As this claim cannot be maintained[18], it follows that the requirements of requisite variety - and hence of hierarchical concepts of regulation - are incompatible with systems characterized by operational closure of their internal interactions.

If we take requisite variety at the more modest level of theory construction as the claim that good regulation requires the more complex to regulate the less complex, there is an important conclusion to be drawn from the argument I have presented: self-regulation must be characterized in contradistinction to the law of requisite variety by processes through which less complex components or subsystems can regulate more complex components or subsystems.

Requisite variety is located merely at the level of the generation of regulating behavior by the component that serves as regulator and that acts as a hierarchical superior due to its *behavioral* variety. But what about *hierarchical organization* as a type of hierarchy that does not necessarily imply the behavioral variety requested by Ashby's law? In everyday experience, the organizational variant of hierarchy appears at least to be the most visible type.

Like any organization, a hierarchical organization is characterized by a pattern of selective interactions between the components of the system. In the case of organizational hierarchy, this pattern of interactions is such that some component-specific behavior or preferences dominate the behavior of the system in a *permanent* and *transitive*[19] manner.[20] In this sense, hierarchical organization can be seen as a functional equivalent to a hierarchical regulator. By its selectivity, it establishes the transitivity of the order in which the components join in the decision-making process resulting in the overall behavior of the system. This participation is such that those taking part earlier define the remaining range of possibilities for those entering at a later point in time. If we disregard what has been said so far about regulation resulting from behavioral differences of components, it is clear that in this case it is the selectivity of hierarchical organization per se that makes components be regulators. If regulators defined in such an organizational way are themselves connected hierarchically, they form a hierarchical system.

This ordering or ranking is in principle independent of any considerations as to the content of the decisions it might produce.[21] However, hierarchical organization, like any organization, is also nothing but a pattern of interactions of components;[22] this pattern of interactions connects components through their outputs and inputs. As follows from the discussion of the different sources of modification of the behavior of components, *inputs do not produce but trigger* (and in this sense of course »select«) *outputs*. As the dynamics of the components keep them mostly within a certain known and expected range of behavior, the result of the selectivity of the organization is *in actual effect* a selection of outputs. This is well known and one reason why it is possible, at least to a certain extent, to influence the course of action of hierarchically organized systems such as, for example, public administrations by changing the personnel occupying the top positions.

Whereas a hierarchical organization is rather well-defined, the same does not seem to be the case for its opposite, the »heterarchical« organization, to use the expression and the model W. S. McCulloch proposed in his 1965 paper »A Heterarchy of Values Determined by the Topology of Nervous Nets.«[23] Of course, that paper does not treat heterarchical systems as such. Rather it discusses the minimal organizational requirements a net of idealized, formal neurons would have to meet in order to generate heterarchical that is, intransitive, choices between different values. Nevertheless the ideas McCulloch advances in this discussion might serve as a starting point for further developments.

McCulloch demonstrates that two different value systems (hierarchical and heterarchical), understood as leading to observable choices between alternative inputs coming from other components of the same system, can be produced as a result of modifications of the organization of the system. In doing this, McCulloch shows that phenomena like transitivity and intransitivity are not restricted to logical analysis, and that the idea of values and value systems does not lend itself only to a verbal type of explanation. Instead he points out that a transitive value system may result exclusively from a hierarchical organization of the system and hence be completely independent of any evaluation of the outcomes.

This allows us to define *heterarchical systems as systems whose components interact in such a way, and hence give rise to such an organization that no component is excluded from the decision process on an organizational basis.* This means that in heterarchical systems every component has organizationally the same opportunity to participate in the interactions of the system and hence in the shaping of its actions. Examples can be seen in all kinds of egalitarian groups or communities. The more or less heterarchical character of organization[24] might be factually or formally established, the main examples being declarations (like the UN declaration of Human Rights) or constitutions that at least stipulate that all citizens should have equal opportunities to participate - via elections, and so forth - in the decision processes of their countries.

At a first glance, it might seem that the organization of a heterarchical system comes close to the case in which all components are connected, thus emptying the concept of organization of its central aspect, its selectivity, which is needed to understand important traits of the behavior of certain systems. Such a position overlooks the fact that the organization is merely one part to be taken into account (the others being the components) if we want to explain the behavior of a system. It is the interaction of both that leads to what we take as the system's activities.

The best way to explore the characteristics of heterarchical systems may be to compare them with hierarchical systems. In a hierarchically organized system, the impact of components on the system's course of action is limited as a result of the selectivity of the organization. Hierarchically organized systems can produce decisions that are consistent with respect to the principle of transitivity. This can be achieved by a sequential participation of the system's members such that at each step all alternatives left over from previous choices fulfill this criterion. Therefore the members of the system are factually restrained in their liberty with the effect that although they choose (according to their own dynamics at the given point in time), the outcome of their selections might be unsatisfactory to more or less everyone except to those who, from in- or outside the system, established its specific hierarchical organization to regulate its behavior. In heterarchical systems, this type of sequentiality does not exist. As a consequence, the organization does not hinder components from contributing to the interactions within the system according to the whole range of possibilities their proper dynamics allow for. Whereas in hierarchical systems creativity and innovations depend on modifications of their organization or on changes in the regulating components, heterarchical systems are, as systems, indifferent to change because they do not prevent it at the organizational level.

But there is another aspect that differentiates hierarchical from heterarchical systems. Hierarchical organization not only assures a specific way to handle events from inside or outside the system; by establishing a certain sequentiality of participation and by limiting the range of choices offered to particular members, it assures a certain efficiency in reaching decisions (if we take »efficiency« to mean merely »rapidity«). Although this aspect is often overestimated, there can be no doubt that systems sometimes encounter situations in which quick decisions are needed. Therefore it seems clear that the lack of the organizational support offered by hierarchical organization disadvantages heterarchical systems. They face the problem of being organizationally unable to decide. This situation, which most of us know from experiences in committee work, results at least partly from the organizational opportunity for all components to participate in decision making. The typical manifestations of this problem are never-ending debates, even when decisions are urgently needed.

The argument of inefficiency is familiar to most of us. Although it is commonly used in defense of hierarchies, it overlooks the importance of the components of a system and of their interplay with the organization (and hence the possibilities the organization offers for their activities). Hierarchical systems are systems in which the freedom or the (relative) autonomy of the components is rather modest; *heterarchical systems are systems in which complex, autonomized components or subsystems interact through a selective utilization of the existing connections between the components*. What does this mean?

Clearly enough, active systems - and this holds especially for heterarchical systems - have to »decide« what the actual situation is and hence what to do. This is the regulation problem. In hierarchical systems, the organization functions as a rather stable regulator with respect to the dynamics of individual components. A deeper analysis thus reveals that the negative aspects of hierarchical regulation are linked not only to assumptions that cannot be met (absolute knowledge), but, as well, to different time intervals serving as references. Following this line of reasoning, problematic effects of hierarchical organization must be derived from the fact that the organization of the system changes much more slowly than individual components. At the same time, and as its raison d'être, the existing organization of a system serves as the precondition for problem solving. This simply means that the organization remains more or less stable, while possibly numerous different problems are solved according to the sequence of choices the organization »selects.« Besides the positive effect of efficiency, the requirement of complete knowledge must produce at some point poor results and discontent. Nevertheless, every decision taken in the presence of competing preferences requires a hierarchical ranking of available possibilities.[25] The problems, then, are a) the stability over time of the hierarchical organization on the organizational level, and b) the required absolute knowledge at the level of the components.

Both problems are solved through the already mentioned »selective utilization of the existing connections between the components«. The lack of organizational support for decision making in heterarchical systems is compensated for by an increased importance of the components that might lead to the formation of *temporary hierarchies*. This allows me to propose what I think to be the *basic theorem of self-regulation*. It states that *self-regulation is only possible in systems with heterarchical organization and which allow the formation of temporary hierarchies.*

The theorem sets forth two conditions for self-regulation: a heterarchical organization *and* the formation of temporary hierarchies. Requiring temporary hierarchies for decision making may seem to overstate the case. But it is not only this issue (which is in fact the subject of this contribution), but more systematic reasons that render this requirement necessary.

One could claim that a sharper distinction is needed between temporary hierarchies of members of a system on the one hand and temporary hierarchies of choices on the other. This would in fact distinguish between choices and their proponents and allow us to put more stress on consensus formation. In fact, heterarchical systems should allow for consensus formation, but they should not be mistaken for communities in the Durkheimian sense. Communities in this tradition are defined as social systems characterized by the absence of internal differentiation, that is, by the presence of a shared construct of reality. It is for this very reason that the more an internally differentiated system approaches the community type, the more problems that need regulation, and accordingly call for self-regulation, tend to disappear.[26]

If we take the case of community, and hence consensus formation, without temporary hierarchies as one extreme of heterarchical systems, we find at its opposite another group of systems in which the formation of temporary hierarchies does not succeed although regulation problems exist and although a heterarchical type of organization - or at least relationships between the individuals concerned which look like a heterarchical organization - has been brought about. These situations seem typical for social systems on the brink of breaking down as composite unities or for sets of individuals who mistake being together for being a social system with respect to the problems they have to solve. In this sense, not only members of the same university but also members of the same faculty might be considered a social system if the university or the faculty comes under attack from outside. But if one looks at certain aspects of collective decision making that call for self-regulation, one can find that much effort is dedicated to preventing temporary hierarchies from forming. The reason is that the individuals concerned often do not constitute a social system in the sense described in the definition above or that, in the name of abstract egalitarian principles, they oppose the formation of *any* hierarchy.[27] We can therefore conclude:

1. The formation of temporary hierarchies is not only a useful but a necessary requirement for self-regulation.

2. Temporary hierarchies presuppose a heterarchical organization of the interactions of the members of the system concerned. Finally both propositions can be connected by the (already drawn) conclusion:

3. Self-regulation depends on heterarchical relations between the components of the system.

How are temporary hierarchies produced and what is their significance beyond being a solution to the decision-making problem of heterarchical systems? If there are, as it is the case in heterarchical systems, no organizational reasons that constitute an a priori hindrance to participation in the group's decision making, the only effective factor of differentiation may be what can be called »attention.« Assuming for the sake of convenience that the participation of the members depends only on their level of attention (including cognitive and physical ability as well as readiness to act), it follows from what has been said about systems and individuals-as-components that this level of attention can be differentiated according to three general groups of causes. They all can trigger changes in those sets of properties through which components of systems are defined. The groups of causes (which sometimes overlap) are:

1. the system's internal functioning and dynamics,
2. influences from the system's environment,
3. individual factors.

The internal dynamics of a social system may well contribute to a differentiation of the level of attention of its members and hence to a difference in their participation in the decision-making process of the system: Not everybody is interested in everything, temporarily varying levels of occupation modulate »free attention« as well as routine, specialization, and habitude.

The cognitive constructs and action programs that exist at the base of every social system necessarily contain definitions of what constitutes the system's environment as seen from the system. These environments are therefore socially defined.[28] Accordingly, there are evaluation schemes and »action programs« that allow them to be handled in a way agreed upon by the system. Everything that happens in the so-defined environment then triggers the corresponding behavior of the system, or, to be more precise, of its members. Although the environment is defined as a class of possible events and actions specified by the system, the system can only partially control the sequence of events that actually occurs. Therefore, and this can be seen as a certain analogy to the influence of the organization of the system, the environment acts upon the system. These actions might accidentally or purposefully lead to differentiations in the system (self-organization) and hence influence its self-regulation.

The last group, individual reasons, is comprised of causes that can be attributed neither to the internal functioning of a social system, nor to its environment in a strict sense. This group includes all those experiences of individuals that participate in the modification of the properties defining them as components of social systems. These causes influence a given system through the ways in which the flesh-and-blood individuals cope with them, eventually changing in this way those characteristics that make them (partly) components of a specific system. They do not belong to the environment of the system because they are not defined as such within the constructs of reality and the action programs that are partly constitutive for social systems.

The Relation between Hierarchical and Heterarchical Systems

From this discussion of hierarchical and heterarchical systems, it may appear that they are mutually exclusive types of systems; that is, that a hierarchical system could include no elements of heterarchical systems and vice versa. Nothing would be more wrong. Constructed as ideal types, both are in fact found in various combinations in our phenomenal world. Even theoretically, there is no reason to disregard the combinations they form.

Since components are defined through the interactive capacities they actually display in a system, the way their behavior is produced is of no importance to the definition. Hence it may as well result from a hierarchical as from a heterarchical type of regulation *in the components* (or subsystems). Here different levels of systems analysis must be distinguished. A system is considered to be heterarchical or hierarchical depending on whether its organization is seen as such at the level at which the system is *actually defined*. Of course, at the empirical level, there appears to be a problem - sometimes difficult to solve - of deciding to which level a given phenomenon belongs. As stated above, systems are analytical tools. Therefore, there is no possibility to find something like »natural« systems in the social domain. It is the scientists who decide, and who must then justify, why they draw the border of what they take as a system in the way they do.

As a result, it may well happen that a system has to be classified at a certain level as heterarchical, although a study of its components, taken themselves as systems, shows that they must be viewed as hierarchical. The contrary can occur as well - for example, a system might be of a hierarchical type but consist of heterarchically functioning components or subsystems. And there are other possible compositions as well. It even seems - at least theoretically - that no combinatory possibility must necessarily be excluded. Empirically, the most frequent case is probably that of a combination of heterarchical processes with hierarchical ones, such that the system changes from one type to the other according to the problems it is working on. The typical examples are those social systems - and they are not rare, especially in science - in which routine work is done in a more or less hierarchically organized way, but where the system changes to a heterarchical mode if problems of innovative research or problems with relevance to the overall politics of the group or institution have to be dealt with.[29] Just as the emphasis on internal versus external explanations overstates the case, the postulate of a mutual exclusion between hierarchy and heterarchy holds only with respect to a specific system and its treatment of a particular problem at a specified point in time. Moreover the theory-dependent answer to the question into which category a system should be classified is overwhelmingly empirical and can only be decided through an investigation of the way it works.

3. Conclusion

Two types of self-regulating processes have been identified: self-regulation from self-organization and self-regulation from the activities of components in a heterarchical organization. Seen in a Durkheimian and historical perspective, self-regulation through self-organization is the »older« and more »archaic« way to produce changes in the behavior of systems. But it is a necessary condition for the generation of a heterarchical type of organization. Whereas the stability of social systems results to a large extent from their organization, it is through heterarchical self-regulation that they generate a variety of different realities to which they adapt. Social change might then be understood as the self-organized generation of autonomized components or subsystems (be they individuals or social systems) that, through their interactions patterned by the actual state of the system's organization, generate realities to which the system adapts in a heterarchical process of self-regulation.

Notes and References

1. The theoretical model presented here resulted from and owes much to continuous discussions I have had over recent years with G. Roth, A. Barsch, S.J. Schmidt, and many colleagues from different disciplines. Yet, without the patient help of R. Hunt, whom I have the pleasure to thank particularly, it would have been very difficult to produce this text.

2. P.M. Hejl, *Durkheim und das Thema der Selbstorganisation*, LUMIS - Publication No. 18, University Siegen, 1988.

3. See for more details and for the relation of the approach to traditional system theory or to holism, P.M. Hejl, *Self-Regulation in Social Systems: Explaining the Process of Research*, LUMIS - Publication, in print, University Siegen, 1989.

4. See for more details P.M. Hejl, »Towards a Theory of Social Systems: Self-Organization and Self-Maintenance, Self-Reference and Syn-Reference«, in H. Ulrich, G.J.B. Probst (eds.), *Self-Organization and Management of Social Systems, Insights, Promises, Doubts, and Questions*, Berlin, Heidelberg, New York: Springer, 1984, p. 60-78; *op. cit.*, 1988, note 2.

5. See on the used understanding of constructivism E.v. Glasersfeld, »The Concepts of Adaptation and Viability in a Radical Constructivist Theory of Knowledge«, in I. Siegel, R. Golinkoff, D. Brodzinski (eds.), *New Directions in Piagetian Theory and their Application to Education*, Hillsdale, New Jersey: Erlbaum, 1980, p. 87-95; »An Interpretation of Piaget's Constructivism«, *Revue Internationale de Philosophie* **36**, 4 (1982), 612-635.

6. This definition of the properties of members or components of a system is in no way trivial as a comparison with the influential definition of A.D. Hall, R.E. Fagen, »Definitions of Systems«, *General Systems* **1** (1956), 18-25 demonstrates. They defined components through properties *and* interactions, establishing a distinction that unavoidably leads to confusion.

7. See for instance N. Luhmann, »Lob der Routine«, in N. Luhmann, *Soziologische Aufklärung. Aufsätze zur Theorie sozialer Systeme*, Vol. 1, Opladen: Westdeutscher Verlag, 1970, p. 113-142; *Funktion und Folgen formaler Organisation*, Berlin: Schriftenreihe der Hochschule Speyer, Vol. 20, 1972.

8. There is the possibility that minor causes might produce effects that seem to be out of proportion. Effects of this kind are normally observed when the system reaches, for whatever reason, an unstable state, allowing some modification in the behavior of a member or some unexpected result etc. to trigger a whole series of modifications.

9. See for a more detailed explanation of the concept of self-organization to be developed in this part Hejl, *op. cit.*, 1988, note 1.

10. See D. Crane, *Invisible Colleges. Diffusion of Knowledge in Scientific Communities*, Chicago: University of Chicago Press, 1972.

11. In fact we must distinguish two extreme cases. If there are only very few members who change their relevant cognitive constructs, the system will be affected only in a negligible way. If, on the contrary, an overwhelming majority undergoes changes, there might be cases where the existing organization remains invariant even though the defining reality constructs and action programs have evolved.

12. This definition takes into account the fact that no explanation can avoid the use of unexplained assumptions. For this reason and because the explanation would go beyond the aim of this contribution, the question of the source of events that lead to self-organizing processes is left open. For the present purpose it may suffice to point at the implication that the definition does not exclude successful external influences on the system, even if it must be thought as operationally closed to the extent that it is self-organizing and self-regulating.

13. In system theoretical terms, »coercion« might be understood here as an unacceptable gap between the dynamics of components and the inputs they continue to receive from the social system to which they belong.

14. The formation of hierarchical static systems such as classificatory or norm systems are good examples. From the point of view of systems theory, cognitive hierarchies are situated at the level of components.

15. See N. Luhmann, »Politische Planung«, in N. Luhmann, *Politische Planung. Aufsätze zur Soziologie von Politik und Verwaltung*, Opladen: Westdeutscher Verlag, 1971, p. 69.

16. W.R. Ashby, *An Introduction to Cybernetics*, London: Methuen, Routledge Chapman & Hall, 1965, p. 207.

17. R.C. Conant, W.R. Ashby, »Every Good Regulator of a System Must Be a Model of That System«, *International Journal of Systems Science* **1**, 2 (1970), 89-97.

18. In technical contexts it is less problematic because regulators are constructed to meet the requirements of predefined events. If the engineer speaks of »self-regulation«, using the term not merely as a »facon de parler«, he refers to environments and hence possible events of either limited variety, or it is understood that there are limits to regulation beyond which breakdowns might happen.

19. The logical theorem of transitivity establishes a linear order of preferences. Applied to a set of for example, three choices (A, B, and C), it stipulates that if A is preferred to B and B to C, A is also preferred to C.

20. Of course one might think of hierarchically organized social systems, in which modifications of the system's behavior are brought about through decisions from top of the hierarchy, that allow intransitivity as well. This case, which is not taken into account at the level of the general discussion of this contribution, can be understood as a historically produced combination of both types of hierarchical regulation discussed here.

21. Of course social hierarchical organizations are often, if not mostly, designed to assure a specific output as results of a preselected sequence of activities of the components of the particular system. Nevertheless, as a type of organization, hierarchies and their functioning are independent of the activities

they structure as well as of considerations of which particular component should participate in which way.

22. And to avoid any mystification, it should be underlined that the organization »does nothing«. The only active agents in a system are in fact the components, the role of the organization being precisely that of a social pattern of interactions, which for this very reason is not at the disposal of singular individuals who participate as components in the system's activities.

23. W.S. McCulloch, »A Heterarchy of Values Determined by the Topology of Nervous Nets«, in W.S. McCulloch, *Embodiments of Mind*, Cambridge, Massachusetts: Cambridge University Press, 1965, p. 40-45.

24. It has to be stressed that we are concerned with ideal types. Clearly enough, observable systems normally present characters which make them for good reasons partly hierarchical and partly heterarchical. See below, section 2.4.

25. The minimal case being the hierarchical ranking of »decision« and »nondecision«!

26. Not surprisingly, heterogeneous membership in a research group is associated with its superior performance. See C.G. Smith, »Scientific Performance and the Composition of Research Teams«, in *Administrative Science Quarterly* **16** (1971), 486-495.

27. An attitude that may easily turn out to be in practice one way to establish a hierarchy: see below the discussion of factors leading to the formation of temporary hierarchies.

28. As a result, events that do not belong to the so-defined environment of the system are normally not taken into account, or, if they are, are considered to be merely »individual problems« of the members of the system.

29. See C.G. Smith, »Consultation and Decision Processes in a Research and Development Laboratory«, in *Administrative Science Quarterly* **15** (1970), 203-215, who gives evidence of this »organizational mix« in relation to the type of problems to be dealt with on the basis of an analysis of Decision Processes in a R. and D. laboratory. Smith of course uses the traditional opposition, »hierarchical vs. decentralized organization«, and hence decision, which could not answer the question of how decentralization was coupled to the overall unit. See as well T. Shinn, »Scientific Disciplines and Organizational Specificity: The Social and Cognitive Configuration of Laboratory Activities«, in N. Elias, H. Martins, R. Whitley (eds.), *Scientific Establishments and Hierarchies*, (Sociology of the Sciences. Yearbook VI), Dordrecht, London: Reidel, 1982, p. 239-264, for a demonstration of a specific relation between task specificity and type of organization.

Systemic Therapy
A Particular Drift between Systems Theory and Psychotherapy

KURT LUDEWIG
University of Hamburg

Abstract

As an applied science, psychotherapy has grounded its concepts and methods to a great extent upon the contemporary prevailing understandings of man and nature in the basic sciences. During the last decades it has additionally incorporated the concepts of self-organization, cybernetics, and systems theory and their epistemological and ontological consequences into its conceptual repertoire. This paper gives a succinct account of the historical development of paradigms in psychotherapy by reconstructing step-by-step the introduction of systemic thinking into this field. Then, it presents an outline for a clinical theory that intends to interrelate the different aspects entailed in clinical (psychotherapeutic) work into a comprehensive whole that is sparing enough to provide for practicability. The paper ends by briefly presenting the results of first outcome studies made within this conceptual perspective that suggest its efficiency.

1. Historical Remarks: The Converging Drift Between Psycho-Therapeutic Practice and Concepts of Self-Organization

The beginnings of psychotherapy as an autonomous discipline, derived from both medicine and psychology, date back to the last quarter of the 19th century. Keeping step with its sources in medicine as an applied natural science in the Zeitgeist of the 19th century, personal suffering was conceived of as a symptomatic manifestation of inner psychic disorder, as a coming to the surface of improperly balanced psychic energies. It was assumed that if the process of intrapsychic binding of inborn energies did not succeed through adequate interconnections with the immediate environment, it would leave traumatic defects in personality that would then unconsciously interfere with the realization of conscious goals in adult life, thus eliciting suffering. The therapist's task was correspondingly conceived of as providing for a relational setting in which these »defects« could be analytically disclosed and worked through. Psychoanalysis and its corresponding psychopathology modeled itself within the paradigm of mechanics. Its main concern was with the inner functioning of the psychic apparatus (psychodynamics); its main purpose was to find ways to enter it through undermining its resistance against disclosure in order to make it analyzable.

During the first quarter of the present century, a partial shift of orientation took place. The experimental approach to learning had rendered a mechanistic vision of behavior (conditioning) in terms of in- and outputs mediated by a not-yet-explainable mechanism (the organism). The concern with psychic processes was deemed mentalistic and of no interest for scientific inquiry. All that mattered was the observable, that is stimuli and reactions (= behaviors). The application of learning principles to healing

W. Krohn et al. (eds.), Selforganization. Portrait of a Scientific Revolution, 128–142.

began, though, only after World War II, as large groups of veterans in need of psychological treatment overwhelmed existing treatment capacities. This new branch of therapy, behavior therapy, claimed that, in comparison with psychoanalysis, it could considerably shorten the length of treatment and scientifically prove its effectiveness. Essentially, the psychopathological assumption of this approach was that suffering stems from either learned, maladjusted behavior or unlearned, adjusted behavior.[1] Problematic behavior was no longer regarded as a symptom of an abnormal apparatus, but as the illness itself; the application of the medical model to behavioral phenomena could thus be abandoned. The goal of this therapeutic orientation is, basically, to promote the learning of the missing and the unlearning of the inadequate.

By the time behavior therapy began to expand in the 1950s two other main orientations in psychotherapy stemming from practice appeared as well. The first, the so-called humanistic or phenomenological approach, gained importance during the 1960s; the second, family therapy, during the 1970s. In the work of Carl Rogers[2], a main exponent of the humanistic trend, a central consideration is paid to the person of the client in his or her totality. The psychopathological problems to be dealt with in therapy are regarded as an outcome of disturbing experiences that restrain personality from otherwise normal development (unfolding its possibilities). The work of the therapist consists of providing for an interpersonal setting that allows the client to experience him- or herself more fully and thus be freed from hindrances. For this to occur, the therapist needs to center him- or herself in the client, refraining from alienating diagnostical and interventionist attitudes and behaviors. Helping to become fully a self-actualizing person is the goal of therapy.

As may be inferred from the above, the short history of psychotherapy moves slowly from analogies from medicine and physics (mechanics) toward psychologically based models. Whereas with regard to psychopathology, the location of its origins is shifted from the inner psychic to the interpersonal (learning, experiencing), with regard to therapy, its proposals change from the notion of a socially detached analyst to that of an active participant in a social relationship. The next step in this course expands its focus to the social network of which the person is a part. It is taken by family therapy.

This new practice of psychotherapy is emerging simultaneously at different locations in the USA and in Europe. It responds to pragmatic difficulties arising especially in the treatment of young schizophrenics and juvenile delinquents. As we learn from Harry Goolishian, one of the innovators of the time, the teams experimenting in what later became »family therapy« engaged in transgressing the limits of accepted standards (i.e., individual treatment) without having either support from their institutions or from other teams.[3] These practitioners had observed that most young so-called schizophrenics would recover during inpatient treatment, but that many of them would come back with serious relapses after having been released to their families. More in terms of trying to understand the dynamics entailed in these relapses than of pretending to expand therapy into the family, they noticed regularities in intrafamilial behavior that, if they could be triggered to change, would result in an increase in the probability of lasting recovery. This mainly pragmatic by-product of practitioners' inquiry led to an increasing repertoire of new therapeutic techniques that were, at first, not backed by consistent theoretical concepts. It was Gregory Bateson who, having engaged together with J. Haley, D.D. Jackson, and others in trying to understand schizophrenic communication at a Veterans Administration Hospital in Palo Alto, California, began influencing theoretically the field by bringing in cybernetics as a means to describe and explain the functioning of families in terms of complex, circularly organized behavioral patterns. The notion of self-regulative feedback loops that maintain homeostatical constancy in families became central. This work was continued at the Mental Research Institute in Palo Alto by Paul Watzlawick and colleagues.[4]

From this point of view, the family is seen as an open system committed to conserve the stability of its structure in the face of environmental strains. Symptoms are conceptualized as self-regulative mechanisms maintaining the system's homeostasis, that is, as fluctuation-reducing, negative feedbacks. Symptoms are thus implicitly functional. That they are nevertheless termed dysfunctional results obviously out of deviance from normal expectation about how family members should operate. Suffering stems from the impossibility of stopping such »dysfunctional« processes. Correspondingly, therapy is modeled on the need to reregulate the system's functioning through interventions from the outside. Depending on the orientation of the therapist, the goals of therapy vary between restructuring the family, interrupting their dysfunctional communication patterns, and logically tackling their inappropriate world views.[5] The development in family therapy reached its peak in 1978 in terms of field-wide acknowledgement when a Milan team reported their impressively effective work with families of schizophrenic youngsters.[6] Their concepts integrate the thoughts of cyberneticians and communication and systems theorists with practice into a coherent and effectively applicable form. In 1980 the Swiss psychiatrist Gottlieb Guntern proclaimed finally »The Copernican Revolution in Psychotherapy«, meaning a radical change from an analytical paradigm based on Newton's physics to a systemic one relying on modern systems theory and related disciplines.[7]

During the late 1970s an increasing interest in systems theory, self-organization, and their epistemological consequences spread into the field. Although many of the leading family therapists adhered to the terms of ad hoc explanations of their impressive therapeutic success (mostly in terms of rather simplistic analogies to systems dynamics, e.g., mobiles), a few others, in search of more coherent explanations, made a marked shift toward the work of Prigogine, Jantsch, von Foerster, Maruyama, and other scientists involved in studying self-organizing systems.[8] The historical accomplishment of family therapy is to have drawn attention to notions like patterns and processes instead of elements and states, and to complexity, interpersonal dynamics and circularity instead of traits, intrapersonal dynamics, and lineal causality. Systemic therapy adds to this, essentially, a shift in orientation from objectivism to constructivism. It thus implies an epistemological shift.

2. Systemic Therapy

During an international conference in 1981 in Zurich, Paul Dell thoroughly tackled the theoretical relevance of almost all that had become indispensable as a foundation of family therapy, for example, the notions of homeostasis, information, open system, dysfunctionality, strategic and paradoxical intervention, and so forth[9] A little later, the 1982 issue of the journal *Family Process* included a few papers that would initiate a momentous shift in theory and practice.[10]

The advanced propositions challenged therapy theory right at its base. They were: structure-determinism and operational closure of all biological and thus cognitive and social systems with its epistemological and ontological consequences.

In one of these papers, Paul Dell calls attention to the unsuitability of a therapy theory based upon homeostasis and stability that fails to explain its main goal: change. Drawing on Maturana's writings he proposes to do away with the notion of a family as an open system embedded in larger systems that determine it. Instead, he proposes to think of family in terms of a structure- determined entity that operates in closure, that is, only with its own states, and that therefore may not be determined from the outside. From this position, the notion of pathology becomes obsolete, since all systems are at any time functioning at their best. The proposition of operational closure poses a tremendous challenge to therapy theory, since its right to exist is grounded precisely on the possibility of deliberately changing (healing) individuals and groups. Bradford Keeney, following mainly Bateson's thinking, calls the attention of the therapist to the epistemological understanding that all knowledge, and thus also all knowledge concerning the functioning of families and of therapists, is bound to distinctions drawn in the cognitive domain of observers and hence not to an independent, transcendent reality. Since this is the case, the therapist is no longer an objective authority entitled to make decisions about his or her clients. Thus, ethics rather than objectivity should become the foundation of therapy, and therapy should be understood as applied epistemology. From this it follows that therapy may not be understood adequately if conceptualized only in terms of pragmatic utility. It needs to be implemented in a realm defined by the complementarity of pragmatics and aesthetics, that is of technological craftsmanship and poetical artistry. Finally, Steve de Shazer contends that clients by joining a therapist configure together with him or her a common social system that becomes the relevant entity. The therapist may no longer be seen as an intervening agent from the outside. From this point of view, he contends that a theory of therapy should deal with the specific mode of cooperation between clients and therapist and lay aside such traditional notions of psychotherapy as »power« (on the part of the intervening therapist) and »resistance« (on the part of the client). The assumption that clients ultimately always cooperate with the therapist leads the latter to assume responsibility for behaving in a manner that is likely to meet the specific ways of cooperating of each clientsystem. This notion reflects more adequately the self-referentiality of the therapy system; it also defines a more ethical setting by allowing the therapist to devise procedures that are better fitting to the client and thus less painful and time-wasting, hence more efficient.

The 1980s are the era of systemic therapy. As a further development out of family therapy it relates to its origins in an equivalent manner as does 1st to 2nd order cybernetics.[11] Its theoretical background has incorporated especially Maturana's theories of autopoiesis, cognition, and sociability, but also von Foerster's cybernetics of observing, von Glaserfeld's radical constructivism, and Luhmann's sociological systems theory.[12] Within this framework, the notion of systemic therapy may be reserved for the therapeutic practice that takes place whenever the therapist reflects upon and guides his or her activities from a systemic point of view. That is, whenever he or she conceives him- or herself as an active, constituting member of a social system - the therapy system - in which he or she conjointly with his or her clients bring forth a therapeutic reality that is propitious for eliciting change. Since theoreticians of systemic therapy have not yet attained a consensus with regard to their subject (apparently an unavoidable aspect of discourse in psychotherapy), the following restricts itself to the position of our Hamburg team.[13] It presents an outline for a clinical theory based on systemic thinking that comprises the relevant literature and orders it into a comprehensive and yet condensed, thus workable, unity.[14]

3. Outline for a Clinical Theory

A systemically conceived, comprehensive theory of psychosocial healing-practices needs to entail coherently the following: 1) a definition of the »systemic«, that is, a demarcation of the domain of knowledge to which it pertains; 2) a description of social systems that is compatible with 1); 3) a specification of problemsystems, that is, of social systems centered on a problem; 4) the formulation of criteria of evaluation for therapy that are congruent with 1); and 5) a methodological solution to the problem of therapy and its evaluation in accordance with 1) - 4).

A Definition of »Systemic«

The term »systemic« has been used in the field of therapy in a very loose manner. In fact, it has traditionally served mainly as a distinguishing mark from »nonsystemic« therapies but without a consensually accepted meaning. For the purposes of this paper, the well-established adjective »systemic« will be specified in terms that bindingly delimit the entailed domain of discourse.[15] In a first gross approach, »systemic« is defined as the specification of a general method of thought, indeed a »way of seeing«[16] concerned with systems. In order to make this overall definition more precise, six propositions seem to suffice which, in the framework of this publication, do not require further explanation as they refer to the core of thinking entailed in this book, for example, bio-epistemology, 2nd order cybernetics, self-organization, and modern systems theory. They are: (1) Everything said is said by an observer.[17] (2) An observer is a languaging living being who draws distinction in language and thus brings objects into existence. (3) Everything said is brought forth by observers in conversation with other observers. (4) It follows that reality becomes realities-in-parenthesis, that is, arguments in communication; and (5) since a system is a unity that was brought forth by an observer who regards it as a composite unity (i.e., composed of elements and their interrelations). (6) »Systemic« refers to a »way of seeing« concerned with systems (as specified in (5)).

Propositions (2) and (3) define the terms included in the universal proposition (1): the observer and what is said. Proposition (4) follows from (1-3). Proposition (5) specifies the term »system« in accordance with the foregoing propositions, and finally (6) follows from (1-5) substantiating the general definition of »systemic« given above. This set of propositions intends to follow closely what Humberto Maturana has called the explanatory path of objectivity-in-parenthesis.[18] By this is meant that if the observer accepts that he or she is a human living system, he or she must also accept being inescapably embedded in the biological phenomena that constitutes him or her as an observer. In this sense, an observer cannot do anything else but constantly realize his or her condition as a living system inlanguage. As an autopoietically organized unity, the living system is described in modern biology as a structure-determined, operationally closed, and thus autonomous system. It follows that an observer has no operational basis to make any statement or claim about objects, entities, or relations as if they existed independently of what he or she does. Consequently, the notion of a transcendental reality becomes irrelevant, and objectivity, in terms of an accurate representation of the world, becomes biologically unattainable. It is the observer who constitutes existence through the operations of distinction he or she makes while living in a domain of languaged coexistence. Observers realize and conserve their basic organization as human living systems as long as they conserve adaptation (compatibility) with their environment. The organism, as a totality, does not interact with its environment; it may do so only through its components (e.g. receptor cells). Occurrences in the environment may have a perturbing effect upon components of the organism and therefore may trigger their functioning, but they never determine the course of the reaction of the organism, which is at all times determined by its own structure. An observer who focuses on the organism in its process of living may chose to speak of »instructive interactions« (i.e., determining inputs). By doing that, though the observer overlooks deliberately or inadvertently the structure determinism of the organism. No environmental agent may purposively determine the behavior of a living being. However, a thorough understanding of the structure of a living being may enable an external subject to provoke in it expected reactions with some degree of predictability. For this, he or she must either »properly« trigger its functioning (within the range delimited by its structure) or configure marginal conditions in which expected reactions become probable. In both cases, even though the behavior of the living being may appear to an observer as »controlled« by these circumstances, it is the affected living being that determines the course of its reactions.

Social Systems

Presently, two differing conceptions of the social system dominate the domain of systemic therapy. One of them is more attached to biological thinking, the other to sociological theory. Biologically grounded explanations define social systems as composed of (bodily) human beings and their interrelations (structure couplings). Social interaction occurs embodied in the organism of the participants.[19] From a sociological point of view, Luhmann[20] defines social systems as composed of communicational actions (elements) and the adhering of these elements to each other (relations). In the sense that they produce their own elements (communications), social systems are understood as autopoietic; the human individual is displaced into the environment of the social system. Being social systems composed of sequences of events, they have no spatial existence, they may only be distinguished (inferred) by an observer as temporary unities. In order to make this inference, the observer orders the observed events along a particular criterion of continuity: »meaning« (German: Sinn). Meaning implies a phenomenological process of selection under conditions of complexity that allows us to discern which events pertain to a certain communicational sequence and which do not. In other words, it allows us to order whatever »has meaning« as belonging together, and thus to sort out all other social actions. Meaning delimits the boundaries of a social system (border of meaning). From the perspective of the participants, meaning is what enables them to keep pace in communication (to adhere adequately), and what confers structure to a particular communicational network (through developing expectations).

With respect to therapy, both of these positions pose conceptual and pragmatic difficulties in terms of modeling. If the clients are seen as biological individuals, the therapist's task consists of properly perturbing them as components of the system in order to trigger the system to change. For this to be accomplished, the therapist needs to have attained a deep understanding of the clients in all their individual and social complexity. But, to understand the clients deeply demands engaging in extensive interactions with them (besides having to resolve the epistemological problem entailed in this). The therapist is thus confronted with the difficult task of duly interacting without getting entangled in precisely those behavioral patterns that he or she seeks to change, that is, without becoming conditioned by the clientsystem. On the other hand, the understanding of social systems as mere sociological abstractions composed of communicational networks runs against the therapist's perception of full-sized individuals sitting in front of him or her. Hence, both concepts exceed the cognitive skills of the therapist. Because of this, the following attempts to facilitate the therapist's understanding of his or her task by supplying him or her with a conceptualization of the social system that applies more adequately to therapy. This needs to be a) coherently systemic (see above) and yet b) practicably applicable to therapy, and it should additionally take into account that c) the therapist works in face of flesh-and-blood individuals in d) a conversational, social process that should therefore be formulated in terms of the social domain.

This we have attempted by means of a concept that is concrete enough to fit with the therapist's perceptions and abstract enough to suit the diction of social discourse: the concept of »member«.[21] By a member, we understand a social operator who performs at least one operation that is followed (adhered to) by the operation of another entity of the same class.[22] A member is always embodied by a human being, it is thus able to act and cognize, but its definition does not overlap with that of its »generator«, a human being. The social operator »member« is generated and conserved by a human being whose structure is, in turn, to some extent modulated by the operations entailed in all the memberships it produces. »Member« pertains inseparably to a membership in a particular social system; its definition considers only the (communicational) operations fulfilled while being realized. In this sense, members, communicational acts, and the border of meaning of a specific social system mutually generate and qualify each other and thus reciprocally condition the existence of one another. This is illustrated in the following figure:

(generate)
members → communicational acts → border of meaning
← (qualify) ←

Although the concept of member resembles at first sight that of the role-player, both concepts belong notably to different domains of discourse and should be handled distinctively: Whereas role-player is embedded in role-theory and thus related to structuralist and normative notions, member is defined as an entity that is inseparably connected to a particular social system by co-specifying its dynamics; it pertains therefore genuinely to the domain of (social) systems theory. In the context of membership, the notion of social role refers to a generalized description of a behavioral program as it is fulfilled by a member. As an entity that is distinguished in the social realm by reference to another member, this concept helps as well to refrain from undue individualization, generalization, and reification, and thus to maintain a clear »logical bookkeeping« between the entity and its domain of existence.[23]

With respect to the conceptualization of therapy in systemic terms, the concept of member delivers one main advantage over related concepts: The goal of therapy needs not to be conceived of in terms of having either to understand »deeply« and/or to change a human being, a person; instead it suffices to conceive it in terms of adequately triggering the dissolution of membership in a problemsystem.

The Problem-System

The utility of the above conceptualization of components of social systems becomes more apparent while defining the »problemsystem«, this being a specific social system centered on a »problem«. It is claimed that a problem of living may only become a matter of social concern if it is somehow uttered and communicationally adhered to. A sequence of communicational actions must have established itself that is valued either by the participants or by a noninvolved observer as unpleasant, painful, in short, as needing to change (disappear). In this sense, it may be contended that a problem is always constituted through communication, that is, through a social system that emerges while communicationally dealing with a disliked theme. Such a social system may be termed a »problemsystem«.[24]

In order to explain the coming about of a problemsystem, it is required, as it applies to all other social systems (see above), to consider two domains of discourse simultaneously, being careful not to infringe on the logical book-keeping: these are the domains of the generators (human beings) and of the operators (members) entailed in the emergence of this system. At the level of the participating »generators« the fulfillment of two conditions seems to suffice: 1) at least one of them must negate, to some extent, the right of another to generate an adequate operator »member« in a specific interaction;[25] and 2) at least another one must insist on trying to change this evaluation in his or her terms. If this pattern stabilizes itself, a social system emerges that, at the level of its members, functions just as well as any other social system (otherwise it would dissolve). But, because of the emotional effects this communication has on the persons who generate the members of the system, either through the experience of being negated as a legitimate equal or through comparison with other less troublesome memberships generated by the same person, it is bound to elicit suffering and thus give rise to evaluation as a problem. However, probably precisely because the participating persons wish to quickly (dis-)solve the problem, a great deal of interactions (or variations on the same theme) occur that are bound to keep it going.[26]

The concept of the problemsystem provides a number of theoretical and pragmatic advantages in therapy. They may be summarized as follows: 1) it is coherently systemic; 2) it does away with the necessity to postulate »psychopathology«, »family dysfunctionality«, and so forth, since a problem-system does not differ fundamentally from any other social system; 3) it allows defining the goal of therapy in terms of the dissolution of the problemsystem instead of the repairing or reteaching of a psychic apparatus or a family structure; 4) it allows the therapist to work decidedly with the »resources« of the clients, since they embody many other memberships and thus at any moment use a behavioral repertoire that can be triggered to do »something else« instead of »more of the same«; and 5) it frees the therapist from having to be an expert in matters of life, since all he or she needs to do is to provide favorable conditions for a shift in the preferences of his or her clients. In sum, this concept liberates therapist and clients from having to engage in the traditional game of negatively connotating each other (diagnosing, treating, resisting, and so forth) while trying to find and give help.

Criteria of Evaluation

The acceptance of the above advanced systemic premises leads basically away from the traditionally acknowledged criterion of therapy evaluation, that is, the assessment of its efficiency in terms of having »objectively« removed suffering through the application of a suitable treatment. A systemically founded clinical theory must therefore develop criteria for evaluation that are coherent with its premises. Faced with this problem in the domain of science, constructivists have reoriented their understanding of science from the search for truth toward the utility of knowledge, that is to the viability of scientific solutions. Unable to validate scientific knowledge against an independent reality, all scientific utterances necessarily fall back upon their authors, and their validity depends on the attained plausibility within the scientific community. For this to occur, knowledge needs to prove useful or viable, that is, it must allow orienting actions effectively without running into hindrances. Knowledge may not ever »match« reality, but it serves to surpass its constraints through »fitting« with them.[27]

The concentration on viability in terms of effective operationality as the main criterion of validity for a therapeutic orientation proves to a great extent coherent with the systemic view. It disregards though, in the last consequence, the central importance of the observer as the source of all knowledge, since the cause of viability is posed ultimately on the »treated«. Conversely, within the path of objectivity-in-parenthesis it is the observer who constitutes existence through languaging, and who therefore may at any time reflect about this process of reality construction. Insofar as the observer has this capability of reflecting, he or she unavoidably takes responsibility for the world he or she lives. Following this line of thought, the consideration of viability as the sole criterion of acceptability for a method appears insufficient; it should be complemented by criteria that bind the activities of the performer to his or her personal responsibility. Otherwise the discourse is in danger of falling back into objectivism.

Taking into account that therapy is a social undertaking at the core of human interests - the overcoming of suffering that permits no »trivialization«[28] -and with the intent of modeling it without abandoning the path of objectivity-in-parenthesis, we propose to complement the criterion of utility (viability) with two further criteria: beauty of form and humanity.[29] In this sense, for a therapeutic process to be valued as acceptable, it must not only prove effective in terms of solving a problem by whatever means, it must also meet aesthetic and ethical considerations. These additional criteria allow therapists and/or therapy theoreticians to keep in touch with the systemic premises without having to dispense with the personal responsibility entailed in therapeutic processes. Since these essentially subjective considerations escape nomological ruling, they refer inescapably back to their proponents. The observer (theoretician and practitioner) who is willing to abide by these criteria must therefore make explicit how far his or her choices regarding theory and methodology as well as their application will render a beautiful, human, and useful therapeutic process. In Maturana's words, the knowledge of knowledge obliges.[30]

Methodology of Therapy

In the light of all that has been stated above, therapy may be defined formally as the conversational process occurring while generating, conserving, realizing, and terminating a social system - the therapy system - under the following conditions: 1) at least one of the persons involved must define his or her membership in it as that of the therapist; 2) at least another one must define his or her membership accordingly as that of the client (customer), that is by making out of his or her participation as a member of a problemsystem the subject matter of the conversation with the therapist; and 3) their interactions must be im- or explicitly limited in time. If these conditions are fulfilled, all that can happen, from the point of view of an observer who acknowledges them as such, is therapy. All other conditions that might be imposed upon therapy, especially in normative terms, are not substantial since they may vary from place to place and from time to time.

If this formal definition of therapy is to serve the therapist as an orientation for defining him- or herself as such in terms of attitudes and activities bound to elicit effective cooperation with his or her clients, it needs to be operationalized. A number of alternative options have traditionally been used for this purpose. They are: 1) the effects of therapy; 2) the interactions between therapist and client; 3) the duration of therapy; 4) the structure of the client; and 5) the structure of the therapist. For the following reasons, we have opted for the latter: for therapy to be operationalized in terms of its success, it must be assumed that clients may be deliberately »instructed« or determined (see above); the utilization of the interactions between therapists and clients requires their standardization and thus an unnecessary a priori restriction of the scope of possible actions in therapy; the choice of time duration as the base for a methodological decision proves senseless prior to each therapy; the assessment of the client's structure, for instance, in terms of severity of illness or of adequacy for therapy (predictability of success that outweighs the efforts) is grounded on the assumption of objectivity (without parenthesis) and, in addition, it dispenses with any ethical substantiation. Having dismissed all these options, what is left for the purpose of outlining a method is the structure of the therapist's membership.

The methodological definition of the therapist's structure may be attained by proposing a more or less generalized program of actions, that is, by defining the role of the therapist. This definition needs to take into account both that 1) whatever the therapist does during therapy is to be done »as a therapist« and 2) that the therapist is embodied by a person in the only possible manner this person can do it. The definition of the therapist needs therefore to be both task- and person-oriented. Furthermore, therapeutic activities should be conceptualized according to the criteria stated above, that is, in terms of their expected utility (goal-orientation: dissolution of the problemsystem), beauty of form (choice of interactional manners), and humanity of attitude (respectful acceptance of the client). This we have accomplished elsewhere by formulating 10 + 1 guidelines or guide-questions for the therapist that concisely summarize these lines of thought in a manner that has proven helpful both in terms of orienting the activity of the therapist and of providing a frame of reference for supervision.[31] Since they primarily pertain to the practice of therapy, these guidelines will not to be treated here in detail.

4. Outcome Evaluation

In the field of family therapy, as in all other fields of psychosocial therapy, there is a continuing debate going on regarding the efficiency of the applied methods. The contentions of review articles written by prominent family therapy researchers in the handbooks of the last decade point to an overall empirical utility of family therapy.[32] On the other hand, authors concerned mainly with the observance of the rules of research methodology (at times of »methodolatry« or the tendency to let research methods dictate what are proper data) conclude that family therapy is still far from having soundly proved its utility.[33] They claim that most of the positive outcome results stem from either clinical samples (as opposed to a research design providing for matched control groups, pre- and post measurements, placebo groups, etc.) or inadequate methodology of appraisal. Further, they claim that these positive results are contaminated with those of other therapy orientations not properly sorted out from the researched samples. In sum, the discourse about therapy outcome seems to cling to the traditional dispute between clinicians and researchers of the field. Whereas clinicians have actually admitted family therapy into their field by supporting a number of training institutions and professional associations all over the world, researchers remain skeptical about the efficiency of this method.

Interestingly enough, in spite of this discussion, research shows that psychotherapy is, regardless of the applied method, overall of help, that is, of more help than when it does not occur.[34] Obviously, this means that, in the long run, all methods are helpful if they meet, at least to some extent, the expectations and needs of the client. Insofar, a therapy method that claims effectiveness is supposed to at least beat the rate of so-called spontaneous remission.[35] Besides that, there is some evidence that even a one-session-counseling may be all that is needed.[36] Taking these aspects (among others) into consideration, it may be followed that therapy should be practiced within the margins marked by doing too much (too long a time, too much intervention, too much technology, and so forth) and doing too little (too little care, too little orientation toward the client's needs, too little optimism, and so forth).

With regard to »systemic therapy«, understood as an orientation that first appeared in the 1980s (see above), there are still few reports of outcome. This orientation has been more preoccupied with clearing out its own theoretical background than with trying to assert its efficiency empirically. Most of the reported assessments were obtained clinically and not by means of traditional methodology. The Brief Family Therapy Center in Milwaukee reports, for instance, improvement rates of about 80% after an average of 4-5 sessions as assessed by follow-up phone calls (Steve de Shazer, personal communication 1988). These results coincide with those of other institutions of the same or similar affiliation.

In our own studies at the outpatient clinic of the Department of Child and Juvenile Psychiatry at the University of Hamburg we have come to similar results. Since 1982 we have been developing a mode of questioning outcome in accordance with our understanding of systemic thinking and with the criteria of validation expressed above. It concentrates on the appreciation given by the clients regarding three aspects of therapy: utility, beauty of form (fitting of therapist's behavior to the client's expectations) and respect in attitude toward the client. A first major result was obtained in a mail follow-up of all clients who attended therapy sessions with four systemic-oriented therapists in 1983 (42% return rate). It disposes in favor of optimism: 75% of the clients (mostly families) report by the time of questioning (2-4 years after finishing consultation) that the child that had presented the problem is doing well; they attribute this clearly, at least to some extent, to the conversations held with us. The most impressive result is, however, that this outcome was obtained, regardless of the kind of presented problem, after an average of 2.75 sessions (total range: 1-12 sessions).

If these results are substantiated by a current follow-up study of all clients treated between 1984 and 1986 (a first account of 40% of the answers does so), it may be plausibly concluded that a systemic understanding of therapy proves not only to consider duly the autonomy of human beings in theoretical but also in pragmatic terms by efficiently helping to solve problems with the least possible perturbation. As such, it actualizes the criteria of utility, beauty, and respect, qualifying them as adequate conditions of therapy. Additionally, it helps overcome the dilemma sketched out at the beginning of this paper, that is, of having to choose as a theoretician and/or practitioner of therapy between complying with the rules of (uni-versal) science and enacting personal autonomy as a viable means toward the construction of liberating (multi-versical) realities.

Notes and References

1. See, for example, H.J. Eysenck, S. Rachman, *The Causes and Cures of Neuroses*, London: Routledge & Kegan Paul, 1965; Ch. Kraiker (ed.), *Handbuch der Verhaltenstherapie*, Munich: Kindler, 1974.

2. C.R. Rogers, *Client-Centered Therapy*, Boston: Houghton-Mifflin, 1951; *On Becoming a Person*, London: Constable, 1961.

3. H. Goolishian, workshop in Hamburg, 1986. See also H. Anderson et al., »The Galveston Family Institute: Some Personal and Historical Perspectives«, in D. Efron (ed.), *Journeys: Expansion of Strategic-Systemic Therapies*, New York: Brunner/Mazel, 1986; G. Bateson et al., *Schizophrenie und Familie*, Frankfurt am Main: Suhrkamp 1969; L. Hoffman, *Foundations of Family Therapy*, New York: Basic Books, 1981.

4. Bateson et al., *op. cit.*, 1969, note 3. See also P. Watzlawick, J.H. Beavin, D.D. Jackson, *Pragmatics of Human Communication*, New York: Norton, 1967.

5. C. Sluzki (see *Family Process* **22** (1983), 469-476) points out that the three predominant approaches to family therapy focus primarily at either process, structure, or world view of the families. The structural-strategic approach intervenes to restore a (normative) »healthy« family structure, for example, S. Minuchin, *Families and Family Therapy*, Harvard University Press, 1974; J. Haley, *Problem-Solving Therapy*, San Francisco: Jossey-Bass, 1976. The interactional, process-oriented approach points at interrupting »dysfunctional« communicational patterns, for example, P. Watzlawick, J. Weakland, R. Fisch, *Change: Principles of Problem Formation and Problem Resolution*, New York: Norton, 1974. The approach that concentrates on changing the world-views of families (epistemes) stems especially from the work in Milan, see note 6.

6. M. Selvini-Palazzoli et al., *Paradox and Counterparadox*, New York: Aronson, 1978. See also: Selvini et al., »Hypothesizing -Circularity - Neutrality: Three Guidelines for the Conductor of the Session«, *Family Process* **19** (1980), 3-18.

7. G. Guntern, »Die kopernikanische Revolution in der Psychotherapie: Der Wandel vom psychoanalytischen zum systemischen Paradigma«, *Familiendynamik* **5** (1980), 2-41.

8. For example, P.F. Dell, H.A. Goolishian, »Ordnung durch Fluktuation: Eine evolutionäre Epistemologie für menschliche Systeme«, *Familiendynamik* **6** (1981), 104-122.

9. P.F. Dell, »From Systemic to Clinical Epistemology«, in *Zusammenhänge* **3**, Institut für Ehe und Familie, Zurich, 1982.

10. *Family Process* **21** (1982); P.F. Dell, »Beyond Homeostasis: Toward a Concept of Coherence«, p. 21-41; B.F. Keeney, »What is an Epistemology of Family Therapy?«, p. 153-168; B.F. Keeney, D.H. Sprenkle, »Ecosystemic Epistemology: Critical Implications for the Aesthetics and Pragmatics of Family Therapy«, p. 1-19; and S. de Shazer, »Some Conceptual Distinctions are More Useful than Others«, p. 71-84. Regarding the quest of pragmatics or aesthetics in therapy, see also H. Stierlin, »Family Therapy - A Science or an Art?«, *Family Process* **22** (1983), 469-476; K. Ludewig, R. Schwarz, H. Kowerk, »Systemische Therapie mit Familien von ›psychotischen‹ Jugendlichen«, *Familiendynamik* **9** (1984), 108-125.

11. H. von Foerster, *Sicht und Einsicht*, Braunschweig: Vieweg, 1985. With regard to therapy, see L. Hoffman, »Beyond Power and Control: Toward a ›Second-Order‹ Family Systems Therapy«, *Family Systems Medicine* **3** (1985), 381-396.

12. H.R. Maturana, *Erkennen: Die Organisation und Verkörperung von Wirklichkeit*, Braunschweig: Vieweg, 1982; H.R. Maturana, »Reality: The Search for Objectivity or the Quest for a Compelling Argument«, *Irish Journal of Psychology* **9** (1988), 25-82; H.R. Maturana, F.J. Varela, *The Tree of Knowledge*, Boston: New Science Library, 1987; v. Foerster, *op. cit.*, 1985, note 11; E. von Glasersfeld, *Wissen, Sprache und Wirklichkeit*, Braunschweig: Vieweg, 1987; N. Luhmann, *Soziale Systeme*, Frankfurt am Main: Suhrkamp, 1984.

13. The notions summarized here have been extensively elaborated in K. Ludewig, »10+1 Leitsätze bzw. Leitfragen«, *Zeitschrift für systemische Therapie* **5** (1987), 178-191, (engl. in J. Hargens (ed.), *Systemic Therapy. A European Perspective,* Broadstairs: Borgmann, 1989); »Nutzen, Schönheit, Respekt -Drei Grundkategorien für die Evaluation von Therapien«, *System Familie* **1** (1988a), 103-114; »Problem - ›Bindeglied‹ klinischer Systeme«, in L. Reiter et al. (eds.), *Von der Familientherapie zur systemischen Perspektive*, Berlin: Springer, 1988b.

14. The presented concepts are related theoretically to those of the Galveston Family Institute, see H.A. Goolishian, H. Anderson, »Menschliche Systeme. Vor welche Probleme sie uns stellen und wie wir mit ihnen arbeiten« in Reiter et al.(eds.), *op. cit.*, 1988, note 13; and pragmatically to those of the Brief Family Therapy Center in Milwaukee, see S. de Shazer, *Keys to Solution in Brief Therapy*, New York: Norton, 1985, and *Clues: Investigating Solutions in Brief Therapy*, New York: Norton 1988. They differ, though, from them in terms of focal point. Whereas Goolishian centers his interest in understanding meaning systems and in dialogue, and de Shazer focuses on the dynamics of rapid change through fitting interventions, our propositions concentrate on defining the therapist in a manner that is generally propitious for triggering rapid change by means of dialogue.

15. Ludewig, *op. cit.*, 1987, note 13.

16. F. Steier, »On Cybernetics as Reflexive Understanding«, *Continuing the Conversation: A Newsletter of Ideas in Cybernetics* **12** (1988), 7-8.

17. Statements (1) till (5) are closely derived from propositions advanced by Maturana and Varela, *op. cit.*, 1987, note 12.

18. Maturana, *op. cit.*, 1988, note 12.

19. H.R. Maturana, »Biologie der Sozialität«, *Delfin* **5** (1985), 6-14.

20. Luhmann, *op. cit.*, 1984, note 12.

21. Ludewig, *op. cit.*, 1988b, note 13.

22. This notion is to some extent related to that of the individual as discussed by P. Hejl, see, for example, »Zum Begriff des Individuums« in G.Schiepek (ed.), *Systeme erkennen Systeme*, Munich: Psychologie Verlags Union, 1987.

23. Maturana, Varela, *op. cit.*, 1987, note 12. With respect to the domain of therapy, see K. Ludewig, »Von Familien, Therapeuten und Beschreibungen. Vorschläge zur Einhaltung der logischen Buchhaltung«, *Familiendynamik* **11** (1986), 16-28.

24. Ludewig, *op. cit.*, 1988b, note 13. Somewhat similar notions may be found in Goolishian, Anderson, *op. cit.*, 1988, note 14, and in Luhmann, *op. cit.*, 1984, ch. 9, note 12.

25. Here we follow essentially Maturana, *op. cit.*, 1988, note 12, and C.L. Mendez, F. Coddou, H.R. Maturana, »The Bringing Forth of Pathology«, *Irish Journal of Psychology* **9** (1988), 144-172.

26. The intended solution becomes the problem. See Watzlawick et al., *op. cit.*, 1974, note 5.

27. S.J. Schmidt (ed.), *Der Diskurs des Radikalen Konstruktivismus*, Frankfurt am Main: Suhrkamp, 1987.

28. v. Foerster, *op. cit.*, 1985, note 11. He points with this notion to the process involved in functionalizing the human being as if it were a »trivial machine«, i.e. a non selfreferential device determined by inputs.

29. Ludewig, *op. cit.*, 1988a, note 13.

30. Maturana, Varela, *op. cit.*, 1987, note 12. See also »Hinter den Kulissen der Kognition. Mit Humberto Maturana im Gespräch«, *Familiendynamik* **13** (1988), 165-170.

31. Ludewig, *op. cit.*, 1987, note 13.

32. See, for instance, A.S. Gurman, D.P. Kniskern, »Research on Marital and Family Therapy: Progress, Perspectives, and Prospect«, in S.Garfield and A. Bergin (eds.), *Handbook of Psychotherapy and Behavioral Change*, New York: Wiley, 1978; A.S. Gurman, D.P. Kniskern (eds.), *Handbook of Family Therapy*, New York: Brunner & Mazel, 1981; T. Jacob (ed.), *Family Interaction and Psychopathology*, New York: Plenum, 1987.

33. For example, R.A. Wells, A.E. Dezen, »The Results of Family Therapy Revisited: The Nonbehavioral Methods«, *Family Process* **17** (1978), 251-274; H.-P. Heekerens, »Systemische Familientherapie auf dem Prüfstand«, *Zeitschrift für Klinische Psychologie* **17** (1988), 93-105.

34. For example, R.Cohen, »Verhaltenstherapie zu Beginn der achtziger Jahre«, *Psychologische Rundschau* **35** (1984), 1-9.

35. In a series of publications dating back to 1952, Eysenck (for example, *The Effects of Psychotherapy*, New York: Science House, 1969) has thoroughly perturbed the field of psychotherapy research by delivering empirical evidence that some 65% of »waiting-list« patients receiving no treatment spontaneously recovered after some time. A therapy method needs to beat the rate of spontaneous remission in order to be taken seriously.

36. M.Y. Talmon, »Single Session Therapy: Dropouts or Miracles?«, San Francisco, in preparation.

Literary Systems as Self-Organizing Systems

SIEGFRIED J. SCHMIDT
University of Siegen

1. In this paper I want to show how literary studies can profit from a systems-theoretical approach. Accordingly, I shall not deal with the epistemological problem of knowing whether literature *is*, in fact, a system. Instead, I shall consider the applicability of systems-theoretical concepts to »literary matters« and the degree to which these concepts help us to solve problems in the discourse of literary studies.

The application of systems theory to literary studies requires transforming the traditional domain of literary studies. Whereas traditionally the domain of literary studies was defined by a class of objects identified as literary texts, the systemic approach turns towards text-focussing and text-thematizing *activities* as well as to the conditions, contexts, and results of those activities. The dominant goal of literary studies shifts from the interpretation of texts to text focussing, or text-thematizing, their conditions, contexts, and results and the description and explanation of combinations of text-focussing activities.

The basic question that has to be answered is, of course, whether the sociological concept of »social system« can be transferred to literary studies at all. We must also ask how it has to be specified or modified, and how it may be used to organize the empirical material of literary studies.

I shall base my considerations on Luhmann's conceptualization of »social system«[1] at least with regard to its basic features (e.g. environment, borderline, complexity, selection, and dynamic stability). Special attention will, however, be devoted to a more specific characterization of the relations between the components of social systems because I do not share Luhmann's assumption that social systems can be modelled in terms of autopoietic systems in the same way that Maturana and Varela have modelled living systems.

2. In what follows I shall try to look at literature as a social system. I take *literary activities* to be the basic components of a literary system. Literary activities are those activities that follow the basic conventional regulations (macro-conventions) of a literary system.[2] Literary activities focus on those types of phenomena (mostly texts) that people deem literary according to whatever aesthetic/poetic criteria and values they have acquired during their literary socialization. Literary activities are manifestations of *acting roles*, which, since the late 18th century, have become professionalized and institutionalized, viz. the production, mediation, reception, and post-processing of literary texts. The relations between the acting roles define the structure of a literary system.[3]

In order to define the type of system that literary systems are theoretically subsumed under, the relations between the components of literary systems have to be specified. For this purpose I refer to Roth's definition of two notions that undoubtedly belong among the most crucial concepts of both biological and sociological systems theory: self-reference and self-organization. Roth has offered the following definitions of these two concepts. »*Self-referentiality*: Self-referential systems are systems the states of which

W. Krohn et al. (eds.), Selforganization. Portrait of a Scientific Revolution, 143–153.

cyclically interact with each other such that any state of the system contributes to its next state in a substantial way. Therefore, self-referential systems are internally state-determined systems. ... *Self-organization*: Self-organizing processes are such physico-chemical processes that reach a specific ordered state [...] under a (more or less extensive) domain of initial conditions and constraints. [...] Arriving at a given state of order is not (or at least not *essentially*) imposed on the process from outside, but is the result of the specific characteristics of the components involved in the process. The state of order is achieved ›spontaneously‹«.[4]

What I now wish to consider is whether these biological concepts can be applied to literary studies in a non-metaphorical way.

3. It is evident that literary activities are necessarily related to each other (e.g. literary reception is related to literary production, literary post-processing is related to literary reception, and literary production is related to literary production). Moreover, literary activities are *exclusively* related to each other as a result of the efficiency of certain macro-conventions. Thus, the literary system constitutes a *closed organization*. Literary activities result from the respective state of the literary system and contribute to its next state in a substantial way. This is certainly not true of any single action, but it applies to the instantiations of the action roles seen from a social point of view, that is, at the level of communication.

The history of modern literary systems clearly reveals that other social systems (e.g. the political, religious, scientific, and economic system) never succeeded in exercising a long-term external control over the literary system. Like all self-referential systems, the literary system cannot be managed, intentionally and causally, from the outside. Internal evaluations of literary activities are always provisional and have to prove their mettle; also the evaluation of »test norms« remains internal to the system.

The order arising in self-referential literary systems can be called *self-organizing* in Roth's sense: it is achieved spontaneously and results from the specific features of the components involved in the process of the production of order.

Applying Roth's concept of self-organization to literary systems becomes even more plausible when we follow Hejl's proposal to establish a close connection between self-organization and *self-optimalization*: »In self-organizing systems, evolutionary and/or developmental changes lead to a relative optimalization of interactions between subsets of components, within the system. This can be regarded as the development of functionally differentiated and specialized subsystems«.[5]

A brief glance at the emergence of modern literary systems in the 18th century[6] as well as at present literary systems shows that internal differentiation is indeed the main feature. The differentiation in question has to do with the action roles, the formation of literary discourse (e.g. the genre system), or the self-reflection of literary systems in terms of (ever-changing) aesthetics and poetics.[7]

4. To characterize literary systems as self-referential and self-organizing systems we must take into account some additional aspects of systems. Not every differentiation of a system is self-organizing (e.g. the deliberate foundation of a sports club). Nor are all processes in social systems self-referential or self-organizing. To grasp the difference we must take into account processes of control and hierarchization (e.g. the hierarchies in a firm). In a self-organizing social system, the »central« function is no longer attributed to one specific subsystem but can in principle be exercised by every subsystem. Thus a self-organized social system is characterized by a heterarchical form of regulation or »control.« Although no single subsystem exercises control over the next, the self-regulation of the system as a whole is nonetheless maintained.

In regard to the *closure* of the social system of literature, we must take note of Klüver's[8] suggestion that social systems cannot be modelled as closed systems, like biological systems, but that they also exhibit degrees of openness. In my view, this problem can be solved by including *actors* (= living systems) in the systems-theoretical framework while also making use of Hejl's concept of *synreferentiality*. According to Hejl, living systems constitute a social system by establishing a common model of reality and with it a domain of meaningful actions and communications. As long as the actors mutually interact with regard to this domain they »belong to«, or act in, the social system in question.

In functionally differentiated societies, actors adopt various roles at different times, thereby acting in various social systems. In addition to this role-playing, they act in a non-specialized or private sphere (the »everyday world»). By virtue of this multiple »membership« in various social systems, they contribute, on the one hand, to the integration of the set of social systems into a unity called »society«, while on the other hand, they import everyday relations, ideas, and acting potentials into the respective social systems in which they take part.

This proposal entails a distinction between *two different levels*: a macro-level (that of the literary system) and a micro-level (that of the literary activities). On the micro-level, literary systems may be conceptualized as relatively open but on the macro-level they are modelled as closed. As Klüver puts it, the system as a whole tends towards a permanent restitution of its self-referentiality by compensating and thereby closing the microstructural openness on the macrostructural level.

Accordingly, literary systems can be conceptualized as systems the behaviour of which is determined by their *internal states*, that is, by the mutual interaction of the four basic types of action roles. This is the case because any change of the potential activities belonging to one role brings a shift in those of the other basic roles in the literary system.

Let us examine an example. The formation of the book market as a capitalist organization in the late 18th century was due, on the one hand, to the significant growth of the reading public and to a corresponding increase of literary productions. On the other hand, it created the economic conditions for bourgeois writers to become freelance writers. Accordingly, the action role of the literary producer was split, for it could be filled either by professionals or by amateurs. The internal differentiation of the role of the producer contributed to an internal differentiation of the role of the reader (naive vs expert) as well as to a differentiation of textual categories (the hierarchy of high vs trivial forms of literature). This development was, in turn, synchronized with the emergence of the role of post-processor, our systems-theoretical term for the literary role comprising those various activities that one may associate, for example, with literary criticism. The emergence of this role was the result of internal differentiations of producers and receivers, etc.[9]

5. The *boundary* of the literary system, that is, the difference between the system and its environment, can be theoretically established by the two macro-conventions I already mentioned above. These macro-conventions single out from all the activities of the actors in the social system of literature all and only those activities that are deemed adequate in literary discourses (= components of the system internal communication).

These conventions, which have held a sway since the late 18th century (at least in Germany), can be specified as follows:

Aesthetic Convention

It is common knowledge in our society that all actors in literary systems must be willing and able

- to extend their action potential (or the action potential of other participants in the literary system) beyond the usual criteria of true/false or useful/useless and to orient themselves towards expectations, norms, and criteria which are deemed aesthetically relevant in the respective literary system;
- to designate communicative actions intended as literary by appropriate signals during production and to follow such signals during reception;
- to select as a frame of reference for expressions in literary texts not just the socially-established world model he/she is accustomed to in his/her respective social group but virtually all constructible frames of reference and
- to de-emphasize the fact convention.[10]

Polyvalence Convention

It is common knowledge among all actors in literary systems in our society that:

- text producers are not bound by the monovalence convention;[11]
- text receivers have the freedom to produce different readings of the same text at different times and in different situations (= the weak version of the polyvalence convention hypothesis) or during a single reading process (= the strong version of the polyvalence convention hypothesis) and they expect others to do likewise;[12]
- text receivers evaluate the different cognitive, emotive, and moral reading results obtained at different levels of reception in terms of their needs, abilities, intentions, and motivations, although the reasons behind these evaluations may differ in function of the participants and situations and
- text-mediators and text-post-processors should not overtly counteract these regulations.

These macro-conventions establish the most basic distinguishing feature (or communication code) of the literary system in terms of a basic dichotomy between literary and non-literary activities.

The stability of the literary system is dynamic and capable of integrating within itself all sorts of conflicts. An aspect of this dynamic stability is manifested in the fact that the system's basic distinguishing feature is itself open to historical as well as social interpretations: »literary« means whatever actors believe it to mean according to their norms, values, needs, and knowledge. Yet as soon as this dichotomy is established and operates as a selection mechanism (working via a negative definition following which A = non B), the system can maintain its identity on the macro-structural level in spite of the controversies that may occur at the micro-level.

What actually *is* deemed literary emerges from the intrasystemic interactions between the four basic action roles. Accordingly, the identity or *autonomy* of the literary system results from the self-referentiality of literary activities through which all states of order in the system arise.

Whereas the discussion of autonomy during the last decades has exclusively centered on the autonomous work of art, a systems-theoretical approach allows us to make some crucial distinctions. In fact, we have to consider three kinds (or aspects) of autonomy:

- the autonomy of literary works;
- the autonomy of aesthetic experience;
- the autonomy of the literary system as a social system.

These distinctions make us realize that the autonomy of literary systems is not at all dependent upon the (historically contingent) idea of autonomous literary works, that is, it is independent of the kind of poetics actors in the literary system favor *as long as* the boundary of the literary system successfully selects those activities deemed adequate to the literary system. Instead, the system's autonomy solely depends on a full-fledged state of self-referentiality regarding literary activities.

The autonomy of the literary system should not be mistaken for a lack of *function*. Instead, as the emergence of modern literary systems in 18th-century society reveals, the autonomization of the literary system started with a re-specification of the function of literature. Literature was transformed from being a source of pleasure and diversion for the aristocracy into an instrument with which the rising bourgeoisie could criticize feudal society and explore its own needs, values, and interests. Literature was then supposed to become a means of enlightening the rising bourgeois public and of improving its morality. At the end of the 18th century, the institutionalized and professionalized activities in the four acting roles became self-legitimating and the autonomized[13] literary system built upon these roles gradually rejected pragmatic tasks which other systems (religion, politics, education, philosophy etc.) tried to impose on literary activities. Even so, the function of the literary system for society as a whole and for the acting subjects was quite clear. Let us look at just one modelling of this function as it was developed in so-called classical literary theory (Karl Philipp Moritz, Goethe, Schiller, Fichte).

According to this theory, the subjective, everyday production of knowledge and meaning which had been neglected by the rise of science and philosophy was taken over by the literary system which undertook to engage in forms of communicative cultivation. In other words, a multiplication of realities in the imagination and an innovative anticipation of new forms of experience, feeling, and self-expression came to be the central topic of literature's cognitive domain, which thus promised to re-establish a continuum between everyday life and culture.

In the moral domain, the literary system developed ways of coping with the consequences of the functional differentiation of society from which it originated, namely, the problems of personal identity, conflicting norms, purposelessness, and various phenomena that may be associated with the word »alienation«. The literary system promised to reconcile the dispersed and fragmental human faculties by means of the utopian figure of the »perfected human being«.

In other words, according to the classical approach, the literary system was supposed to be the only social system that could reconcile the basic divisions between individuals and between individuals and society. The reconciliation was to be achieved through aesthetic *play* (Schiller) thought to bring forth and develop cognitive, emotional, and moral dimensions of the individual's subjectivity in equal parts, thus perfecting human kind.

The classicist specification of the function of »literature« touches but one functional aspect that has to be taken into account by a systems-theoretical approach, viz. the function of literary systems for society, for there remain two others (1) the function of the literary system for other autonomized social systems, and (2) its function for itself. In regard to the first, the emergence of an autonomized literary system redefines or respecifies the environment of other systems, thus establishing new potential activities (which other systems may or may not select according to their internal organization). In regard to the second, the literary system functions in terms of self-reflexion, mostly embodied in the form of aesthetic or poetic doctrine.

6. The gradual change in the definition of the system's function clearly reflects an increase in self-reference and self-organization. The more the literary system established and strengthened its relations to other social systems emerging from the overall process of functional differentiation, the more it became obvious that literary discourse could only define, transform, and legitimate itself by recourse to literary discourse itself (a fact that was reflected in contemporary poetics). In the process, literary discourse proved to be enormously flexible; nearly all topics could, in principle, become taken up - as soon and as long as they could be subordinated to their contribution to the basic dichotomy »literary vs non-literary«. That is to say, *intersystemic interaction* was regulated by the self-referentiality of the systems involved.

A good example is the interaction between the literary, the juridical, and the economic systems with regard to the regulation of the notion of intellectual property and the respective royalty regulations between 1760 and 1780.[14] Other examples are trials concerning allegedly pornographic literature or the interaction of literature and education in respect to questions of curriculum and canon formation.[15]

7. As soon as the literary system reached a mature status of autonomization, the search for a consensual *legitimation* of the opposition »literary vs non-literary« began. At the end of the 18th century an increasing number of works on aesthetics and poetics appeared.[16] In them literary discourse specified in an overtly self-referential way what was accepted as literary and how literary phenomena had to be ranked according to their degree of poeticity. This poeticity was, in turn, increasingly specified in terms of the formal organization of works of art, and finally, came to be defined as the anticipation of bourgeois self-redemption.

8. How can *fictionality* be explained in a systems-theoretical approach? Here again I shall refer to the process of social differentiation in the 18th century.
In feudal societies, highly consensual and homogeneous models of reality provided (nearly) all members of society with an ontological, moral, religious, and pragmatic orientation. The administration and preservation of this model of reality guaranteed that state and church held a monopoly on the production of meaning.

Through the emergence of functionally specified, autonomized social systems, the ordo-model (which proposed a global interpretation of the world) was shattered. Each social system now developed and administered its own partial model of reality, but these various models no longer added up to only one compulsory, over-arching ortho-model. In the long run, each system (and within it every individual cognitive system) had to face the task of coping with the self-referentiality of cognition, emotion, and communication which was inevitably becoming apparent (as is testified by the tremendous success of transcendental epistemology in the 18th century).

The differentiation of world models caused by the functional differentiation of society did not result in a generalized social schizophrenia (although it did produce many cases of individual neurosis) for the sole reason that the economy had taken over the »leadership« of society. The objectivity of money, which made all goods and all kinds of labour quantitatively commensurable, responded to the crisis created by the ontological differentiation into types of world models. Money became the guarantee of reality and the objective measure of goods, labour, and human beings. Money became, as it were, the general mode of equivalence.[17] Based on an economy that functioned by means of monetary mechanisms, the state could afford to pass the »ontological question« along to society and to its various functional subsystems. The need to interpret »world and life« and to explain disappointments were now delegated to social subsystems as well as to individuals. In addition, the developed juridical system allowed for the coexistence of contradictory world models without jeopardizing state and society, because repressive claims for universalistic world models - as evidenced by the religious wars of the 17th century - could no longer be enforced.[18]

Consequently, the bourgeoisie could realize its recently detected subjectivity and creativity (incorporated in the figure of the genius) in terms of the communicative freedom of subjective »world-and-life-interpretation« - long before it achieved its political freedom. Yet this freedom was more or less restricted to the private sphere and to special discourses, for example, the philosophical and the literary ones. The poets (at least since Klopstock) insisted on their ability and vocation to develop alternative models of (social) reality thus becoming profane priests of the bourgeois public.[19] These alternative models were no longer expected to be reflexions of a compulsory ortho-model of reality. Accordingly, literary communication in the autonomized social system of literature gradually became *fictionalized*. Fictionalization, that is, the orientation towards possibility instead of truth or probability, became the »engine« of a thematic and stylistic differentiation of literary productions, resulting gradually in an exclusive orientation of literary activities towards *literary* norms and values. In this process of aesthetization, the literary code (i.e. literary vs non-literary) and nothing else served to select literary activities.

At the end of the 18th century, the literary system can be described as becoming progressively autonomized (self-referential and self-organizing) and literary communication as being fictionalized. As a result, the literary system and its literary discourses were sharply separated from other social systems and their discourses, as well as from daily life. Any attempt to achieve an immediate application of »literature« to »life« became an illusion.

This fictionalization and aesthetization of literary discourse gave rise to a proliferation of literary semantics, since all thinkable modes of referentiality that could be contrived and expressed were admissible, thus remarkably increasing the complexity of the literary system.

9. In a survey of the development of the notion of self-organization in various disciplines Küppers, Krohn and Paslack[20] compiled a list of the distinctive features that seem to be consensual in discourses on self-organization. As all these features can be shown to be pertinent to the conceptualization of literary systems that I have proposed, it would seem all the more appropriate to follow this theoretical direction. The following synopsis of the basic features of self-organizing systems, together with exemplary applications to a systems-theory of literature, may thus be taken as an additional confirmation of my approach.

(a) Models of self-organization assume a self-regulated differentiation of autonomized systems into complex subsystems that form dynamic networks and are open to reorganization. (Example: The differentiation of the literary system into the subsystems of the four action roles).

(b) Self-organizing systems are conceived of as exhibiting dynamic stability. All relevant properties of the system are based on non-linearity. (Example: The relation between the literary system and the acting roles as subsystems has undergone remarkable modifications, but no substantial change since the 18th century).

(c) The environment acts upon social systems in a non-specific way; moreover, highly developed self-organizing systems actively influence the environment, that is, manipulate, to a certain degree, the environmental conditions necessary for their survival. (Example: the interaction of the literary, the economic, and the juridical systems in matters of intellectual property or in the transformation of the notion of incest).[21]

(d) Self-organizing systems are endowed with the ability to create their own identity and to preserve it through the purposeful control of their boundaries. (This is clearly the case in what I have described as the reflexivity of the literary system).

(e) In the analysis of self-organizing systems process comes to the fore (see Luhmann's concept of »temporalized complexity»). (My insistence on literary activities as the units of the literary system corresponds to this emphasis).

(f) Complex processes in self-organizing systems are characterized by circular instead of linear causality, that is, outputs can serve as inputs in the same system. (As an example, I may point to the formation of the genre-system in the complex relationships between the subsystems of the literary system and its interaction with other social systems).

(g) Social systems are characterized by a system-specific relativity of time, that is, the system coordinates its internal processes according to its eigen-time (as is the case with the coordination of divergent literary schools or approaches).

Finally, Probst's and Ben-Eli's remarks on self-organizing systems also seem to match my model of literary systems as self-organizing systems. According to Probst and Ben-Eli, »self-organizing« means »operating simultaneously on a number of different levels.« They add the following points about the operation of self-organizing systems:

- The operation consists in discrete activities, involving recent pertinent events (e.g. literary works relate to literary works).
- The operation brings about the mutual adjustment of all singular activities so as to produce a meta-systemic whole (e.g. singular literary production activities generate the subsystem of literary production).
- The system's operation operates recursively on itself, thus improving on a previous norm of self-organization (e.g. the relation between poetics, literary criticism, and literary production).[22]

10. Let me sum up. I think that conceptualizing »literature« in terms of self-organizing social systems offers some tangible advantages:

- the *literary field* is theoretically conceived of in such a way as to integrate agents, texts, and contexts; micro- and macro-sociological aspects; and social as well as semantic dimensions;
- the *rise* of the literary system is explained as resulting from functional differentiation without relying on notions of causality or intention, or on notions borrowed from the philosophy of history (like progress, rise, decline etc.);
- the mechanism of that emergence is explicated in terms of self-referentiality, self-organization, self-optimalization, and the three main categories of the system's dynamics (internal differentiation, intersystemic interaction, and reflexivity);
- the system-theoretical approach allows us to respecify and reorganize the historical material in a plausible and sometimes innovative way, thereby integrating it into a theoretically coherent model.

In a word, it does what a theory is expected to do. It opens up a fresh view of history as well as of the present; it solves problems; and it, of course, produces a lot of thrilling new ones - in a word, it keeps scientific communication going!

Notes and References

1. N. Luhmann, *Soziale Systeme*, Frankfurt am Main: Suhrkamp, 1985.

2. See section 5 below.

3. For details see S.J. Schmidt, *Die Selbstorganisation des Sozialsystems Literatur im 18. Jahrhundert*, Frankfurt am Main: Suhrkamp, 1989.

4. G. Roth, »Selbstorganisation - Selbsterhaltung - Selbstreferentialität«, p. 154f, in A. Dress, H. Hendrichs, G. Küppers (eds.), *Selbstorganisation - Die Entstehung von Ordnung in Natur und Gesellschaft*, Munich: Piper, 1986, p. 149-180.

5. P. Hejl, *Systemdefinition*, unpubl. paper, 1987, my translation.

6. See Schmidt, *op. cit.*, 1989, Note 3.

7. See G. Jäger, *Die Avantgarde als Ausdifferenzierung des bürgerlichen Literatursystems*, Paper presented to the Colloquium *Modelle des literarischen Strukturwandels*, University of Passau, 1986.

8. J. Klüver, »Auf der Suche nach dem Kaninchen von Fibonacci oder: Wie geschlossen ist das Wissenschaftssystem«, in W. Krohn, G. Küppers (eds.), *Selbstorganisation - Aspekte einer wissenschaftlichen Revolution*, Wiesbaden: Vieweg, 1990, p. 201-229.

9. For details see Schmidt, *op. cit.*, 1989, Note 3.

10. Fact convention: It is common knowledge in our society that communicative objects, especially texts, should permit reference to the world model accepted in that society such that people can decide if the assertions conveyed by the text are true and what their practical relevance is.

11. Monovalence convention: It is common knowledge in our society that: (a) text producers are expected to shape their texts in such a way that different people at different times can assign them one and the same reading, and (b) text receivers are expected to strive for the assignment of a single reading to the texts.

12. For a detailed discussion of the strong and the weak version of the polyvalence convention, see N. Groeben, »Methodologischer Aufriß der empirischen Literaturwissenschaft«, *Spiel* **1** (1982), 26-89.

13. According to Hejl, autonomized systems are systems the autonomy of which exceeds the autonomy of its subsystems.

14. G. Plumpe, »Kunst und juristischer Diskurs«, in J. Fohrmann, H. Müller (eds.), *Diskurstheorie und Literaturwissenschaft*, Frankfurt am Main: Suhrkamp, 1988, p. 330-345.

15. F.A. Kittler, *Aufschreibesysteme*, Munich: Fink, 1985.

16. In 1799 a first bibliographic raisonné of poetics appeared: Joseph von Koller, *Entwurf zur Geschichte und Literatur der Ästhetik, von Baumgarten bis auf die neuste Zeit*, Regensburg: in der Montag und Weißischen Buchhandlung, 1799.

17. See M. Fontius, »Produktivkraftentfaltung und Autonomie der Kunst«, p. 489, in G. Klotz, W.Schröder, P. Weber (eds.), *Literatur im Epochenumbruch*, Berlin-Weimar: Aufbau, 1977, p. 409-529.

18. See N. Luhmann, *Grundrechte als Institution*, Berlin: Duncker & Humblot, 1974.

19. See J. Schönert, *Professionalisierung der Schriftsteller? Zu Praxisformen und Reflexionstypen des Schriftstellerberufs zwischen 1850 und 1920*, unpubl. paper, University of Hamburg, 1988.

20. H. Krohn, G. Küppers, R. Paslack, »Selbstorganisation - Zur Genese und Entwicklung einer wissenschaftlichen Revolution", in S.J. Schmidt (ed.), *Der Diskurs des Radikalen Konstruktivismus*, Frankfurt am Main: Suhrkamp, 1987, p. 441-465.

21. M. Titzmann, *Literarische Strukturen und kulturelles Wissen: Das Beispiel inzestuöser Situationen in der Erzählliteratur der Goethezeit und ihre Funktion im Denksystem der Epoche*, manuscript, University of Munich, 1987.

22. G.J.B. Probst, M.U. Ben - Eli, »The Way You Look Determines What You See or Self-Organization in Management and Society«, p. 283, in R. Trappe (ed.), *Cybernetics and System*, Dordrecht: Reidel, 1986, p. 277-284.

Chekhov's Letter: Linguistic System and its Discontents

CHRISTOPHER S. BRAIDER
Harvard University

In speaking here for language and literature (and I do speak, not »on« or »about,« but precisely *for* them: on their behalf and in their name) I face an impossible task. It is not just the vastness of the topic; there is also the enormous body of scholarship, either itself systems-theoretical or of a sort congenial to systems theory, dating back in Saussure and the Russian Formalists to the earliest decades of this century. Ideally I would present some kind of report summarizing views, methods, and findings, delineating common assumptions and problems, and hinting at profitable lines for future research. Yet given the dimensions of this paper, any such inevitably shorthand report is out of the question. I must take a quite different tack. I shall of course name some of the most important names and say something of what they stand for. But my general method will be suggestive rather than descriptive, less a conventional exposé than a series of what Wittgenstein would call thinking-points. I shall present a thesis of sorts, grounded in the interpretation of a certain story; and the fact that I ground my remarks in the analysis of a particular literary text already tells you something about how we do things in my neck of the woods. But this thesis is meant to serve merely as an occasion, a place for thought; the actual *work* must come later, in your own subsequent musings on the points it raises.

Mind you, one of these points is precisely that it is of the essence of the systematic process at work in language, and especially literature, to exert its most decisive impact only after the fact: in the way what we hear or read claims a second life not just in what, but in how we think. Literature (and the point applies to lyric poetry or the essay as much as narrative) deals preeminently in *fictions*. We thus meet here with a case of what systems theory calls »reality constructions« - the more unmistakably in that the reality so constructed is in some sense a pure invention. Yet we recall how in the first great post-classical realist, William of Ockham, the word »fiction« means what we today call a *picture*.[1] For Ockham, we know the world by framing pictures of it, now »mental,« now »verbal,« the whole art of philosophy consisting simply in making them as clear and faithful as we can. A point all the more intriguing when we remember the link between the German word for picture, *Bild* (a relative, incidentally, of the English »build«) and the word for formation or education, *Bildung*. To conceive something is to form a *Bild*, or build a fiction - what we call »reality.« And when we commit this *Bild*, fiction, or reality to language, commending it to an interlocutor, the act of picturing becomes a *Bildung*. It forms its hearers or readers by *informing* them with the picture it makes. But the picture's true force lies less in what we see in the instant of formation itself than in how we see in its aftermath. And what is true of philosophy or science is even more true of literature, where the picture-making or *Bildung* proceeds by the curious detour of a »fiction« in the ordinary sense. By our participation in constructing the fictive world it presents, literature induces us to reconstruct the *non*-fiction world in which the *Bildung* itself takes place: if the fictive *Bildung* has done its job, we picture the world of ordinary

W. Krohn et al. (eds.), Selforganization. Portrait of a Scientific Revolution, 154–168.

experience in some new way. The literary fiction, the literary text and the singular communication it bears, *make* something of me: someone who, picturing differently, restructures the outer, unreconstructed world that forms the constant horizon of all we think, see, or do.

So to the extent that my remarks on language and literature project into the future of your own subsequent meditations, I invite you to see my method as an instance of its own object. This sort of recursiveness is moreover endemic to language and literature both. The original working-title of this paper, »the languages of language,« though abandoned in the upshot, remains apt from several points of view. For it is a distinctive faculty of the thing called language to be capable at any moment, like the proverbial serpent, of swallowing itself by its own tail. And it lies in the logic (and thus in the systematic character) of the situation in which we find ourselves relative to language that language is itself our only means of coming to terms with it - to the point in fact where there is room for asking whether the very question of language is not an expression of language as such, a phase in the self-governed, self-organizing, »autopoietic« evolutionary life of the system, language, itself.

You will also find my method representative not just because it reflects something of the character of its object and our analytic situation relative to that object, but because it images certain *styles* of interpretation and analysis characteristic of the current state of my discipline. You will sense in the way I go about things more of the »post-structuralism« of Derrida, Lacan, de Man, or the later Barthes than of the high semiotic scientism of Jakobson, Lotman, Riffaterre, or Eco. In an evolution in literary and linguistic studies likely to seem familiar to those engaged in recent efforts to redirect classical systems theory, the structuralists of the first half of this century proceeded on the implicit premise that by analyzing a given language, culture, or text either as itself being or as forming part of a global functional system, they obtained an unassailably »scientific« point of view *on* it by which it comes to be defined as a genuine *object of knowledge*. What distinguishes subsequent approaches, generally designated as »post-structuralist,« is a critique of the hidden assumption underlying structural method: the view that there exists, relative to the »object,« a privileged *subject* wholly external to and uncompromised by the system on which it reports. Like »post-classical« physics since relativity and quantum mechanics, and like systems theory since its acknowledgement of the need to factor into its equations not just the properties of the object, but the observing scientist or analyst as well, post-structuralism asserts that subject and object coexist as equal partners in the system of knowledge itself. Given the nature of the systems (the »objects«) with which linguists or cultural scientists are concerned, whatever we do as »subjects« expresses the very thing we study since it is only *within* a language or culture that such a thing as a science of language or culture is conceivable. Small wonder, then, if I adopt here a distinctly self-conscious, self-referential attitude toward language and literature. We are always problematically *implicated* in the system of which we speak or write. And to this extent any critique of language is simultaneously a critique of its *own* language, and thus of its own conditions of possibility. There is no rock-solid Archimedean vantagepoint outside the world of language from which it could be levered out of its customary orbit: the work of understanding must be pursued from within and as it were *against* language itself. In the very movement of self-consciousness we therefore discern something of the nature of both the phenomenon at issue and the activity required to comprehend it.

To speak the »language of language« or, as we say in the trade, to use the »metalanguage« in which the »target« or »object« language can be spoken, we must stem the tide of our own subject, resisting the system of which we are a part and to whose function and growth our analyses actively contribute. Thus, properly understood, the style or, better, the moment of thought called »deconstruction« is nothing but a certain strategy of resistance intended, if not to escape the order of language, at least to create within it, and by the only means at our disposal, namely language itself, a bit of *play*, some room for maneuvre. True, this will not and cannot secure our unconditional release from the system of language. But the gesture by which we attempt to dismantle the constructions to which language compels us in the very forms in which we speak of it constitutes a necessary step in the process of apprehending what they are, how they work, where they tend; a process without which we remain condemned simply to rehearsing what language has already said without us.

The need to resist the system of language as a precondition of understanding it more fully and authentically has however a telling corollary: it is just when we *do* resist that the system becomes most clearly visible in all its coercive power. This is why, practising what he preaches, Roland Barthes inverts the famous tag associated with the Romantic myth of Original Genius, that the essential aim of literature is to »express the inexpressible,« urging instead that it is just the reverse, to »inexpress the expressible.«[2] The problem for any writer trying to say, and *only* to say, just exactly what he or she means is far less one of saying what has never been said than one of saying differently, with the accent of *truth*, what has already been said far too often. Language is always already there fully formed, putting words in our mouths, and never more conspicuously than when, like the Romantics, exemplary in this of all of our avant-gardes from Mannerism to Surrealism and the various shades of the »anti-aesthetic« characterizing our contemporary »post-modernity,« we try to position ourselves at its frontier. For what indeed (echoing Jonathan Culler)[3] has become more obvious, even embarrassingly stereotyped, than the Romantic rhetoric of Spontaneity and Sincerity, the »true voice of feeling« by which the Romantic poet is constituted and signalled as such? But the fall into stereotype (which is also, as Barthes shows, a fall into history: what is more *dated* than the various ways in which we »write for the Ages?«)[4] is not simply an accident of Romanticism or even literature at large. It represents an unavoidable condition of language itself. A point Saussure makes when he observes, as the foundation on which the whole of the new science of language rests, that for language to exist, i.e., for language to *happen* in the particular act of communication, requires the prior existence of an entire structure of rules and conventions, a systematic code, the grammar of the *Langue* of which any given *discours* or *parole*, any given utterance, is merely a functional part.[5] Language as such is either the entire self-organizing *system* of the language, defining the terms available for individual acts of meaning, the laws according to which these terms are used, and the play of differential relations by which they identify themselves, or it is nothing at all, mere inarticulate sound. Either everything has already been said in the conditions governing meaningful saying as such, or nothing is said, for nothing will in fact be sayable.

What this means in practical terms emerges more sharply if we place ourselves for a moment, not at the level of the system of the *Langue* itself (German, English, French) but at that of the sub-systems Saussure calls *langages*. Under this heading stand the various dialects, sociolects, ideolects, parlances, trade languages, and argots that characterize the system of the *Langue* at any given moment. Each *langage* is distinguished first of all by a specialized vocabulary, syntax, and grammar. This is indeed a major part of what the Russian Formalists had in mind when they spoke back in the '20s of the object of a genuine science of literature as being neither the individual work of literature nor even the entire set of works of literature, but *literarity*: the common properties of idiom and form by which a given text designates itself *as* literature. But this alone fails to tell the whole story. For a *langage* also encodes, at the level of the system as a whole, an extra meaning in which we detect as it were the signature of the system itself; and this signature in turn bestows its peculiar flavor on everything it comes to encode. Science, for example, has a specialized vocabulary (its usages and terminology) and syntax (e.g., that of the »scientific paper« as defined at the level of diction for instance, in English at least, by a pervasive use of the passive voice and, at the level of composition, by the pattern of »scientific argumentation and experiment«). What we recognize in such things is precisely the *scientificity* of scientific discourse. »Science« in this sense, in addition to being a certain kind of activity (itself, as T.S. Kuhn has shown, densely coded)[6], is a certain *meaning* (more exactly, a connotation) embedded in the structure of its idiom. A »proper« scientific paper is one that is properly constructed, properly *parsed*; the parsing is indeed one of the marks or signs by which we learn that it is in fact (or purports to be) a scientific as opposed, say, to a journalistic, literary, or philosophical paper. But with this example we also see the self-organizing (i.e., the *systematic*) character of the language of science. To *be* science, a given address must present patents of authenticity in the form of the requisite syntax. But this syntax functions not simply as a mode of composition. It implies a whole structure of thought: that of problems properly stated (»theoretical introduction«) leading to hypotheses properly framed (»thesis« and »rationale«) in turn dictating lines of experiment (the »protocol«) organized in such a way as to provide a meaningful answer to the questions outlined in the original hypotheses (»analysis,« »evaluation,« or »assessment«), winding up with »findings« or conclusions as often as not wrapped in a rhetoric of modesty asserting their »tentative« nature.

In stressing the degree to which science as self-referentially imaged in the scientific paper is (or is at least characteristically associated with) a certain linguistic stereotype, I do not mean to devalue its unique epistemological power. However far it strays from what scientists actually do in their offices and laboratories, the syntax of the scientific paper has proved an immensely potent and fruitful tool. The point is simply to illustrate how far the syntax by which we recognize science defines not only its form but its content. Not just how it is expressed, but how it is thought and done. And what is true of science is equally true of literature. It is in this spirit that the Russian Formalists not only insisted on the primacy of »artistic« verbal form over the presumed content of the literary text, but focused attention on what Boris Eichenbaum called »the very *dynamics* of literary form,« the generative energies unleashed in the movement whereby a given content *takes* form. And it is in this same spirit that the Formalists asserted that »the central problem of the history of literature is the problem of evolution without personality - the study of literature as a *self-formed social phenomenon*«.[7] For it is of the essence of the code by which the literary text comes to be, as it is indeed the essence of any code as such, that it exist as a social (and intensively socializing) entity binding at least two persons together in a common act of communication and understanding. The language of literature not only implies, but is itself, *in* itself, a social institution, and to speak or write it is therefore not only to become a part of that institution, but to reflect and reproduce, propagate and prolong it.

But this also defines the peculiar place and unique demonstrative value of literature. As Jakobson has pointed out, one of the things chiefly distinguishing literature is the fact that, in the literary text, language is not simply a means, but an end: not just that *with*, but that *on* which the writer works.[8] We thus encounter in literature, in a more pronounced and critical form than elsewhere, a tension, conflict, or dialectic common to all acts of language as such. The system, defining the conditions under which the individual utterance or text is possible, tends at the same time to define its *content*, contriving thereby, in the very movement of individuation by which utterance or text is produced, to reproduce *itself*, its *own* character and content. And yet it remains the system's constitutive aim to enable particular acts of communication to take place - in literature, to make room for the individual text and the peculiar *experience* (of pleasure, discovery, ritual enactment, wish-fulfillment) each text provides on the occasion of reading it. Which is why, from the Platonic dialogue to the present, literature has characteristically entertained a certain anxiety about its own nature and status: an unbroken series of texts, of works *of* the system, in which the system curiously *turns* on itself, aiming thereby at its own undoing[9], yet in such a way as to make itself more palpable, pointed, and present in the process.

It is worth recalling in this context where Saussure got the name he gave the general science of signs and systems of signification of which the sciences of language and literature are themselves but two complementary faces. This general science is of course »semiology.« But the interesting thing is that this word derives from a sometime medical term, »semeiology,« meaning what we would today call *symptomatology*: the art of interpreting the diverse clinical signs (sniffles, fevers, rashes, »referential« pains) by which a patient unconsciously signifies the underlying pathology to which he or she is subject - or more accurately, by which the underlying pathology *signifies itself*, pronouncing its own name, appending its own signature. From here it is but a step not just to Freud, but to Nietzsche and that »symptomology« by which the human comedy is diagnosed and denounced in his writings. And indeed, if both Freud and Nietzsche exert such a powerful influence precisely in literary circles, the reason lies in the curiously etiological and forensic character our proceedings naturally take as soon as we turn our eyes to the intersection of the particular phenomenon (the given text, the individual work of literature) with the system of its conditions of possibility in the social evolution of language and form.

But this in turn explains why the time has come to pass from general analysis to the particular case: from the system to its simultaneous *mise en oeuvre* and *mise en cause* in an individual text. The text I have chosen is a story by Anton Chekhov - a very short one, barely six-and-a-half pages in the edition I have used.[10] Yet short as it is, the reading I offer is by no means exhaustive. The number of things for which this text is a sign (the number of things giving off signs in this text) is practically infinite - all the more so in view of its own deliberate self-questioning, the way it implicates itself in its own anecdote. I shall, then, focus on a single thread. The story tells of a certain *letter*: how it is written, and by whom; how it reaches its destination, and to what effect. We shall simply follow this letter through the different incarnations by which it migrates through the text from its inception to its eventual arrival in the hands of the person for whom it is intended. But before we begin, I ought to confess that I neither speak nor read Russian and have therefore entirely to rely on an English translation. This fact does nothing, however, to invalidate the reading. On the contrary, given the nature of the fate awaiting the letter in question, a fate which is also that of the Chekhov's text itself, the risk of translation is one of the things the story is *about*. That we should have to read it in translation, as the form in which it finally reaches us, is precisely the *point*.

The story (the text) begins with its title: »At Christmas Time.« In the peculiar reversible time of composition, the recursive time not just of writing, but of revision, which is also that of *reading*, the title is at once the first and last thing: first in the time of what Derrida calls the »trace,« the sequence of graphic signs as we meet them on the page, but last in the time of the Text, the form these signs compose in the gesture of committal by which the original hand (now lost) commends them to their posterity. With the title, we cross a certain threshold: the boundary between the world (the experience) in which the text appears and the singular series of events that is our experience as we read it.[11] At this threshold, as we cross it, the title sets a mood or tone, puts us in mind. The text has a certain topic, a certain *topos*: we enter, as the etymology of the word *topos* indicates, a certain »space« or »place,« which happens in this case to be a certain time or date, a »season.« The title thus makes a *difference*: as Saussure teaches, the functional basis of meaning itself. It differentiates what is (will be, can be) »in« the story from what is not. It marks out and marks off certain possibilities, and in the as yet dimly perceived pattern of these possibilities we discern a *theme* experienced as a certain potential or *energeia*, as Aristotle would say. It is Christmas time, the long, dead mid-winter cold. But it is also the feast of the Child, and of the Family, a celebration of light in darkness, of fertility and renewed growth in the womb of the dead earth. Most of all, it is the feast of the Christ child whose Nativity it commemorates, the season, too, of the Epiphany, the radiant light of the Incarnate Word.

In all of this we find as it were the story's origin, both what it is about and the very forms (coded, symbolically charged) by which it is produced. By virtue of all that it encodes, the title plays a *causal* role - and indeed, when the text explains the writing of the letter with which it is concerned, we read: »But now it was Christmas time, and Vasilisa could bear it no longer.« (p. 434) Mind you, all of this is the *function* of a title, what it is *for*. And in this sense, we already find ourselves in the presence of a coercive dynamic: a certain machinery of language, convention, form has locked into place, framing a semantic system governed by specific and very powerful laws binding the various elements of its environment into itself. And among the elements of the textual environment are of course ourselves as readers and the special, densely coded activity to which we are called and by which we, too, are defined, namely the reading. In picking our way through the different possibilities formed by the title, we are *engaged* in the system of the text, taking on a role assigned us in the writing itself as such. To read the text is to become a part of it, to enter into the play of its forms and signs - as is in fact, at the end of the story, the lot of the person to whom (for good or ill) the letter is actually written and sent.

But in the movement by which it differs, the title also *defers*.[12] It postpones the moment at which the text will (as it ultimately must) face up to and embrace its own difference. The title proposes many possibilities, but which of them apply and in what sense remains to be seen. The title has thus a double function: to define or delimit a space of meaning, a »world« in relation to which a certain question is raised and to hold that question open, leading the reader on. The moment comes however when the text moves out of the realm of mere potential into that of actual meaning; when, like the Word announced in its title, it must bow to the logic of its own incarnation, consenting to surrender the curious purity of relative in-difference, underdetermined, free, to become instead wholly (indeed, tragically) overdetermined, differentiated Flesh, as *this* child, *this* body, *this* text. A choice has finally to be made, foreclosing even as it fulfills the possibilities with which the title teems - a fact reflected in the opening words of the text itself proper: »What'll I write?«

Whose question is this? True, set off by quotation marks, it issues from inside the fictional world; and no sooner does it register than we get an attribution: »›What'll I write?‹ asked Yegor, and dipped his pen in the ink.« (p. 434) With the name »Yegor« we get indeed an irrevocably differentiating, incarnating Word. The writing in question, like the question itself, is henceforth Yegor's, and his alone - as we shall see, a fact of some consequence to what actually gets written. Further, the moment it names the author of the question, the text goes on to tell the rest of the story - that of a certain letter. Mindful of the season (»But now it was Christmas time«), the old peasantwoman, Vasilisa, and her husband Pyotr want to send word to their daughter Yefimya who, having married and moved to Petersburg four years ago, »sent two letters home, and then disappeared without leaving a trace.« (p. 434) But unlike the story, the text does not begin with the old couple. Like writing itself, it opens *in medias res*[13], with the decision to write and the actual physical as well as formal act of writing itself. In the instant before the name »Yegor« commits the text to the tale, we hesitate, or at any rate find room to hesitate, an opening, an occasion; and in that instant the question can in fact be turned back on the text itself as the trace of *someone else*, its Signatory, Chekhov, *his* question: »What'll I write?« For here too we meet a need to communicate, to send word to absent loved ones. By its initial question, Chekhov's text ties its fate to that of the letter it speaks of, its own destiny reflected in what happens to the missive that circulates through its pages. Yegor's question, dipping his pen in the ink, preparing for dictation, is the question of authorship itself; and in what becomes of the writing that issues from it, it mirrors that of its parent. For what *does* Chekhov mean by this story? What is he trying to tell us, what is he trying to *say*? Given all the steps intervening between the willto-write and the text's final arrival at its destination some ninety years later and in translation, given all the chances and changes attending composition and transmission, the drafting, revising, submission for publication, distribution, reception, and scholarly commentary, who can be sure?

Only look at all the forces that come into play with the writing of the letter itself, this further incarnation of the word. To communicate with their distant daughter, Vasilisa and Pyotr must write a letter: a written must do what no mere spoken word can, make the journey through space and time required to reach Yefimya, or ourselves. But Vasilisa and Pyotr (and note everything inscribed in that name: Pyotr, or Peter, the holder of the keys; the rock, *petrus*, on which the Church is built; an unlettered peasant, a fisherman, become a Fisher of Souls, a tender of nets and snares intended to catch us as we pass) are illiterate, cannot write; there is a necessary intermediary skill they have not learned, a code they never mastered: the art of turning (but *what* exactly? words? wishes? feelings?) into graphic symbols. At which stage, enter precisely *Yegor* as an expression of the logic of the situation, its systemic character. To write Yefimya, the old couple must seek out Yegor; already, even before it has been written, the letter begins to travel, and *change*.

Nor is merely the *art* of writing at issue here. With the need to write, a whole society comes to bear on the question of this text, its »What'll I write?« Society as such after all, in the time of Chekhov's writing as in that of our reading, is largely organized in accordance with the laws of the text. The law itself is a written code, closed to who cannot read. Writing is also the foundation on which the modern bureaucracy is based; indeed, as an historical phenomenon, writing is exactly coeval with bureaucracy, originating in the work of the ancient scribes, that dubious double priesthood whose office was not only to keep, copy, and interpret the Sacred Writings, but to maintain statistical records of transactions and supplies.[14] Further, writing is a principle of social differentiation, social *stratification*; literates and illiterates stand in a hierarchy, literates on top - a hierarchy which, as a determined adversary of the Pharisees, those *other* technocrats of the written word, Christ himself tried to overturn by naming precisely the illiterate fisherman, Peter, his successor. Writing is thus a form of social power, a form by which power projects itself and shapes its environment - as we see in the ensuing development of the anecdote.

Unable to write the letter for themselves, Vasilisa and Pyotr are compelled to seek a deputy to write it for them. The search represents a hesitant ascent up the social ladder mediated by a whole series of modes of social relation: class, family, gender, and above all *money*. They go to the village teahouse, for they have heard that the proprietor's brother-in-law, recently back from the army, »could write a fine letter if he were properly paid« for it. »At the teahouse Vasilisa had a talk with the cook, then the proprietress, and then with Yegor himself. Fifteen kopecks was the price agreed on.« (p. 434) The mediations writing imposes do not stop here, however. There comes the moment when Vasilisa and Pyotr meet with Yegor to dictate the letter itself, arriving at the »scene of writing« as such.[15] The experience is a distressing one:

> Yegor was sitting at the table, pen in hand. Vasilisa was standing before him, thoughtful, an expression of care and grief on her face. Pyotr, her husband, a tall, gaunt old man with a brown bald spot, had come with her; he stood staring fixedly ahead of him like a blind man. On the range a piece of pork was being fried in a saucepan; it sizzled and hissed, and seemed actually to be saying: »Flu-flu-flu.« It was stifling.
>
> »What'll I write?« Yegor asked again.
>
> »What?« asked Vasilisa, looking at him angrily and suspiciously. »Don't rush me! You're not writing for nothing; you'll get money for it. Well, write: »To our dear son-in-law, Andrey Hrisanfych, and to our only beloved daughter, Yefimya Petrovna, our love, a low bow, and our parental blessing enduring forever and ever.«
>
> »Done, keep going.«
>
> »And we also send wishes for a merry Christmas, we are alive and well, hoping you are the same, please God, the Heavenly King.«
>
> Vasilisa thought for a moment and exchanged glances with the old man.
>
> »Hoping you are the same, please God, the Heavenly King,« she repeated, and burst into tears.
>
> She could say nothing further. And yet before, when she had lain awake at night thinking of it, it had seemed to her that she could not get all she had to say into ten letters. Since the time when her daughter had gone away with her husband much water had flowed under the bridges, the old people had lived like orphans, and sighed heavily at night as though they had buried their daughter. And during all that time how many events had occurred in the village, how many weddings and funerals! What long winters! What long nights! (p. 434-435)

A number of things are at work in this passage. First we note the gestural confirmation of writing's social dimensions: Yegor sits while the old peasants stand; the veneration that the younger man would normally be expected to show his elders is replaced by a petty bureaucrat's impatient sense of superiority. It is due in part to this social dimension that, once past the salutation, Vasilisa is struck dumb. Writing is not simply a neutral instrument of communication, a mere copy of the words Vasilisa would like to say, contingently reserved for certain classes of society (*petite bourgeoisie*, lesser functionaries) as opposed to others (rural peasantry, urban proletariat), it is a medium of expression geared to the idiom and thus the *ideology* of certain classes as against certain others. As we shall see in a moment, despite Vaslisa's silence, a letter does in fact get written. Like aristotelian nature, language abhors a vacuum. But it is written precisely by Yegor and, through him, by the literate institutions of which he is an extension. The letter is indeed written *by the book*, straight out of the pages of the »Statutes on Disiplinery Fines and Criminal Laws of the War Department,« the book of »Militery Regulashuns« (p. 436) - a grotesque homily exhorting Yefimya's husband (like Yegor, a former soldier) to live up to his former exalted military calling, and above all to shun strong drink, reinforcing him in this pious resolve by reminding him that; as stated in volume five, »Soldier is a common name and an honorable one. The Topmost General and the lowest Private is both called soldier.« (p. 436) Needless to say, this is pure humbug; but the point is that the humbug expresses the false consciousness of the literate classes by and for whom such a thing as a book of Military Regulations (however spelt) exists - to the point where, to see through the cant (and get the *joke*), you have to be able to read and write, and thus be conversant with the medium of expression and with the modes of thought it encodes. Vasilisa falls silent because she is quite literally out of her element, a fish out of water. Writing is not a »natural« mode of expression, still less a mere mechanical transcription of the spoken word. What Vasilisa means to say simply cannot be *said* in writing - an impossibility socially and psychologically as well as formally determined.

Look again at Vasilisa's salutation, with its »low Bow,« »parental blessing,« and ritual invocation of the »Heavenly King.« A great deal of research, largely stimulated by the »Homeric question« as solved by the American classicists, Milman Parry and Albert Lord, has been done since the 1930s on oral as opposed to literary modes of composition and, stemming from this, on the systematic contrasts between the modes of consciousness conditioned by the media in which oral and literate cultures respectively encode, preserve, and transmit knowledge and information.[16] One discovery to which this has led concerns the vital role played in oral cultures like that of our illiterate peasants by the *formula*. For once you get beyond the immediate needs and usages of everyday verbal communication, once it is a question of more than ordinary requests, commands, and exchanges of news referring to the immediate existential context, oral cultures encounter remarkable difficulties unknown in literate ones. How for example, in the absence of written records, and *a fortiori* of devices like our contemporary computerized information banks, is basic knowledge of whatever sort, technical, historical, moral, philosophical, to be preserved from one place and generation to the next? Where there are no books to remember for you, where you cannot refer your interlocutor to »volume five of Militery Regulashuns,« you have to do the remembering yourself - and one of the central preoccupations of the entire culture becomes the business of memorization and recall.

The solution (or part of it) is to develop highly ritualized modes of speech and behavior mnemonically geared to the diverse practical situations in which different kinds of knowledge are needed. Such are the proverbial sayings, adages, saws, and aphorisms which, only capriciously and precariously surviving in our post-industrial societies, remain a common coin in abundant circulation in non- and semi-literate cultures. Such too are those immense »tribal encyclopedias«[17] found in the epic poems of the rural African present or our own classical past, giant songs structured by formulaic epithets and the formulaic development of theme and plot encoding and transmitting not only the fund of legendary and mythic tales, but basic information on matters ranging from the collective history, customs, and values of the people who share the language of the songs to such mundane yet vital matters as how to build a boat, skin an ox, or treat an ague. When in the *Iliad*, during his momentous quarrel with Agamemnon, Achilles brandishes his scepter and recites the story of its origins and the nature of its symbolism, we get not just an important episode in the drama of the Trojan War, nor even, over above this, an object-lesson in lines of political authority and a mini-course in political rhetoric; we also learn how to make a scepter - with what sort of wood and where such wood is found; by what steps, with which tools, using what methods. What makes all of this possible, simultaneously ensuring its dissemination and retrieval, is the formula in its various metrical, narrative, and thematic uses. It helps us to remember the many thousands of things we need to know by casting them in a memorable form; memorable because it sings (metrical aspect), because it is embedded in a concrete action (narrative aspect), and because it stands in a constellation of associations with other memorable facts (thematic aspect).

Now, the point of this excursion in Homeric poetics is how it places Vasilisa's salutation. The salutation is itself highly formulaic and for this very reason offers a pocket compendium of peasant mores, customs, gestures, and speech-patterns in turn-of-the-century Russia. But more important is the way it bespeaks the formal basis of a certain *mentality*, that of the culture of an illiterate peasantry. The formulae it so touchingly reproduces are allied with patterns of thought and experience as well as speech quite unlike those inscribed in writing and the structure of literate speech. The formality of address (so distant from our own »Dear Sir« or even »Sehr geehrter Herr Professor Doktor«) implies a formality of *relation* mnemonically imbued with a whole scheme of values and conduct sharply contrasting with the causal rudeness of the semi-literate Yegor. And this too contributes to Vasilisa's silence in the scribal presence. *Everything* in the situation is false. Her silence expresses not just the clash between the artificiality of the written (or indeed spoken) word and the spontaneous emotion that seeks incarnate utterance; the entire pattern of speech, comportment, and thought, her very consciousness is alienated by its mediating encounter with the radically different consciousness encoded in the act of writing. She cannot form the proper terms, cannot frame her thoughts in the idiom dictated by the intervening culture of literacy in the person of the scribbling Yegor. »Don't rush me!« she cries at one point, reacting not only to Yegor's bored impatience, but to a more general pressure for which she has not even a name, but which constrains her no less tyrannically for that.

The machinery is in place, however. And though Vasilisa can think of nothing to say, a letter gets written anyway. More exactly, the letter writes itself - and such a letter! Vasilisa continuing silent, Yegor fishes for information about the people the letter is for, and peasant protocol dictates that Vasilisa mention her son-in-law, as male head of household, first. Once Yegor has learned that this son-in-law was once a soldier like himself and now works in a »hydropathic establishment« in Petersburg, the rest follows almost automatically: »Yegor thought a little and then began writing rapidly.« (p. 436) Vasilisa squabbles with her husband, and thinks of everything the letter ought to say, and won't: how the flour ran out last year even before Christmas, forcing them to sell the cow; how needy they are again this year; how the old man is sick and likely to die soon. (p. 436) And all the while Yegor gets on famously, regurgitating (mangled and ill-spelt) such platitudes as he can recall from his time in the army, all of it addressed to the one person the old peasants mean *not* to write, the man who took their daughter from them:

> The pen creaked, forming flourishes on the paper that looked like fish-hooks. Yegor wrote hurriedly, reading every line over several times. He sat on a stool, his feet spread wide apart under the table, a well-fed, lusty fellow, with a coarse snout and a red nape. He was vulgarity itself, coarse, arrogant, invincible, proud of having been born and bred in a teahouse; and Vasilisa knew perfectly well that here was vulgarity, but she could not put it into words, and only looked at Yegor angrily and suspiciously. The sound of his voice and the incomprehensible words, the heat and the stuffiness, made her head ache and threw her thoughts into confusion, and she said nothing further, stopped thinking, and simply waited for him to cease scratching away. (p. 437)

The sad little story of this misbegotten letter is not over yet. Once Yegor has been paid, the letter must be sent off, and the next morning, after her prayers, Vasilisa walks the seven miles to the train station. In addition to everything else, she must now contend with the postal system and rail network, in fact the entire urban infrastructure of transmission and communication - though, mercifully, no mishap occurs on this occasion. (But I cannot help thinking here of Melville's scrivener, Bartleby, and the Dead Letter Office...) And yet does Vasilisa's letter in fact get through? The first snag at this stage is Yefimya's husband, Andrey Hrisanfych, the doorkeeper in Dr. B.O. Moselweiser's hydropathic establishment, and keeper of other doors as well, in particular the one behind which his wife and children are hidden. To reach Yefimya, the letter, already addressed to him rather than her, must not only run the gauntlet between heart and page and between sender and recipient, it must pass through *him*. And though, indifferent, he does decide to hand it on after scanning a few lines, this cannot be taken for granted. For we learn that Yefimya has in the past given him three or four letters to send home to the village, »but some important business had always intervened; he had not sent the letters and somehow they were mislaid.« (p. 439) In any event, the letter does physically get through, but then a curious thing happens. »In a trembling voice,« Yefimya reads the first lines aloud to her children, »and could read no more«:

> these lines were enough for her. She burst into tears, and hugging and kissing her eldest child, she began to speak - and it was impossible to tell whether she were laughing or crying. »It's from granny, from grandpa,« she said. »From the country. Queen of Heaven, saints and martyrs! The snow is piled up to the roofs there - the trees are white as white can be. Children are out on tiny little sleds - and darling bald old grandpa is up on the stove - and there is a little yellow puppy - My precious darlings!« ... »And little hares hop about in the fields,« Yefimya continued

> mournfully, bathed in tears, and kissing her boy. »Grandpa is gentle and good; granny is good, too, and kindhearted. In the village folks are friendly, they fear God - and there is a little church in the village; the peasants sing in the choir. If only the Queen of Heaven, the Mother of God would take us away from here!« (p. 439)

At one level, this ultimate incarnation simply prolongs and reinforces the sequence of translations and travesties begun the moment the letter was originally conceived. Vasilisa wanted her daughter to learn of her parents' misery and need: how long, cold, and lonely the winters are, how they could use some money to see them through, how badly they miss their child and want her back. But truly to reach its destination, a letter must survive not only the writing and sending, but the *reading*. The act of inscription, of encoding, calls into play a further, posterior act of decoding over which it has minimal and inadequate command. Yefimya never actually reads the letter at all; instead, she projects on it her *own* misery and need reflected in the edenic fantasy she develops for her children. She to whom the letter was intended thus completes the process of estranged miscommunication in which it began. Indeed, as a direct result of the conditions under which it is first written and then read, we get, concerning each of the two chief protagonists in the story of the letter, images of that alienated isolation in the spatial and temporal present the letter is precisely meant to mediate and overcome - Vasilisa, struck dumb in the suffocating heat of the teahouse kitchen, trapped in the presence of »vulgarity itself: coarse, arrogant, invincible,« in the person of the odious Yegor; and Yefimya, just as mute and powerless in the presence of her equally vulgar and odious husband:

> Andrey Hrisanfych returned to his room to have a smoke before another patient arrived, and Yefimya suddenly stopped speaking, grew quiet, and wiped her eyes, and only her lips quivered. She was very much afraid of him - oh, how afraid of him she was! She trembled and was terrified at the sound of his steps, his look, she dared not say a word in his presence. (p. 439-440)

Yet cruel as the work of the letter turns out to be, a shadow of grace hangs over the closing scene. As entirely as Yefimya seems to misread her mother's pitifully deformed letter, has she actually got it wrong? Is there not a sense in which, precisely by misreading it, she writes the letter her mother meant to? Returning now to the startingpoint of our *own* reading of both Vasilisa's letter and its parental double, Chekhov's text, it all depends on what we make of the instigating title, »At Christmas Time.« And yet if only because, in the reversible, infinitely recursive time of the text, the title itself turns to the story to beg a meaning of its own, if only because the sorry yet signal fate of Vasilisa's letter may lead to an *ironic* interpretation of the redemptive promise of the title, suggesting the ultimately unsalvageable character of human suffering, the form of the answer merely reaffirms the body of the question, the story itself, and so on indefinitely, »forever and ever.«

Unless we take this very movement of indecision as our guide, seeing in it less the story's *answer*, solving the terms of its equation, than an *opening*, a way out of the trap that seems so irresistibly to shut, with a snap, at the story's close. When we first discussed the title, implicitly citing Derrida, I noted the way it both *differs* (makes a difference) and *defers*: a double movement of position and de-position, assertion and negation, to which Derrida gives the name *différance*, the French word for »difference« with the difference of an »a« substituting for the penultimate »e«, thereby recalling the participial form of the homonym of the French verb *différer*, »to differ,« i.e., *différer*, »to defer.« We need not follow Derrida through all of the entailments of this move. I want simply to show how far, leaving the meaning (and thus the fate) of the letter open, Chekhov anticipates Derrida. The story is after all meant to *make a difference*. We are to be moved, and changed, and the world with us. Yet this difference is so fragile that the text dare not be too categorical about it. Is it not after all precisely because language in all its forms pushes so inexorably for an answer, a single term withheld from all ambiguity, all differential play, that Vasilisa's letter endures the grotesque transformation the text recounts? But whatever we decide, bleak as may be the fate to which Vasilisa and Yefimya seem condemned in the unrelenting and yet nevertheless unwaveringly loving and particularizing light both of Chekhov's story and of its ultimate pattern in the epiphanic Word of the Incarnate King, the letter that passes between them makes at least this much possible: an insight into the nature of their condition - their life in language itself.

Notes and References:

1. William of Ockham, *Ordinatio Ockham*, D. II, Q. viii, prima redactio; as cited in *Philosophical Writings*, ed. and tr. P. Boehner, London: Nelson, 1957, p. 41.

2. R. Barthes, *Essais critiques*, Paris: Seuil, 1964, p. 15.

3. J. Culler, »Apostrophe,« in *The Pursuit of Signs*, Ithaca, New York: Cornell University Press, 1981.

4. See R. Barthes, *Le Degré zéro de l'écriture*, Paris: Seuil, 1953.

5. F. de Saussure, *Cours de linguistique générale*, critical edition by T. de Mauro, Paris: Payot, 1982.

6. T.S. Kuhn, *The Structure of Scientific Revolutions*, 2nd edition, Chicago: University of Chicago Press, 1970.

7. B. Eichenbaum, »The Theory of the ›Formal Method‹«, in L.T. Lemon, M.J. Reis (eds., tr.), *Russian Formalist Criticism: Four Essays*, Lincoln, Nebraska: University of Nebraska Press, 1965; Bison Book paper edition, p. 136.

8. R. Jakobson, »Concluding Statement: Linguistics and Poetics«, in T.A. Sebeok (ed.), *Style in Language*, Cambridge, Massachusetts, New York: The Technology Press, M.I.T. & John Wiley & Sons, Inc., 1960, p. 350-377.

9. This is, by the bye, an invariant principle of literary evolution. See V. Shklovsky, »Art as Technique«, in Lemon, Reis (eds., tr.), *op. cit.*, 1965, note 8; R. Jakobson, »Du réalisme artistique«, in T. Todorov (ed., tr.), *Théorie de la littérature*, Paris: Seuil, 1965; R. Barthes, *op. cit.*, 1970, note 3.

10. A. Chekhov, »At Christmas Time«, in A. Yarmolinsky (ed.), *The Portable Chekhov*, New York: Viking-Penguin, 1977, p. 434-440. All subsequent references are given in the text.

11. Another feature of the shift from structuralism to post-structuralism bears precisely on the contribution of the *reader*, yielding, for example, the *Rezeptionstheorie* or *Rezeptionsästhetik* of Iser and Jauss (by way of Gadamer), and current American »reader-response« criticism. For a helpful introduction, see S.R. Suleiman, I. Crosman (eds.), *The Reader in the Text: Essays on Audience and Interpretation*, Princeton, New Jersey: Princeton University Press, 1980.

12. See J. Derrida, »La différance«, in *Marges de la philosophie*, Paris: Minuit, 1972.

13. This points to a pervasive theme in Derrida, for whom writing destroys (»deconstructs«) what he considers the central myth of our »logocentric« Western culture: that of an Origin from which phenomena derive all essential reality and truth. See, for example, »signature événement contexte«, in *op. cit.*, 1972, note 14 and, for a general historical and theoretical discussion, *De la grammatologie*, Paris: Minuit, 1967.

14. See, for example, I.J. Gelb, *A Study of Writing*, revised edition, Chicago: University of Chicago Press, 1963.

15. See J. Derrida, »Freud et la scène de l'écriture«, in *L'écriture et la différence*, Paris: Seuil, 1967.

16. For a useful introductory overview to these problems, see W.J. Ong, *Orality and Literacy: The Technologizing of the Word*, London: Methuen, 1982.

17. E.A. Havelock, *Preface to Plato*, Cambridge, Massachusetts: Harvard University Press, 1963, ch. IV.

III. The Appearance of Structure

Concepts of Self-Organization in the 19th Century

MICHAEL HEIDELBERGER
University of Freiburg im Breisgau

Several contemporary schools of thought are participating in the current debate on self-organization. These schools, it seems to me, have not yet reached complete agreement on the question of what self-organization exactly means and what the most central and most basic intuitions of this concept are. In such a situation, it should be fruitful to have a closer look into the history of the most prominent ideas in this debate.

My impression is that two separate sets of scientific ideas dominate our contemporary discussion on this subject. The first set is clustered around the assumption that our cognitive system is self-referential, that is, that it does not mirror an objective external reality but that the subject constructs his reality himself. Formulated in the language of the 19th century, this would mean that our sensory experience is completely subjective. The second set of ideas starts from the thesis that living organisms are self-organized systems such that they structurally determine their own states. Once again in the language of the 19th century, this would mean that living systems follow a built-in tendency to preserve their stability.

In the following, I would like to give a short outline of the histories of these two kinds of ideas in 19th century Germany. I would like to explain why there was little fusion between them and why they had such little impact.

I

The first and most important formulation of the idea that the world as we see it is our own construction comes from the physiologist Johannes Müller (1801-1858). Müller wrote about this as early as 1826. His ideas reached their most mature state of development in the 1838 volume of his *Handbuch der Physiologie des Menschen* (1834-1840). There Müller presented the Law of Specific Nerve Energies that states that sensory nerves, however stimulated, always give rise to the same kinds of sensation. It follows from this law that the nature of our sensation does not mirror the external stimulus giving rise to it but only the state of the nerve being stimulated. If, for example, the sensory nerve of vision is stimulated, we then have a sensation of light, regardless of whether it is caused by a ray of light striking our eyes, by a mechanical pressure on the eyeball, by an electric stimulation, or by anything else. We are immediately aware not of external objects but of the particular states of our nervous system. Therefore, no one-to-one correlation can be established between the stimulus and the sensation. One and the same stimulus can cause different sensations if it excites different sense organs, and one and the same sensation can be the effect of different causes if the same sense organ is stimulated differently. As a consequence, there is no direct knowledge of the qualities of the external objects; we only know the time and the intensity with which these external objects act on our nerves. Neither the modality nor the quality of a sensation tells us anything about the real nature of an external stimulus. In Müller's own words: »Sensation

W. Krohn et al. (eds.), Selforganization. Portrait of a Scientific Revolution, 170–180.

consists in the sensorium's receiving through the medium of the nerves, and as a result of the action of an external cause, a knowledge of certain qualities or conditions, not of external bodies, but of the nerves themselves.«[1]

Müller's law was a cornerstone of sensory physiology throughout the 19th century and it is still regarded as valid. As a matter of fact, I think that Müller's law is still the basis for our conviction today that our cognitive system is *self-referential* and *self-explicative*. By these qualifications, we mean that the brain is not able to »compare« the neural signals it receives by being stimulated from the outside with the external reality. All it can do is compare one electric potential with another one and interpret its own states with the help of internal rules.

If self-reference and self-explicativeness of the brain has already been known for 150 years, then one might ask, why did it take so long for us to think of the brain as a self-organized system, that is, a system that at each step of its development autonomously determines its own structure from the inside? For Maturana, Varela, and their followers this would be the most natural conclusion one could draw from Müller's law. The easiest answer is, of course, that nobody thought of it until our time (or could not think of it for lack of experimental evidence). In fact, this is precisely how Heinz von Foerster explains this. He even denies that anybody in the 19th century ever realized that Müller's principle implies self-reference. He writes: »Although the essence of this principle has already been known for over 120 years, the tremendous consequences of this insight had only been realized in the 1950s. All of a sudden it struck one and one asked how the immense richness of our sensation in each moment, how the multitude of our experience of a colorful, dancing, singing world can be the sole result of the monotonous clicks transmitted by our senses.«[2]

I think that such a view is historically too naive. It was Müller himself who already realized the »tremendous« epistemological consequences of his law. In the second section of this paper, I shall try to show that even the concept of self-organization (under different names, of course) was already available in the 19th century and was even tentatively applied to our own cognitive system.

So we have to search for another explanation. If we take a closer look at the discussion of Müller's law in thc last century we see that it was dominated by a theory that could (at least to the mind of its contemporary audience) successfully cope with its »tremendous consequences«. It did this by combining two contentions that at first sight might seem incompatible, namely, that the brain is *both* a system that interprets itself *and* a system that, at least to a certain degree, represents the external world. Expressed in our contemporary idiom, this would mean that the brain is a self-referential but *not* a self-organizing system. This theory was put forward by the physiologist and physicist Hermann von Helmholtz (1821-1894) and it became immensely successful.[3] For this reason, all theoretical considerations of the idea that the brain is a self-organizing system were very much on the defensive.

So, let us take a look at Helmholtz's theory. Helmholtz makes a crucial distinction between an »image« (*Abbild*) and a »sign« (*Zeichen*). An image always has some similarity with its original, whereas a sign does not need to have any similarity with the object it signifies. All that a sign has to do in order to fulfill its function is always to give rise to the same object for which it stands whenever it is used. In other words, any unequal pair from a set of signs (that are not synonymous) must signify two different objects if the system is to be consistent. Helmholtz now says that although none of our sensations mirror a quality of the external world, some of our sensations can be taken as *signs* of objectivity. His theory of perception is therefore called a »theory of signs« or, as coined by Ernst Mach, »a telegraph-wire theory of sensation«.

Consider now a system of signs whose meanings are unknown to us but from which we know that each of its members stand in a causal relation to the objects they signify. Would this help our interpretation of the meaning of the signs? Helmholtz's answer is that it would. The *order in time* of a series of signs that stand in a causal relation to that which they signify is an image of the *causal order* of their references. If we encounter regular repetitions of sequences of signs, we can try to develop a hypothesis about the causal order of the events signified by these sequences.

Can we apply this to our sensations? We can if we are able to find out which of our sensations are caused by external stimuli and which are not. The outer sensations thus singled out could then be taken as signs for the real objective world. That then means that we could take these outer sensations as a basis for our inferences and start developing hypotheses about the objective causal order of the world.

So the problem is how to distinguish between inner and outer sensations. Helmholtz sees the solution in the consciousness we have of the voluntary movements of our body and our conscious acts in general. If, for example, I am turning my head, some (or all) of the visual sensations I had before change as a result of this. Since I only intended to change the position of my head but not my visual sensations, these changes were not caused by myself and must therefore be the result of the movement of my head, that is, they are caused from the outside. Helmholtz's criterion rests on the capacity of the mind to detect those changes of its sensations that »do not have a psychic relation with the preceding voluntary impulse«[4] to perform a bodily movement. In other words, we are able to separate the outer from the inner sensations if we are able to differentiate between those sensations that can be changed at will and those that cannot, i.e. if we know the lawful effects of our voluntary impulses. This differentiation is the result of an internal process of interpretation by the mind itself; it is not caused by an external stimulus. Although our physiological organization does not allow us to characterize reality or Kant's thing-in-itself in an objective and positive way, »we can, however, attain knowledge of the lawful order in the realm of the real as it is represented in the sign language of our sensory impressions.«[5]

Helmholtz's theory became very attractive and influential, showing how objective knowledge of the world was still possible in spite of Müller's law. At the same time it also provided a new theory of space perception and gave a physiological explanation of the Kantian conception of space as a form of our intuition. I do not want to go into this any further here. The basic idea behind it is to take that which constitutes the external world for us, namely our external sensations, as that which we perceive as spatially characterized.

II

I now come to the second part of my talk that concerns self-organization.[6] Immanuel Kant seems to have been the first thinker to use the concept of self-organization in relation to organic nature.[7] He meant by this that the living organisms are not the kind of systems whose origin is caused from the outside as is the case with the origin of systems of inorganic nature, but that they reproduce themselves from the inside. Organic nature with all her species is not the product of external causes but of an internal capacity of reproduction. Kant's conception was then taken up by F.W.J. Schelling (1775-1854). By assuming that there is a principle of organization built into all »dead« matter, Schelling thought to have found an alternative both to a mechanistic explanation of life (by external mechanical causes) and a vitalistic one (by a *vis vitalis*). He saw the distinguishing feature of life in a certain structure of the relation between cause and effect as it is present in the organism. A system is self-organized if cause and effect make up a closed loop.

Schelling's conception was very fashionable at the time but short-lived because of the later opposition to *Naturphilosophie*. The only »window« through which it was transmitted to later generations of scientists seems to have been the work of the biologist and philosopher Lorenz Oken (1779-1851). In his *Lehrbuch der Naturphilosophie* of 1809, Oken defined the organism as »an individual, unified, self-excited, and self-moved body that is closed in itself.«[8] Following Schelling, Oken pictured the whole universe as a self-organizing system. The universe has two sides that are identical, an ideal (immaterial) and a material (real) one. The form of the ideal side is pure unity whereas the form of the real side is variety, separateness, and dispersion. In this picture, God is to be identified with the universe itself. The development of the world is seen then as the permanent process of God's movement toward self-consciousness. According to the tradition of German idealism, to develop self-consciousness means to put oneself in opposition to the external world, to determine the indefinite absolute into self and object. All mechanical movement, all activity in the world eventually has its source in God's self-organizing process, which is called dynamic. *Naturphilosophie* is therefore seen as the science that deals with the »eternal transformation of God into the world«.[9]

How can such a grandiose and bombastic cosmology contribute anything to an explanation of organic nature? Oken deduces the following consequences from it. Living beings originate as the result of a differentiation of the world's organism into subunits. Since they retain the features of the system from which they came, they are impregnated with divine qualities, that is, they are ideal wholes and autonomous, individual, self-excited, unified, and self-moved subjects capable of developing a self-consciousness. Yet, at the same time, any subsystem is still *part* of the world and thus an object, a lifeless, deadened product of God's movement toward self-determination, externally bound to his activity. Living organisms thus follow two opposing tendencies in their development. »Every living entity is two things at once. It is autonomous and it is immersed into the absolute. Two kinds of processes happen in it, an individualizing, vitalizing one and a universalizing, deadening one. Through the latter, the living being in its limitation longs to become the absolute itself; through the vitalizing process it strives to become the variety of the universe.«[10] This kind of explanation of the dynamics of life is perhaps Oken's most important contribution.

Sixty-four years after Oken's ideas had appeared for the first time, the physicist and philosopher Gustav Theodor Fechner (1801-1887) tried to adapt some of Oken's ideas to the science of his time and published this in 1873 in a book entitled *Einige Ideen zur Schöpfungs- und Entwickelungsgeschichte der Organismen*. According to Fechner's view, life is a special type of movement of the material parts that make up the molecules of an organism. A system of moving parts is either »spontaneous« or »receptive« (or mixed) according to whether the behavior of the system depends on inner or outer forces. The movement of the parts in organic molecules is highly spontaneous, whereas for the inorganic molecules, only the spontaneity of the heat oscillation of its parts remains.

The distinguishing feature of any system endowed with some spontaneity is its tendency to maintain or increase its stability. Fechner states this as the »principle of the tendency toward stability«. This principle, the very heart of Fechner's theory, is his principle of self-organization. Any system that is closed or under constant influence »through the action of its inner forces without retrogression approaches more and more a *stable* state, that is, a state in which its parts periodically return to the same relative position toward each other and to the same conditions of their movement.«[11] Notice that stability here does not mean immovability but periodic repetition. For open systems, the principle states that they try to preserve or raise the degree of stability they have attained by compensating as much as possible the effects of external (receptive) forces.

Fechner wanted his principle to be applied to the universe at large as well as to individual organisms and thereby sought an explanation of the directedness of their development. He was well aware that such an explanation could not be given by purely mechanical means. There is yet another aspect to Fechner's principle in which it also implies a nonreductionist hypothesis on the origin of life. If there is an irreversible tendency in the world to ever higher stability and if the inorganic world is always more stable than the organic one, then life could not have originated from dead matter. One therefore has to assume that before the origin of life the universe was in a very much more unstable condition than it is now, and that the organic world as well as the inorganic one are products of different degrees of stability in this early unstable state. Fechner characterizes the early state of the world as »completely unstable«, »chaotic«, and objectively »random«.[12]

Since Fechner proposed his principle at a time when the reception of Darwinism was almost at its peak in Germany, he was at pains to show that it is compatible with Darwin's view on evolution. He wanted to attain this by introducing another principle, the »principle of mutual differentiation«, according to which the increasing stability of the universe as well as of its subsystems results from their increasing differentiation. Sometimes a system can maintain its stability only by differentiating into two subsystems that mutually supplement each other. The differentiation can appear either as a result of inner forces (i.e., as a self-organizing process) or in an accidental way through variation in Darwin's sense. Natural selection plays a secondary role in correcting these two types of differentiation.

As the founder of psychophysics, Fechner was very much intrigued by the consequences of his ideas for the theory of mind. Since he advocated a strict parallelism of mind and body, he looked for the relevance of his principles to the animate, psychical side of the universe.[13] By transferring the principle of stability to the realm of our conscious life, Fechner even thought that he had found the »cardinal fundament for psychology«.[14] He also believed that he had found a sound basis for ethics with this principle. He sometimes called this the »principle of pleasure«. It states that the physical tendency of an organism to higher stability is the carrier of its psychical tendency to higher pleasure. Any physical increase in the stability of an organism corresponds to a psychical increase in the pleasure felt by it. The more the stability is disturbed, the greater the displeasure.

Fechner's concept of stability bears a relationship to the concept of an »internal environment« (*milieu intérieur*) as introduced by the French physiologist Claude Bernard (1813-1878). There is no indication that Fechner was familiar with Bernard's work. It is true that Bernard used this term from about 1854 on, but it became the center of his theorizing only in the late 1870s, that is, after Fechner's theory had appeared. Bernard differentiated between the external environment, outside the organism, which can vary greatly, and the internal one, which must be stable if life is to be maintained at all. He studied the regulation of this internal constancy and its principles. Bernard's work was further expanded in the present century by Walter B. Cannon (1871-1945) and L.J. Henderson (1878-1942). In 1926, Cannon invented the term *homeostasis* in this very context.[15]

Although Fechner's theory remained almost unknown, it nevertheless had a decisive influence on a small number of important thinkers. The very mention of Fechner's principle of pleasure immediately evokes the name of Sigmund Freud. We have heard of Oken's doctrine of the two kinds of processes taking place in an organism, the dynamic individualizing, vitalizing, and autonomous self-movement and the mechanical, deadening, universalizing tendency to become a lifeless object. Seen through Fechner's eyes, the self-movement of life is a residuum of the universe's early unstable and spontaneous phase, whereas the deadening process is nothing but the tendency to stability. The same duality of the two kinds of processes can again be found in Freud's metapsychology, especially in his *Jenseits des Lustprinzips* from 1920, in which he introduced the distinction between the death instinct and the life instinct. Both kinds of impulses are at work in the ego. The death instinct acts through the pleasure principle that has the function »of terminating all stimulation in the psychical apparatus, or of keeping the lowest possible or an approximately constant level of stimulation.«[16] The decrease in stimulation is felt as pleasure and the increase in stimulation above certain limits leads to displeasure. Freud acknowledged Fechner's influence and called his own pleasure principle a special case of Fechner's principle of the tendency toward stability.[17] In this respect, Freud also talks of the principle of constancy or of the Nirvana principle.

So it seems that Freud has taken up Fechner's suggestion that the reinterpretation of the stability principle for the mental realm can serve as a fundamental basis for psychology.

The American philosopher Charles Sanders Peirce (1839-1914) took over another part of Fechner's theory and turned it into an »evolutionary metaphysics«. He published his ideas in a series of articles from 1891 to 1893. In the same way as Fechner, Peirce characterizes life in terms of spontaneity and instability and even takes these features as the marks of consciousness (»feeling«): »Wherever chance-spontaneity is found, there, in the same portion, feeling exists. In fact, chance is but the outward aspect of that which within itself is feeling.«[18] Fechner's tendency toward stability is now called »tendency toward habit taking«. Peirce conceives of the principle of habit as primarily a mental law and sometimes talks of it as simply *the* psychical law. Peirce greatly surpasses Fechner in saying that the physical world of matter is the product of the tendency in the psychical world of consciousness and feeling toward habit taking. »Matter«, as he claims, »is effete mind, inveterate habits becoming physical laws.«[19]

In fact, wherever we meet uniformity in the world we can, according to Peirce, explain it by the tendency toward habit taking. The same principle enables us to »search out a natural history of laws of nature«.[20] In the end, the development of the universe and everything in it is seen as a grandiose evolution from pure spontaneity and absolute chance to uniformity and stability. Peirce's vision culminates in the following, famous poetic words: a cosmogonic philosophy »would suppose that in the beginning - infinitely remote - there was a chaos of impersonalized feeling, which being without connection or regularity would properly be without existence. This feeling, sporting here and there in pure arbitrariness, would have started the germ of a generalizing tendency. Its other sportings would be evanescent, but this would have a growing virtue. Thus, the tendency to habit would be started; and from this with the other principles of evolution all the regularities of the universe would be evolved. At any time, however, an element of pure chance survives and will remain until the world becomes an absolutely perfect, rational, and symmetrical system, in which mind is at last crystallized in the infinitely distant future.«[21]

Peirce also takes over Fechner's principle of mutual differentiation and calls it »evolutionary love«, advocating it as an alternative to Darwin's »evolution by sporting« and Spencer's and Hegel's »evolution by mechanical necessity«.

So far, we have described some 19th century approaches to how living organisms may be viewed as self-organized systems. The interesting question here is whether anyone thought of extending Fechner's or a similar principle to perception and cognition. I think that such an extension could provide a link between the theory of organisms and a theory of cognitive systems. The discussion of such links is perhaps the most interesting and distinctive feature of our contemporary debate on self-organization.

Fechner himself already gave a first hint in this direction. He observed that »in any material system that is closed or under constant influence, there is a tendency of the parts to group themselves in a regular external form (*gestalt*). This might be connected with the principle of the tendency toward stability. Even the realm of the mind seems to submit to this principle.«[22]

It was the physicist Ernst Mach (1838-1916) who took up this suggestion and tried to develop it further. He was not only influenced by Fechner but also by the physiologist Ewald Hering (1834-1918) who was a student of Fechner for some time and had studied the »automatic control« (*Selbststeuerung*) of physiological processes like metabolism and respiration. Mach was deeply convinced that our thoughts have to be regarded as »expressions of organic life«, as »organic processes«.[23] Following Fechner's psychophysics, he expressed the view from 1863 on that any stable psychical state, that is, a thought or sensation, corresponds to a stable physical state, and vice versa, in such a way that one can argue from one realm to the other.[24] »I am looking for similarities and analogies between those psychical and physical states that correspond to one another.«[25]

Mach also viewed the development of scientific theories, conceived as an ever increasing economy of thought, as a biological process. This view owes very much to Fechner's stability principle: »Continuity, economy, and stability are mutually constraining conditions; in fact, they are only different sides of one and the same feature of healthy thinking.«[26] I think that Fechner's principle is here thought to be responsible for the stability of scientific theorizing whereas Darwin's principle of natural selection is responsible for its economical aspect.

Mach did not elaborate his program of applying the stability principle to the cognitive realm to any great extent. This was done by the *gestaltpsychologists* of our century. The Fechnerian jargon in their theories cannot be overlooked. Wolfgang Köhler (1887-1967), for example, speaks of the tendency of a system toward conciseness of its structure and *gestalt* or of a tendency toward simple formation. A system whose parts have a certain degree of functional connection is subject to the »principle of stationary distribution that arises spontaneously out of its inner dynamics«. Such a system has the tendency »to reach a stationary state in relation to the dominating conditions.« The principle of the tendency toward the stationary distribution is »the only general basis of thought that is used by the theory of physical *gestalts*.«[27]

III

If we look back on our panorama of concepts of self-reference and self-organization, it is striking to see that nobody seems to have tried to bring together these two sets of ideas. That is, nobody seems to have entertained the view that our cognitive system is both self-referential and self-organized - a view that is very popular nowadays. The reason for this lack of awareness lies in the fact that the two kinds of ideas were pronounced by two different schools that were opposed in certain crucial respects. The Oken-Fechner line, leading to the Mach-Freud-Peirce view, rejected (unanimously, as far as I know) the contention that Müller's law gives a physiological proof of the inapproachable nature of Kant's thing-in-itself, that is, of the self-referentiality of the brain. They thought the reasoning behind this conviction to be self-refuting and unscientific. The structure of their argument is as follows: if one talks about the »unknown and unknowable objective reality« or some aspect of it in the style of a Helmholtz-like epistemology, the term »objective reality« must be meaningful if the whole thesis is to make sense. It can only be meaningful if there is empirical evidence from which we can deduce its presence. The thesis itself, however, explicitly denies that such empirical evidence can exist. Therefore the whole enterprise of the thing-in-itself or the self-referentiality of our cognitive system is futile and meaningless.[28]

The Helmholtz tradition of physiological Neo-Kantianism answered these accusations by pointing out their transmechanistic implications. It was the basic conviction of the Helmholtz tradition and the mechanistic worldview in general that mechanics (together with perhaps chemistry) is adequate enough to explain the functioning of organic systems. The attempts of the Fechner-Mach tradition to overcome this fixation on mechanics by introducing nonmechanical principles for the explanation of the organic and psychical world were regarded as sophisticated but still inadmissible relapses into the old, irrational *Naturphilosophie*. The domination of the mechanistic worldview in the 19th century also explains why the idea of self-organization had only little impact on the biological sciences.

An adequate understanding of the history of self-reference and self-organization should help us to see our own situation more clearly. It seems to me that there are two morals to be drawn from the story. All too often today, it is assumed without saying that self-organization and self-reference mutually imply each other. I do not think that this is necessarily the case as our examples show. One should therefore be very careful not to deduce premature epistemological consequences from such assumptions. And second, I do not think we should call the increasing use of concepts of self-organization a scientific revolution. Most often this is done with the sole intention of propagating certain epistemological views allegedly connected with these concepts and of raising their value. I hope, however, to have shown that there is no need for a scientific revolution in this respect when some of the battles have already been fought.

Notes and References

1. Cited according to Boring's translation of this quotation in E. Boring, *A History of Experimental Psychology*, 2nd edition, New York: Appleton, 1950, p. 82.

2. H. v. Foerster, »Erkenntnistheorien und Selbstorganisation«, in S.J. Schmidt (ed.), *Der Diskurs des Radikalen Konstruktivismus*, Frankfurt am Main: Suhrkamp, 1987, p. 133-158, p. 140.

3. Its first formulation is given in H. v. Helmholtz, »Ueber das Sehen des Menschen«, Vortrag gehalten zu Königsberg am 27. Februar 1855, in *Vorträge und Reden von Hermann von Helmholtz*, 2 Vols., 5th edition, Braunschweig: Vieweg, 1903, Vol. I, p. 87-117; its final and classical version in Helmholtz, »Die Tatsachen in der Wahrnehmung«, Rede gehalten zur Stiftungsfeier der Friedrich-Wilhelms-Universität zu Berlin, 1878, in Helmholtz, *Vorträge*, Vol. II, p. 213-247. On Helmholtz in general, start with R.S. Turner, »Helmholtz, Hermann von«, in *Dictionary of Scientific Biography*, Vol. 6, New York: Scribner's, 1974, p. 241-253, and Turner, »Hermann von Helmholtz and the Empiricist Vision«, *Journal of the History of the Behavioral Sciences* **13** (1977), 48-58.

4. Helmholtz, *op. cit.*, 1903. note 3, p. 239.

5. Helmholtz, *op. cit.*, 1903, note 3, p. 242.

6. I have treated this part more explicitly in M. Heidelberger, »Selbstorganisation im 19. Jahrhundert«, in W. Krohn, G. Küppers (eds.), *Selbstorganisation - Aspekte einer wissenschaftlichen Revolution*, Wiesbaden: Vieweg, 1990, p. 67-104.

7. See his *Kritik der Urteilskraft*, B 293 and B 379. For Kant's and Schelling's use of the concept of self-organization, see M.-L. Heuser-Keßler, *Die Produktivität der Natur. Schellings Naturphilosophie und das neue Paradigma der Selbstorganisation in den Naturwissenschaften*, Berlin: Duncker & Humblot, 1986.

8. L. Oken, *Lehrbuch der Naturphilosophie*, 3 Vols., Jena: Friedrich Frommann, 1809-1811, Vol. II, 1810, § 817f, See §§ 906f, 909f.

9. Oken, *op. cit.*, 1809, note 8, Vol. I, § 34.

10. Oken, *op. cit.*, 1809, note 8, Vol. I, § 72. See also § 26f.

11. G.T. Fechner, *Die Tagesansicht gegenüber der Nachtansicht*, Leipzig: Breitkopf & Härtel, 1879, p. 209.

12. Fechner, *op. cit.*, 1879, note 11, p. 221 and Fechner, *Einige Ideen zur Schöpfungs- und Entwickelungsgeschichte der Organismen*, Leipzig: Breitkopf & Härtel, 1873 [repr.: edition diskord, Tübingen, 1985], p. 55, 68, 75f. For the revolutionary features of Fechner's indeterminism in general, see M. Heidelberger, »Fechner's Indeterminism: From Freedom to Laws of Chance«, in L. Krüger, L.J. Daston, M. Heidelberger (eds.), *The Probabilistic Revolution, Vol. I: Ideas in History*, Cambridge, Massachusetts: MIT Press, Bradford Books, 1987, p. 117-156.

13. For Fechner's solution of the mind-body problem, see M. Heidelberger, »Fechners Leib-Seele-Theorie«, in J. Brozek, H. Gundlach (eds.), *G.T. Fechner and Psychology*, Passau: Passavia, 1988, p. 61-77.

14. J.E. Kuntze, *Gustav Theodor Fechner (Dr. Mises). Ein deutsches Gelehrtenleben*, Leipzig: Breitkopf & Härtel, 1892, p. 205f.

15. For an overview of this tradition, see R. Dubos, »Environment«, in P.P. Wiener (ed.), *Dictionary of the History of Ideas*, New York: Scribner's, 1973, p. 120-127.

16. S. Freud, »Jenseits des Lustprinzips«, in A. Freud et al. (eds.), *Sigmund Freud, Gesammelte Werke*, Vol. 13, 7th edition, Frankfurt am Main: S. Fischer, 1972, p. 3-69, p. 67f.

17. For the relation between Fechner and Freud, see H.F. Ellenberger, »Fechner and Freud«, in *Bulletin of the Menninger Clinic* **20** (1956), 201-214.

18. C.S. Peirce, »Man's Glassy Essence« [1892], in C. Hartshorne, P. Weiss (eds.), *Collected Papers of Charles Sanders Peirce*, 2nd edition, Cambridge, Massachusetts: Harvard University Press, 1960, Vol. 6, p. 238-271, p. 265.

19. C.S. Peirce, »The Architecture of Theories« [1891], in Hartshorne, Weiss, *op. cit.*, 1960, note 18, p. 25.

20. Peirce, *op. cit.*, 1891, note 19, p. 12.

21. Peirce, *op. cit.*, 1891, note 19, p. 33.

22. Fechner, *op. cit.*, 1873, note 12, p. 32.

23. E. Mach, *Die Principien der Wärmelehre. Historisch-kritisch entwickelt*, 3rd edition, Leipzig: Ambrosius Barth, 1919 [1st edition 1896], p. 382, 390.

24. E. Mach, *Compendium der Physik für Mediciner*, Vienna: Wilhelm Braumüller, 1863, p. 234; *Die Analyse der Empfindungen und das Verhältnis des Physischen zum Psychischen*, 9th edition, Jena: Gustav Fischer, 1922 [1st edition 1886, 2nd edition 1900], p. 41, p. 268.

25. Mach, *op. cit.*, 1922, note 24, p. 305.

26. Mach, *op. cit.*, 1922, note 24, p. 268, p. 273.

27. W. Köhler, »Gestaltprobleme und Anfänge einer Gestalttheorie«, in *Gestalt Theory* **5** (1983), 178-205, 193, 201, 198f. (First in *Jahresberichte für die gesamte Physiologie und experimentelle Pharmakologie für das Jahr 1922* **3** (1925), 512-539).

28. »Objectivity« always means »objectivity to us«. It is logically impossible to say that there is a feature of the world that cannot be mirrored by our cognitive system. If it is true that there is such a feature then it *must be* mirrored by our system (there is no other way of knowing empirically the existence of something). If such a feature does not exist then the thesis is also false.

Cognitive Systems as Self-Organizing Systems

MICHAEL STADLER AND PETER KRUSE
University of Bremen

1. Introduction

The theory of self-organization is beginning to penetrate into the entire spectrum of scientific disciplines. The initial ideas that led to self-organization theory can be traced back into the historical tradition of the social and natural sciences. The theory of self-organization in a narrower sense, however, is usually said to have begun with the works of Manfred Eigen, Heinz von Foerster, Ernst von Glasersfeld, Hermann Haken, Humberto Maturana and Ilja Prigogine.[1] If one considers these works as the basis, two main conceptual points of reciprocal causal relationship may be distinguished: the emphasis on the purely theoretical, system-related aspect and the emphasis on the epistemological aspect of self-organization.

The latter emphasis is necessarily made when the idea of an autonomous creation of order is applied to the area of *cognition*. Every experienced or reconstructed reality, according to this view, is considered to be the result of a self-referential attribution of meaning and thus constructed in a radical sense. This epistemological consequence reacts upon all approaches of the theory of self-organization in a theoretical recourse.

In the following this general epistemological consequence will not be our main focus of interest.[2] Instead, we would like to outline the special significance the idea of autonomous order formation may have for the concrete psychological investigation of the functioning of cognitive systems. Arguing first that self-organization theory is a possible answer to a basic problem of psychology and second that this theoretical approach has a long, well-elaborated but nevertheless nearly forgotten tradition in *gestalt psychology*, we will point out some methodological consequences of the self-organization concept for psychological experimentation. In the last part of our contribution some already performed experiments will be demonstrated to illustrate the theoretical conclusions.

2. The Significance of the Tradition of Gestalt Theory

The contradiction between the experienced directness and the analytical complexity of mental phenomena is and has been a central challenge and a touchstone for theory construction in psychology. In consideration of cognitive processes, uncountable possible system states are opposed to the self-evident stability of emergent states of order in everyday life. The difficulties of the predominant scientific reductionism reveal problems in psychology more than in other research domains. On the one hand, there seems to be the methodological necessity to reduce the occurring complexities to the behavior of elementary units; on the other hand, the holistic view that new qualities of behavior are emerging with increasing system complexity has too much evidence to be generally

W. Krohn et al. (eds.), Selforganization. Portrait of a Scientific Revolution, 181–193.

ignored.[3] Psychology is confronted with the problem to parcel out complexity without losing its peculiarities, to *reduce without reductionism*.

At a first glance the attempts to maintain a certain level of complexity might be looked for in the field of social cognition and language-processing.[4] But this is certainly not the case. It is much more evident that in everyday life, and consequently in the history of psychology, the organization of perception especially in the visual system plays a major role in the preoccupation with cognitive phenomena.[5] Therefore we will develop our line of argumentation mainly using the facts and theories of this special field of research.

The *information processing approach*, which still prevails in wide parts of perceptual psychology, states that environmental data are analyzed by the activity of feature detectors and are depicted in a succession of ascending integrations. The structural similarity of this psychological model to the neurophysiological concept of a convergent hierarchy of stimulus processing seemed to bridge the gap between the elementarism of nervous excitation patterns and the holistic complexity of perception.[6] This expectation, however, has been by no means realized and there are strong arguments that this is for fundamental reasons.[7] The range of validity for models of hierarchic information processing has been limited to simple perceptual processes.[8] And such models lead inevitably to the infinite regress of a homunculus concept.

An alternative to the reductionist theory of perception is offered by Gibson's *ecological approach*.[9] In his perspective, the perceptual world is not built up via integrative processing or cognitive inferences but is taken over »directly« from the given environment in the sense of situative affordances. The contradiction mentioned above between the phenomenal directness and the analytical complexity of mental states is not solved but avoided: the experienced directness of perception and the process of perception are regarded as equal. As a consequence, the problem of reduction disappears. But on the other hand the attempt to reconcile physiological and psychological models is discarded. Despite the creative power Gibson's concept undoubtedly has, this renunciation is hard to accept in the light of the history of perceptual research. Furthermore, the approach does not satisfactorily explain the self-active status of perceptual processes as demonstrated, for example, by multistable phenomena such as reversible figures.[10] An adequate solution of the described paradox seems to require a theory which includes the functional autonomy of the cognitive system and at the same time avoids the dissection of the perceptual process.

In this situation, the tradition of gestalt theory might indicate a viable way of modelling. For gestalt theory, the origin of singular order out of the synergy of elementary multiplicity was the explicit starting point of research.[11] The concept of gestalt quality is defined as emerging from elementary interactions but not as completely reducible thereto.

Confronted with this immanent obstinacy, gestalt theory has identified a number of fundamental principles of order formation in the cognitive system which anticipate in detail some of the concepts developed in the theory of complex nonlinear systems.

Certainly the earliest formulation of a self-organization theory of perception within the framework of gestalt theory, and perhaps the most consistent overall, may be found in Wolfgang Köhler's works.[12] The idea that perception must necessarily be understood as a process of autonomous creation of order runs through all his work. Köhler made it his primary task to design and test a model of brain function in which the phenomenal organization of the perceptual world is not explained as stimulus-dependent but as the result of the perceptual system's organizing factors with the stimuli acting as boundary conditions.

The starting point of Köhler's considerations was the fact that biological systems tend to reach the same final state in different ways and starting from different initial conditions. This system-immanent equifinality was the reason for him to reject an elementary determinism as stated in the machine models of his time. But instead of postulating an immaterial organizing life factor as the vitalists did (see Bergson: »élan vital« or Driesch: »Entelechie»), Köhler referred to the existence of physical systems in which complex structures originate spontaneously from the system's own inner dynamics.

As a matter of fact Köhler was, in the development of his model, mainly oriented to the observations and the theoretical concepts of contemporary physics. He was fascinated by the fact that spontaneous creation of order, i.e. order arising without any external control, may even be shown in inanimate matter. In accordance with the assumptions of *linear thermodynamics*, the general systemic tendency towards final equilibrium was considered at Köhler's time to be the only basic principle of self-organization. Consequently, Köhler applied the model of physical fields, striving independently to a balance of forces, directly to the way the visual system functions. In 1925 Köhler wrote: »The somatic processes underlying static visual fields are stationary equilibrium distributions developed from the inner dynamics of the optical system itself«.[13]

Based on the phenomena to be explained, this idea proved to be extremely successful, though hardly tenable and almost provocatively contradicting the findings of neuroanatomy and physiology. On the basis of the field-theoretical model of perception, the brain is not seen as a complex network of many different interacting neurons but as a homogeneous conductor (in an extreme sense comparable to a container full of water).

If one analyses Köhlers statements more closely, it becomes evident that - contrary to the vehemence with which he defended his field theory against various attempts of falsification - his central aim was less the favored concretization than the aspect of *self-organization* behind it. Thus Köhler was fully aware of the fact that the general principle of development of linear thermodynamics was rather unsuitable for application to biological systems as long as it was oriented exclusively towards the final equilibrium. In 1955 for instance Köhler criticized the principle of final equilibrium as follows: »Although this is a perfectly good principle, it cannot, in its present formulation, be applied to the organism. For the organism is obviously not a closed system; moreover while the direction indicated by the principle may be called ›downward‹, the direction of events in healthy organisms is on the whole surely not ›downward‹ but, in a good sense, ›upward‹.«[14]

In view of such and similar formulations, it may easily be supposed that if the knowledge about the creation of order far from equilibrium obtained in the last two decades had been available to Köhler, he would surely have been led to a completely different, rather network-oriented development of his application of the principle of self-organization on perception.

Whatever today's evaluation of Köhler's psychophysical field theory of cognitive processes and brain functioning may be[15], the general emphasis of autonomous order formation makes gestalt theory an important precursor of the currently fast developing theory of self-organizing dynamic systems of Haken and others. Analogously to a fundamental assumption of self-organization theory, Köhler already distinguished clearly between a *microscopic* and a *macroscopic level* of system behavior. This opened the possibility to search for laws of complex order formation independently of the incalculable degrees of freedom and the stochastics of elementary state changes on the microscopic level. For the gestalt psychologists this gave reasons for the methodological primacy of analyzing the macroscopic level of system behavior from the *phenomenological* side. So they collected a great number of examples for the autonomy of cognitive systems and, moreover, also described general principles of phenomenal order formation.[16] Nevertheless, it was always regarded as necessary to ask for the neurophysiological basis of these phenomenological descriptions and principles as Wertheimer did in the first phenomenological-experimental work of gestalt theory.[17] But Köhler's controversial field-theoretical specification of such processes remained the only integrative attempt of gestalt theory to find a solution of Wertheimer's problem of psychophysical isomorphism.[18] Up to now, attempts to develop a neurophysiological explanation of complex processes of gestalt formation in perception beyond simple feature detection seems to prove unavailing because of the incompatibility of the analytical and the holistic system view.[19]

The theory of self-organization may be considered as a necessary methodological expansion for modelling inner-systemic interactions on the microscopic level, and gestalt theory may serve as a corresponding fund of descriptions of macroscopic behavior. Therefore a linkage between gestalt and self-organization theory may help to bridge the gap between neurophysiological and psychological aspects of cognition, it may help to solve the problem of reduction without reductionism.

3. Experimental Investigation of Self-Organizing Processes in the Cognitive System

For the investigation of the cognitive system as an informationally closed self-organizing system being determined by external stimuli only in the sense of energetic boundary conditions, there seem to be at least two methodological approaches:

1) The boundary stimulus conditions of the cognitive system are kept constant (or varied continuously in only one parameter), i.e. a stable stimulus pattern is exposed to the visual system. Observed are *fluctuations* and *multistable states, phase transitions* which show the self-organizational activity of the cognitive system in search of order.

2) The boundary stimulus conditions given to the cognitive system are either of high complexity or at random. These are exposed to the subjects for a short time. The immediate reproductions are set as new stimulus conditions. This is done in a continuous *recursion*. Observed is the process of *order formation*.

Multistability

A promising example for the development of a linkage between microscopic and macroscopic modelling may be seen in the perceptual phenomenon of multistability. Indeed there are already first attempts to simulate multistable perception on the basis of self-organizing neural networks.[20]

In the view of self-organization theory, systems are characterized by a continuous energy flow which causes continuous inner-systemic *fluctuations*. During these fluctuations a system may be brought into an entirely new state of order by means of self-amplifications *(autocatalysis)*. Usually the system leaves stability, passes through a chaotic phase of instability and reaches one of two or more possible stable states, and so on. Some stimulus patterns as in figure 1 are unstable enough to directly perceive the continuous fluctuations said to be the basic activity of perception. During fixation of this pattern circular rosettes are created in perception at various locations which are decomposed in the next moment.

Fig. 1. Stimulus pattern that produces continuous fluctuations in perception[21]

Hermann Haken, the founder of synergetics, considered perceptual multistability an outstanding example of phase transition and symmetry breaking and therefore of self-organization in cognitive systems.[22] In gestalt psychology multistable - or in case of only two alternatives - reversible figures have played a central role in the theoretic discourse from the beginning. Köhler wrote in 1940 outlining his neurophysiological field theory of perception: »Stationary visual percepts, a tree, a stone, or a book are as a rule extremely reticent as to the nature of the neural events which underlay their existence. We may hope to learn more about brain correlates if we turn to instances in which percept processes seem to be in a more active state. This is the case with ›reversible figures‹.«[23]

Reversible figures are usually demonstrated by two dimensional projections of three dimensional bodies, as shown in figure 2, which have to be fixated for some time. These phenomena are, however, by no means restricted to such situations. Three dimensional bodies such as wire cubes also tend to reverse when observed over a longer period and change their front-rear distribution in the process. Because of the optical projection conditions, this results in distortions and, in this case, the regularity of the cube no longer holds.

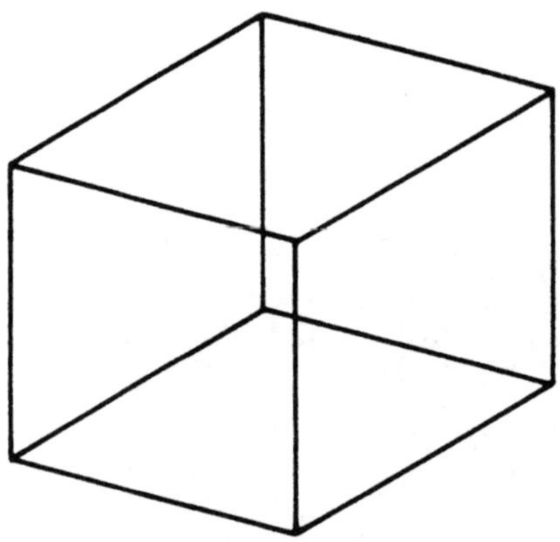

Fig. 2. Reversible figure: The Necker cube.

Even purely two dimensional patterns, such as the Maltese cross in figure 3, are perceived as bi-stable. This is principally the case in every two dimensional contour-pattern, that is, every contour may fulfil its limit function towards one side or the other. The gestalt laws of figure formation, however, make in each case one version so much more probable that continuous symmetry breaking is prevented in every day life.[24]

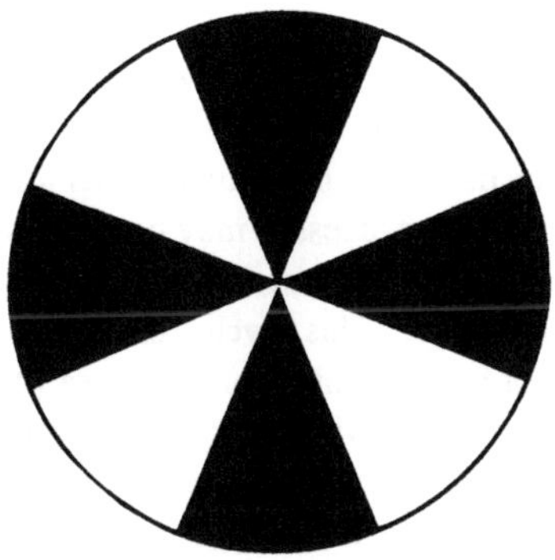

Fig. 3. Reversible figure: The Maltese cross.

Following self-organization theory one could assume that every percept is principally multistable. There are only differences in the time the system needs to decide and in the duration of the elicited alternative. So one may not be astonished to learn that symmetry breaking not only occurs with perceptual phenomena but also for example in actions. In the case of the types of conflict first described by the gestalt psychologist Kurt Lewin[25] - the appetence-appetence conflict, the aversion-aversion conflict and the appetence-aversion conflict - the acting person is confronted with two equally attractive or two equally aversive alternatives. The actor finds himself in a delicate balance between these two alternatives. And he is only able to resolve such a conflict by achieving a symmetry breaking through a creative act of decision. Very generally the theory of self-organization postulates that the behavior of things, just as the acting of human beings, is the more unpredictable the more equivalent the alternatives are. A ball that falls exactly on a razor's edge is subject to an unpredictable symmetry breaking. At the same time the principle of »small cause having great effect« applies. Just as the ball demonstrates a great effect with regard to preference of one side during the slightest bending of the razor's edge, a minimum valence of a goal alternative may decisively influence the direction of action in a conflict situation. In today's cognitive social psychology the so-called »balance theories« or »dissonance theories« which are based directly on Lewin's psychological field theory[26] are of great importance.

The Bartlett Scenario

We have seen in the last paragraph that multistability is an interesting way to demonstrate and analyze self-organization processes in the cognitive system. But there are two problems with this phenomenon making experimentation more difficult: first the process of iteration from one stable state to another happens very fast, nearly instantaneously, and second it is taking place inside the cognitive system and is so never accessible to an external observer. So it would be convenient to find a method to prolong the phase transition and to externalize the process at any given moment. Both criteria are fulfilled in a procedure which we would like to call the »Bartlett-scenario«.

Self-organization is critically bound to innersystemic fluctuations and the existence of autocatalysis. Autocatalytic or self amplifying processes may be understood as simple feedback loops in which the same operation is carried out repeatedly, the output for one operation being the input for the next one. This cycle of iterations has been demonstrated to be a powerful principle of self-organization. The use of iterations as a methodological tool for the demonstration and analysis of order formation is a relatively new idea in psychological research having only one precursor: Sir Frederic Bartlett, who worked with such a methodological principle in the thirties.[27] He had the idea to investigate slight systemic tendencies of remembrance by making the reproduction of a previous perceptual experience of one subject the starting point of the following reproduction of the next subject and so on. These serial reproductions which may also be taken from one and the same subject, may be regarded as data for the deeper analysis of the self-organization process in cognitive systems. Bartlett's methodology has been used for instance in the linguistic analysis of text understanding, acknowledging its sensitivity to symmetry breaking, that is, restructuring of the context, change of frame of reference, conflict resolution etc.[28]

4. Some Experimental Results

In the following we shall demonstrate some results from a recursive experimental procedure using visual material. An 8x8-matrix (a Reversi-game) filled with chips randomly (see upper left display in figure 4) was shown to a subject for 5 seconds then taken away and an empty Reversi-board was presented to the subject who was asked to attempt to reproduce the original layout of the chips as previously perceived. This reproduction was shown to the next subject for 5 seconds and so on. The experiment was broken off as soon as the same structure was reproduced consecutively three times. This structure was taken to be one of several possible attractor states of the initial stimulus condition.

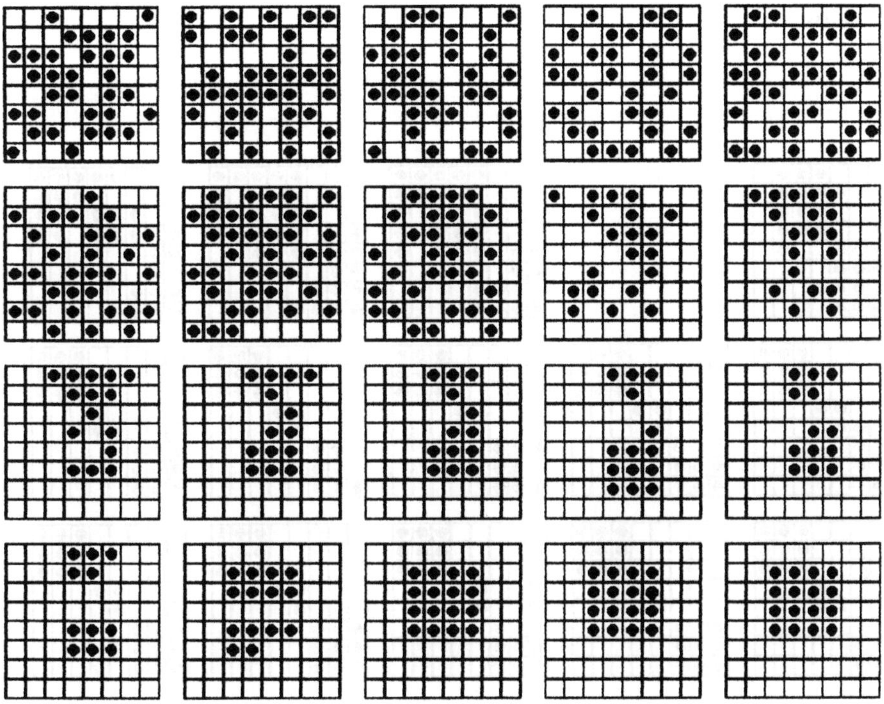

Fig. 4. Serial reproduction of a random dot matrix; see text.

Figures 4 and 5 show the results of two reproduction series (the figures have to be read line by line). In figure 4 we find random fluctuations in the reproductions 1 to 6. In the seventh reproduction (third display second line) we see the beginning of order formation: one column remains empty. In the next we see one column and one line free of chips and then two columns and one line and so on. The tenth reproduction (first display third line) already shows the tendency towards a closed figure. In the next reproductions we first observe a bifurcation of this figure into two groups which may be seen as a preparative act for the final square figure. In the fourteenth reproduction (right display third line) we see a phenomenon which would have been called »requiredness« by Wolfgang Köhler[29], for the upper group seems to fit the lower one. The final square structure is in the Manhattan-geometry of the underlying 8x8 matrice the most simple two dimensional figure possible. So we can conclude that the attractor for a random dot distribution in self-organizing cognitive systems is a simple figure - or as the gestalt psychologists would say, a figure of high salience (Prägnanz).

This result was confirmed in other series of reproduction with the same initial pattern - even if the final figure was not always the same and in different positions on the matrix.

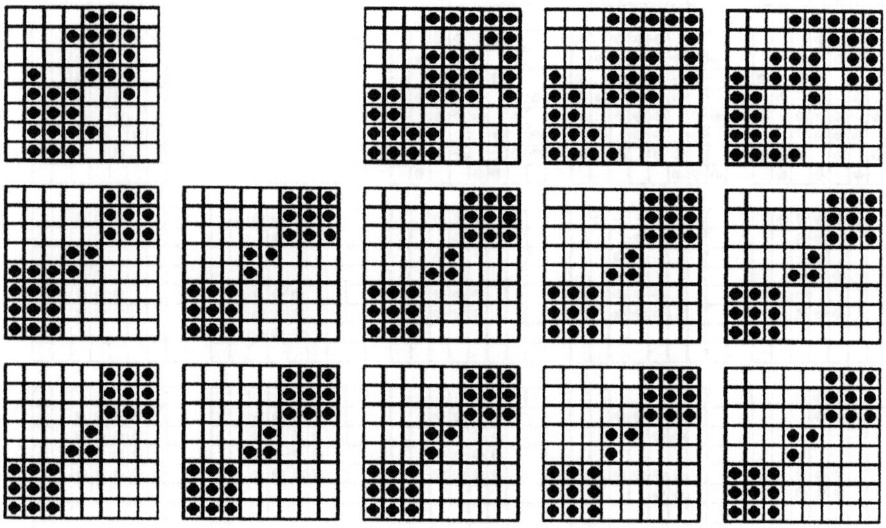

Fig. 5. Serial reproduction of a complex symmetric matrix; see text.

In the next experiment presented in figure 5 we used somewhat different initial conditions: the pattern (upper left display) had a complex symmetric structure. In the first reproduction this symmetry is already broken and it is followed by three more non-symmetric reproductions. Then another simpler symmetric structure is found which has the character of an attractor for it shows stability for the rest of the trials. Only the small triangle in the middle flaps around more than one time without disturbing the symmetry of the attractor state. This example shows that the way from a complex initial structure to the final attractor need not be characterized by a linear increase of order. On the contrary, chaotic phases may be passed through during the change from one ordered state to another.

5. Summary

In this article some fundamental arguments are presented for the investigation of cognitive systems as self-organizing systems. In the theoretical part the significance of gestalt tradition for this approach is emphasized. Gestalt psychologists like Wolfgang Köhler have demonstrated the importance of the autonomous creation of order in perceptual systems in their theoretical and experimental work. But Köhler already expressed his discontent with the only principle of self-organization known in his time: the tendency towards the final thermodynamic equilibrium did not hold for the increase of order in the development of biological systems.

In the second part of the paper, experimental evidence for self-organizational processes in cognitive systems is presented. Two kinds of methods are proposed: the investigation of multistability by keeping constant the boundary stimulus conditions of the cognitive system and the so-called Bartlett-scenario which is based on a recursive re-input of stimulus patterns. Experimental results show a general tendency towards simple and symmetrical structures in a nonlinear process of order formation.

Notes and references

1. See M. Eigen, »The Hypercycle«, *Die Naturwissenschaften* **64** (1977), 541-565; 65 (1978), 1-41, 341-369; H. von Foerster, »On Self-Organizing Systems and their Environment«, in M.C. Yovits and S. Cameron (eds.), *Self-Organizing Systems*, New York: Pergamon Press, 1960; E. von Glasersfeld, *Wissen, Sprache und Wirklichkeit*, Braunschweig: Vieweg, 1987; H. Haken, *Synergetics*, New York, Berlin: Springer, 1977; H.R. Maturana, *Biology of Cognition*, Urbana: University of Illinois, 1970; I. Prigogine, *From Being to Becoming*, San Francisco: Freeman, 1980.

2. For the epistemological discussion see M. Stadler, P. Kruse, »Gestalttheorie und Theorie der Selbstorganisation«, *Gestalt Theory* **8** (1986), 75-98 and P. Kruse, M. Stadler, »Radikaler Konstruktivismus - psychologische Überlegungen zu einem philosophischen Zweifel«, in M. Amelang (ed.), *Bericht über den 35. Kongress der Deutschen Gesellschaft für Psychologie*, Vol. 2, Göttingen: Hogrefe, 1987.

3. See M. Bunge, R. Ardila, *Philosophy of Psychology*, New York, Berlin: Springer, 1987.

4. See for instance K. Lewin, *Field Theory in Social Science*, New York: Harper & Brothers, 1951 and M. Stadler, W. Wildgen, »Ordnungsbildung beim Verstehen und bei der Reproduktion von Texten«, *Siegener Periodicum zur Internationalen Empirischen Literaturwissenschaft* **6** (1987), 101-144.

5. Actually most theoretical battles in psychology were fought in the field of perception: The Hering-Helmholtz controversy in the late nineteenth century as well as the struggle between behaviorism and gestalt psychology in the first half of this century; also in the last twenty years theoretical revolutions of cognitivism, ecological psychology and critical psychology were preceded by books on perception (see U. Neisser, *Cognitive Psychology*, New York: Appleton-Century-Crofts 1967; J.J. Gibson, *The Ecological Approach to Visual Perception*, Boston: Houghton Mifflin, 1979; K. Holzkamp, *Sinnliche Erkenntnis*, Frankfurt am Main: Fischer-Athenäum, 1973).

6. S.M. Anstis, »What does Visual Perception Tell us about Visual Coding?«, in M.S. Gazzinaga, C. Blakemore (eds.), *Handbook of Psychobiology*, New York: Academic Press, 1975; H.B. Barlow, »The Twelfth Bartlett Memorial Lecture: The Role of Single Neurons in the Psychology of Perception«, *Quarterly Journal of Experimental Psychology* **37A** (1985), 121-145.

7. O.D. Creutzfeldt, *Cortex Cerebri*, Berlin: Springer, 1983.

8. K. Nakayama, »Biological Image Motion Processing: a Review«, *Vision Research* **25** (1985), 625-660.

9. Gibson, *op. cit.*, 1979, note 5.

10. See P. Kruse, »Stabilität - Instabilität - Multistabilität, Selbstorganisation und Selbstreferentialität in kognitiven Systemen«, *Delfin* **6/3** (1988), 35-57 and R.N. Shepard, »Ecological Constraints on Internal Representations: Resonant Kinematics of Perceiving, Imaging, Thinking and Dreaming«, *Psychological Review* **91** (1984), 417-447.

11. M. Wertheimer, »Untersuchungen zur Lehre von der Gestalt« II, *Psychologische Forschung* **3** (1923), 301-350.

12. W. Köhler, *Die physischen Gestalten in Ruhe und im stationären Zustand*, Braunschweig: Vieweg, 1920 and »Gestaltprobleme und Anfänge einer Gestalttheorie«, *Jahresberichte für die gesamte Physiologie und experimentelle Pharmakologie* **3** (1925), 512-539.

13. Köhler, *op. cit.*, 1925, note 12. From the english translation »Some Gestalt Problems«, in W.D. Ellis, *A Source Book of Gestalt Psychology*, London: Routledge and Kegan Paul, 1938, p. 55-70.

14. From »Direction of Processes in Living Systems«, *Scientific Monthly* **80** (1955), 29-32.

15. See note 1 and Stadler, Kruse, *op.cit.*, 1986, note 2.

16. See Wertheimer, *op. cit.*, 1923, note 11 and W. Metzger, *Gesetze des Sehens*, 3rd edition, Frankfurt am Main: Kramer, 1975.

17. M. Wertheimer, »Experimentelle Studien über das Sehen von Bewegung«, *Zeitschrift für Psychologie* **61** (1912), 161-265.

18. W. Köhler, *Dynamics in Psychology*, New York: Liveright, 1940.

19. In the seventies K.H. Pribram propagated his holonomic theory of perception based on Fourier coding of cerebral oscillations as a third way between specific neural interaction and the global field-theoretic view (*Languages of the Brain*, Englewood Cliffs: Prentice-Hall, 1971); see also M. Stadler, »Feldtheorie heute - von Wolfgang Köhler zu Karl Pribram«, *Gestalt Theory* **3** (1981), 185-199.

20. See P.K. Kienker, T.J. Sejnowski, G.E. Hinton, L.E. Schumacher, »Separating Figure from Ground with a Parallel Network«, *Perception* **15** (1986), 197-216 and T. Poston, I. Stewart, »Nonlinear Modelling of Multistable Perception«, *Behavioral Science* **23** (1978), 318-334.

21. From M. Stadler, F. Seeger, A. Raeithel, *Psychologie der Wahrnehmung*, Munich: Juventa 1977.

22. Haken, *op. cit.*, 1977, note 1 and *Erfolgsgeheimnisse der Natur*, Frankfurt am Main: Ullstein 1984.

23. Köhler, *op. cit.*, 1940, note 17, p. 67f.

24. Bi-stability is utilized systematically in anagrams, i.e. puzzle pictures in which the background, on closer observation, may obtain the function of a figure. Artists like Salvador Dali or Maurits C. Escher are famous for such pictures.

25. K. Lewin, *A Dynamic Theory of Personality*, New York: McGraw-Hill, 1935.

26. See M.E. Shaw, P.R. Costanzo, *Theories of Social Psychology*, New York: McGraw-Hill, 1970.

27. F.C. Bartlett, *Remembering*, Cambridge: University Press, 1932.

28. M. Stadler, W. Wildgen, *op. cit.*, 1987, note 4.

29. W. Köhler, *The Place of Value in World of Facts*, New York: Liveright, 1938.

IV. The Selforganization of Science

Self-Organization and Autopoiesis in the Development of Modern Science[1]

RUDOLF STICHWEH
Max-Planck-Institut für Gesellschaftsforschung

I

The following considerations try to analyze the interrelationships between the autonomy and self-referentiality of modern science as a social system. Such interrelationships are obvious. Even if one only considers the literal sense of the Greek word autonomy one will realize that it says that something - an entity or a system - is regulated by rules that are generated by itself.

Recent new paradigms such as »*self-organization*« and »*autopoiesis*« - are taken here as conceptual devices that allow a specification of the classical notion of autonomy. The new theories underlying them will not primarily interest us here as conceptual structures in themselves. Instead I shall try to present an analysis of the decisive *developmental steps of the social system of science between the 16th and the 20th century* in terms of these two paradigms. The self-organization and autopoiesis of science are described as two successive phases in this developmental process. This presupposes that »self-organization« and »autopoiesis« *differ* with regard to the criteria of autonomy they imply. Furthermore, the criteria of autonomy are not only different but also sequentially ordered. This means that science first had to be a self-organizing system before it could become an autopoietic system.

II

I shall start with an elementary definition of self-organization. Theories of self-organization - as they are found in the writings of Heinz von Foerster or Henri Atlan among others[2] - refer to systems in their environments. These systems constitute an internal order and structure in the system by disorganizing or destructuring the environment relative to the system. This implies that systems are built against a surrounding disorder of much higher probability. It is obvious why such a process can only be a process of self-organization. No system and no observer external to the self-organizing system possesses a superior knowledge of the organizational needs and opportunities in the developing system. Only the system itself can select the next step or item in its ongoing process of organization. Each progressive state of the (self-)organizing system is therefore the most complete »description« of the system available up to now. Undoubtedly, there are systems that can - and perhaps must - be described before being organized. They are manufactured by teleology - i.e., made according to a plan that includes an anticipated final state of the system - and it is obvious that systems manufactured by teleology are non-self-organizing systems.

Before entering into a discussion of science as a social system, I would like to mention the interrelations of self-organization and two different but related scientific

W. Krohn et al. (eds.), Selforganization. Portrait of a Scientific Revolution, 195–207.

theories. First, there is an obvious interrelation of *self-organization* and *evolution*. Self-organization as a theory describes demands and processes that in present-day theories of evolution are called *structural* or *internal selection*.[3] This can serve as an indication that even the scientific theory which has most convincingly demonstrated the shaping of systems by external »forces« must now take into account a certain prevalence of internal structural demands. This is not obliterated by the fact that in certain cases when studying extremely long periods one may be allowed to disregard structural selection or self-organization, since evolution has had enough time to shape the self-organization of a system by an infinity of steps and phases of selection via external forces. It may be supposed that an evolutionary theory that primarily studies the evolution of different types of self-organization and of different types of autopoiesis would be of considerable scientific interest.

A second remark refers more specifically to the history and systematics of *sociological theory*. If self-organization is based on the *duality of order/disorder*, one may note that the *sociological theory of differentiation* (theory of structural differentiation) - being perhaps the dominant sociological theory of the formation of systems - is concerned with a *relation of order to order*, and is primarily interested in the reciprocal elaboration and differentiation of types of ordering reality. *Articulating order* against the menace of disorder or *individualizing order* in comparing or relating it to other existent possibilities of ordering reality seem to be two rather distinct activities and imperatives. Any theory of system formation and of system/environment-relations should be general enough to include these two types of organizational imperatives.

III

In what sense has science been a self-organizing system and what does this mean for the autopoiesis of science as the next step that occurred in the evolution of the social system of science? In the first part of the following exposition, I shall argue that the history of European science from the 16th to the second half of the 18th century is most plausibly interpreted *as a process of self-organization of science*. From then on and in the 19th century a *transformation from self-organization to autopoiesis* takes place.

Early modern science or learning is an enterprise that primarily deals with *received knowledge*. »Received knowledge« means that there are huge quantities of comparatively new knowledge not produced by early modern science and learning itself. This received knowledge comes from a plurality of sources: an ever growing corpus of ancient scientific, rhetorical, and literary texts and papyri; information from extra-European scientific traditions (the Islamic world, China); the appropriation and production of knowledge in manifold contexts such as the mechanical arts, mining, and the descriptions of travellers. From the perspective of institutionalized learning - or the perspective of the »republic of letters« as a social system - this knowledge, in most of its types, is fortuitous knowledge, since learning has neither controlled its production nor the accidents of its advent.

Under these specific circumstances, early modern science is primarily a practice of *organizing received knowledge*. The urgency of the problem of organizing an incoherent diversity of knowledge may be seen in a passage by Leibniz found in one of his projects for an encyclopedia written for Louis XIV. Leibniz thinks that contemporary knowledge is »une grande boutique ou magazin ou comptoir sans ordre et sans inventaire: car nous ne savons pas nous en servir au besoin ... *nous sommes pauvres au milieu de l'abondance*«.[4]

This knowledge without internal organization is not only useless in the sense of not being readily available for intellectual operations, since these cannot always search the whole stock of knowledge. There exists an even greater risk of available knowledge getting lost again. In early modern texts on learning, on power and the state, on war, on population and other forms of desirable plentitude, there is an almost universal apprehension of not being able to maintain existing states and patterns, and of decline and decay as the probable fate of all things existing in time.

Conservation and *organization* constitute therefore the two main problems of early modern learning and science. Additionally or more precisely, one may identify at least three central values or ideals of early modern learning that are operative in achieving whatever conservation and organization is achieved: *imitation* functions as the social ideal for learned knowledge, *preservation* as its temporal ideal, and *order* as the material value for knowledge.[5]

It is obvious that at least two of these ideals are intimately connected to *language* and *text* as the primary media of early modern learning. Imitation normally means the imitation of classical styles and classical authors. In this respect, it is remarkable that imitation as an intellectual exercise in schools and universities presupposes a certain shift from oral exercises - primarily disputations - and the adversarial culture of medieval schools and universities to classroom work with texts.[6] This may be seen as a shift from local knowledge, which can only be transferred through the spatial mobility of persons, to a more impersonal, less interactional view of knowledge, which presupposes elements or entities of knowledge - texts - that can move easily over temporal and spatial distances. The imitation of classical styles in Latin and sometimes Greek brings with it at the same time a new understanding of language. Latin is no longer the colloquial language of clerics that it was in the European Middle Ages. It becomes an artificial and resurrected classical language system, which means that it is a rather difficult foreign language, that has to be studied for a long number of years.[7]

But there are wider societal ramifications of imitation as a social ideal. In the society of estates (»ständische Gesellschaft«) of early modern Europe, imitation as a behavioral ideal is a means of maintaining the identity and separateness of the estates. Imitation that transgresses the boundaries of estates should therefore be sanctioned. Christian Thomasius, in a text published in 1729, conjures this up in rather harsh words: »Niedertracht« (baseness) is the word for imitation of a lower estate, and »Hoffart« (haughtiness) is the inverse »Unanständigkeit« (indecency) if you should desire to imitate a higher estate.[8] It would be interesting to compare the benefits of textual imitation with its inherent risks. While socializing certain essential social competences, textual imitation offers the student the opportunity to imitate any style whatsoever. The debate on Jesuit theater in »collèges« is an interesting case in point. There was the Jesuit position that was obviously motivated by an interest in the flexibility of role-playing, and there was what might be called the Augustinian-Jansenist position that says that in dramatizing vice the scholars may actually become evil.[9]

Imitation as a social ideal of conserving knowledge has its temporal equivalent in *preservation*. Preservation is the activity of organizing received texts by using philological methods and of collecting diverse bits of knowledge in texts and books. It can be interpreted as being already more inclusive and more scientific than pure imitation. Imitation is restricted to a few select passages from a small number of authors. But his command of the three - sometimes more - classical languages enables the early modern professional scholar to read any text, or appropriate any particle of knowledge humanity has ever known and written down. There exists an obvious correlation between the learning of languages as the prime activity of early modern schools - and occasionally universities - and the preservation of knowledge as an essential ideal of early modern learning as a scholarly activity.

Ordering knowledge as the material or substantial ideal is the most extensive activity of early modern learning. Ordering or organizing knowledge neutralizes both socially coded differences between levels of style and temporal distances between authors. Every bit of knowledge becomes contemporaneous in an order of knowledge. In this respect, it is significant that the early modern notion of *history* does not make a conceptual distinction between the spatial distribution of the facts of natural history and the temporal distribution of the events of human history.[10] However, the essential effect of the ordering activity of premodern science and learning results in the *discovery of structures* and in the *perception of connections* that were not at all obvious in the received traditions. To me, this achievement seems to be the most important *self-organizing property* of early modern science. Whereas elements of knowledge could intrude upon science from anywhere, there was no external place or institution that could generate any idea, power, or competence in order to control the immense work of ordering knowledge as it was achieved in the European encyclopedias and classifications of science from the 16th to the 18th century.

Between ordering and preserving, manifold points of transition exist. Even the formal prerequisite of early modern learning, translating every bit of knowledge into Latin as the language of learning, implies systematization and an adaptation of knowledge to the conceptual structures that were available and had been developed in learned Latin prose. Regarding the philological tradition of transforming unstructured papyri into texts structured by paragraph and page[11], of editing these texts, and commenting upon texts, it may be argued that these activities never have been simple preservations of received stocks of knowledge only, but have always included the construction of a new order of the text. This constructive component is intensified by criticism as the explicit method of philology. Printing is another case in point. In the first instance, it means only a new reliability of preservation and a new level of accessibility of texts in their printed form[12]. But the three formal universals of *printing*, *Latin*, and - whenever necessary - *philologi-*

cal criticism of texts imply such a massive presence of all knowledge ever known that this combination can be considered the most important stimulus for the activities of ordering knowledge. The formal organization in which these universals of early modern learning come together is the *library*. It is remarkable that even in present-day empirical research on universities, the quantity of library holdings relative to the magnitude of personnel is one of the best predictors of the scientific quality of a university.[13] Libraries in early modern science are - if I may misappropriate the word - macropaedias of those progressively extensive encyclopedias that are the final results of the ordering activity of early modern learning. A rather close conceptual and factual affinity of »bibliography« and »encyclopedia« was established in Europe in early modern times.[14]

In finishing this first part of my argument I would like to introduce an additional analytical concept from systems theory. Niklas Luhmann distinguishes three system references for any social system.[15] Luhmann calls them *function*, *performance*, and *reflection*, and, in the case of the social system of science, this is thought to mean: 1. The production of truth as the *function* of science in *society* (society being the most extensive social system integrating all communication); 2. the application of scientific knowledge in nonscientific social contexts as the *performance* of science for specific *other functional systems*; and 3. the *self-reflection* of science on its unity as a system. When these distinctions are applied to our description of early modern science and learning as a self-organizing system, it becomes evident that the self-organization of science as a process gravitates towards a certain *primacy of self-reflection*. Science ensures its nonexistent unity as a system by *presupposing unity*, i.e., by making use of *order* and *connection* as the guiding premises of all its operations. Under these circumstances, the production of new truths and performances (or services) for other systems had to be deferred for the time being.

IV

Self-organization in early modern science is not only an ordering activity. Two aspects of self-organization can be distinguished that seem to me to be of general import and to be relevant for the transformation from self-organization to autopoiesis. If self-organization is described as a relation of *externally produced elements* and *internally generated structures*, two ideal types of self-organizing systems can still be distinguished. At one extreme there is a self-organizing system that accepts any intruding element and concentrates all its internal energies on the formation of structures of highly integrative capacities. In many respects, early modern science is near this pole of self-organization. At the other extreme there are self-organizing systems with little internal structure but which have a maximum capacity for distinguishing between the external elements they internalize and other intruding elements that are rejected at the boundaries of the system. The paradigm for this type of self-organizing system is the thermodynamic example that presupposes two containers with particles of different speed and a Maxwellian demon sitting at the opening between the two containers. Of course, in the real world the system needs a great deal of energy to feed its demon, and this energy may not be available for the additional formation of internal structures. The homogeneity of elements has to function as a substitute for structure. In another theoretical language, the two systems might be comparatively described as consisting of a heterogeneity of elements with *strong couplings* between them versus a system with relatively homogeneous elements that are *loosely coupled* among themselves.

It seems to be attractive to rewrite the history of early modern science as being characterized by a slow displacement from the one type of self-organizing system to the other. Perhaps the most important indicator for this shift is the development of *skepticism* and the higher prominence of what might be called the *critical spirit*, especially in philology. This implies a certain *differentiation of philology and encyclopedia*, which takes place in the 17th and 18th century. Philology discontinued the practice of transforming any text by accumulating annotations into a nonalphabetical encyclopedia.[16] Encyclopedia on the other hand relativized the primacy of the author as the ordering principle for knowledge. In the process of systematization, knowledge also gained a certain distance from the normative primacy of Antiquity as a source of knowledge.[17] Moreover, one can observe in the 17th and 18th century the separation between science and certain intellectual traditions that had initially been an integral part in the reception of knowledge: alchemy, astrology, hermetism, and physicotheology. Furthermore, a growing distance begins to emerge between »history« and »philosophy« in the early modern sense of these two terms: history being the knowledge you know by the testimony of another person, and philosophy being the knowledge of something that you know by the knowledge of its causes.[18] Even in the arguments that explicitly give preference to *preservation* over *novelty*, a shift occurs from *tradition* to *examination* as the critical standard. In William Temple at the end of the 17th century you still find the argument based on tradition and consensus: »... taking Knowledge to be properly meant of Things that are generally agreed to be true by Consent of those that first found them out, or have been since instructed in them«.[19] In a German text from the end of the 18th century by Johann Georg Schlosser in which he explicitly polemicizes against *curiosity* (»Neugierde«) as the search for the new since he prefers *thirst for knowledge* (»Wißbegierde«), the argument advanced for old knowledge is that it has probably survived a number of tests.[20] The establishment of a critical and selective awareness in contradistinction to the earlier preference for structure formation and ordering can be observed in these and similar instances.

In the text by Schlosser just cited an additional argument is contained to the effect that older knowledge may have the advantage of being *applicable*. The obsession of the 18th century with *usefulness* of knowledge becomes a further motive for selecting which knowledge is to be received. Only one passage from the article on »Polyhistorie« in Zedler's encyclopedia from 1741 will be cited here for illustrative purposes. Even in this alphabetical encyclopedia one no longer finds an appreciation of »polyhistoria« as the individual activity of ordering knowledge. The objection is that polyhistorians are dispersing their forces into too many activities, and therefore *deprive themselves of being able to serve other men*.[21] While there may be a pietist undercurrent in this argument and while it is possible that polyhistoria is associated with human pride, the essential point is that

»service« and »usefulness« support a critical-selective spirit in science. It can be argued that the shift from the formation of structure to the selection of elements in this phase of self-organizing science went parallel with a shift from self-reflection to performance in the relative prominence of the three system references of science.

The question might be asked if the value attached to *»usefulness«* is compatible with the concept of self-organization of science. However,it must be seen that in the 18th century »usefulness« is by no means a perspective external to science. It is willingly adopted after preservation and ordering of knowledge are no longer seen as being seriously endangered. »Usefulness« offers a kind of universalist perspective of the world that prepares the ground for new types of learning. In the 1780s Joseph von Sonnenfels, for instance, could claim that for the new *cameral sciences* »usefulness« as a universalistic perspective is a functional equivalent to »justice« as the basic theme of classical legal learning.[22] Perhaps the best institutional embodiment of the compatibility of usefulness and progressive self-organization of science was the University of Göttingen, where Schlözer could firmly state that *»being very learned and working for the common weal are pure synonyma.«*[23]

V

Up to now, the first type of the autonomy of science as a social system based on self-organization has been discussed. At times, stable intermediary equilibria in this process of self-organization occur. Without doubt early modern European Latin learning is an impressive symbol for the autonomy of science.[24] But, as we have already seen, the self-organization of science is an ongoing process that not only consolidates progressively more inclusive structures of learning but is equally characterized by an internal shift of primacy to the selection of elements as a mode of self-organization. This shift of primacy prepares and initiates a second type of autonomy of science that I would like to subsume under the notion of *autopoiesis*. The transformation to autopoiesis rests partly on the successes of early modern science in ordering and systematizing knowledge and reflecting upon knowledge. In other respects, this transformation has to be described as a discontinuity. Before analyzing the development of science at this stage in greater detail I shall bring to mind the notion of autopoiesis.

The common ground covered by the concepts of self-organization and autopoiesis seems to me to lie in the interest both theories attach to the capability of systems to *neutralize causalities that are external to the system*. These external causalities are, undoubtedly, somehow involved in all processes occurring in the system. Any system is susceptible to manifold influences. External influences, however, will never be able to rule out the self-referential constitution of the system, which means that only internal rules of connectivity control the way in which the system will be influenced in its operations by external causalities and imports.

The way in which external processes or elements are handled distinguishes self-organization from autopoiesis. We have already seen that in principle a self-organizing system can be built from any material, although in the process of system building there is an evolution of selective capacities that favor some imports and categorically exclude other materials. Which form of autonomy does such a self-organizing system have? The development of its autonomy has already been described as a process of integrating diverse elements that can only be controlled by the self-referential informational capacities of the system. The criterion of autonomy in a theory of autopoiesis is more demanding in certain respects. It implies such a high degree of homogeneity of the elements of the system that this criterion can only be met *if the system itself produces the elements by which it is constituted*. This means that no element and no operation can be introduced into the system through external sources. The system may be stimulated or disturbed by any event in its environment, but under no circumstances are these disturbances internalized as *elements* in the ongoing processes of the system. Disturbances cannot do more than to stimulate the synthesis of a new element in the system itself that will always be produced according to the rules of the system.

This organizational form of autopoietic systems is normally called *operational closure*. The dynamic process of the system will always be structured in such a way that externally induced changes of states of the system will have to be strictly connected to its earlier states. When applying these notions to *social systems*, some additional properties are needed. According to Niklas Luhmann, the most important point is that the *elements* of social systems are *events* that can be localized at a precise point in time[25]. Events disappear in the same moment in which they occur. It is obvious that this event-structure of social systems induces a dynamic of changes of states that hardly can be controlled by normative structures of social systems. A second property of social systems may be seen by specifying the notion of event. Events in social systems are *communications* or *social actions*, which means that only those entities in the world that can be adapted to the *formal constraints of communication and action as elements* may be elementary operations in a social system.

VI

How does science become an autopoietic system? The argument up to this point has emphasized two steps: the formation of structures in a system that accepts elements of knowledge from a great diversity of sources, and - based upon the previous successes in the activity of ordering and perhaps after perceiving redundancies - the genesis of a system of increased selectivity thanks to a self-constructed Maxwellian demon who distinguishes between acceptable and unacceptable imports. The novel aspect of an autopoietic system of science seems to be that the system starts to produce all its elements itself and, after having done this for some time, changes its identity: *the self-production of new elements seems to become the essence of science*. In Luhmann's language of function, reflection, and performance, this implies once again a change of primacy. In the reflection on the unity of the system or performance, values such as »usefulness« are no longer of prime scientific relevance. Instead, the functional view, which prefers the production of truths and the addition of novelty, establishes its rhetorical and factual supremacy.[26]

The first step in this transformation of science has an intuitive economical plausibility. The more time and energy your demon must spend to find the right elements flying along, the more economical it becomes for the system to produce necessary additions to its stock of elements itself. But there is an obvious inherent difficulty. If it is to take the risk of self-production of its elements, science must no longer fear variation and it must no longer fear self-referentiality. A system of science that *receives* its elements of knowledge is not without variation, but it can always externalize the responsibility for variation. If you make your own observations with your own self-constructed scientific instrument and if you describe these observations with your self-generated theoretical notions, you can no longer escape the closed circularity of scientific cognition. You will have to institutionalize scientific methods that will at least allow you to invent internal differentiations in cognitive procedures which simulate external controls.[27]

But before being able to understand self-referentiality, science has to accept *variation*. This is in many respects prepared by the phase of self-organization, since ordering knowledge means transforming an intransparent corpus of knowledge into a structured whole, which itself, in turn, begins to posit new problems that can no longer be solved by integrating supplementary external traditions. We have pointed to the interrelation of increased selectivity and the self-production of elements. These developments are accompanied by a *secular shift in early modern theories of knowledge from preservation and order to variation and innovation.* One can observe this, for example, in the debate on the relative supremacy of the Ancients over the Moderns, and among the arguments for the Moderns one may find references to the bizarre audacity of French Gothic architecture,[28] but also references to the extraordinary achievements of men like Galileo or Newton as an element of proof.[29] Whereas in the 16th century, theology could still be regarded as a privileged subject in its possibilities of progress for thought - and this in contradistinction to such subjects as logic, physics, or medicine[30] - , in the 18th century the natural sciences are established as the paradigm of the possibility of »new inventions«, and other fields are exhorted to imitate this progress.[31]

No attempt at an exhaustive historical analysis of this transformation can be made here.[32] Let me briefly mention three additional aspects. 1. The rather slow transfer from Latin to the national languages. In principle, this transfer carries the risk of variation, since one cannot simply translate from Latin but has to invent new words and concepts. Huge quantities of established knowledge get lost in this switch of language, because they will be excluded from translation. This circumstance is an additional proof for the radical change in the preferences of science. Preservation of knowledge begins to be of no crucial interest anymore, since there will be too much new knowledge anyway. The remarkable disjunction of Latin and vernacular learning may be illustrated by the physics course taught in French »collèges« of the 18th century. At many collèges, the course in theoretical physics was given in Latin up to the end of the Ancien Régime. The separate experimental demonstrations, which were given by itinerant lecturers at the beginning of the century, were later performed by the physics professor in separate sessions open to the general public. This latter course was now offered in the vernacular.[33] Another implication of the transfer of learned discourse to the national languages is that for persons involved in learning and science Latin could be experienced as a foreign and restrictive language system. In the educational literature of the 18th and early 19th century, one can find explicit discussions of the fear that the creative liberty of intellectual expression might become endangered by the exclusive use of Latin.[34] 2. A second aspect already referred to is the extensive legitimation of empirical and experimental cognition in the 18th century. In contradistinction to rational, philosophical (i.e., knowledge of causes), or logical truth, the status of empirical cognition had always been insecure in the European tradition of thinking about science. Kant was the first to consider *systematic unity* to be the distinguishing criterion of *science*, and that means that the *definition* of science no longer discriminates between rational and empirical truth.[35] The acceptance of empirical and experimental knowledge means at the same time the acceptance of the self-production of the elements of knowledge and a confrontation with self-reference. The observation of »undisguised« nature was only briefly the paradigm of the empirical search for knowledge. Synthetic substances, the institutionalization of laboratories, and the dependence on self-constructed instruments confronts science with the reality of self-referential circularity. Science has to accept artificiality and correspondingly delegitimizes knowledge in natural situations.[36] 3. Apart from this differentiation of knowledge production, there is a third aspect. The 18th century is probably the last period in which science is exposed to a massive influx of external knowledge and experience.[37] This results from the economical, professional-technical, and territorial expansion of European society. The *unexpected growth of knowledge* that goes with it discriminates against old knowledge and suggests that one can do without it. It brings pressure to bear on science to process it. In the 19th century, Alexander von Humboldt is perhaps the last representative of a view of science that aims at collecting a diversity of geophysical, geological, meteorological, and ethnological-historical knowledge worldwide.[38] But in all fields of knowledge, the alternative view that the specialities and disciplines of science should exclusively work with self-produced knowledge asserts itself. This shift to an exclusive reliance on self-produced elements of knowledge seems to me to be the point of transition to science as an autopoietic system.

For this transition, two further changes are of decisive relevance that I shall only name here without analyzing them in any detail. They concern *subsystems* and *elements* of the system of science. For the system/environment difference of science, it is decisive that it is repeated continuously as an internal system/environment difference in science by the means of *disciplinary differentiation*. Any scientific process - in whatever discipline or subdiscipline - is now situated in an internal scientific environment constituted of other systems of disciplinary communication, which means that the possibilities of internal self-stimulation of science grow indefinitely.[39] In analyzing the *elements* of the social system of science, we can observe a specific type of operational closure of the system that rests on constituting and differentiating two complementary types of elementary events. There is the system level of research acts, which in many respects may be described as the microlevel of science. Doing research is the way of doing one's daily work in science. This level of research acts has operated since the early 19th century under the imperious demand of producing novelty. Apart from it, a second and macrolevel in science exists that processes elementary communicative acts.[40] In processes of scientific communication that make use of publications as their elementary and universalist form, truth claims and research descriptions are transmitted and worked upon. Any scientific idea, discovery,or truth claim whatsoever must be transformed into this form of communication by publications which have become the universal medium of communication for science.

Notes and References

1. I am indebted to Wolfgang Krohn and Helga Nowotny for detailed comments on two earlier drafts of this paper.

2. See H. v. Foerster, *Observing Systems*, Seaside, California: Intersystems, ^{2}1984; »Disorder/ Order: Discovery or Invention« in P. Livingston (ed.), *Disorder and Order*, Stanford Literature Studies 1, Saratoga, California: Anma Libri, 1984; »Entdecken oder Erfinden. Wie läßt sich Verstehen verstehen?«, in *Einführung in den Konstruktivismus*, Munich: Oldenbourg, 1985; »Erkenntnistheorien und Selbstorganisation«, in S.J. Schmidt (ed.) *Der Diskurs des radikalen Konstruktivismus*, Frankfurt am Main: Suhrkamp, 1987; H. Atlan, *Entre le cristal et la fumée. Essai sur l'organisation du vivant*, Paris: Seuil, 1979.

3. For these concepts see D.T. Campbell, »Neurological Embodiments of Belief and the Gaps in the Fit of Phenomena to Noumena« in A. Shimony and D. Nails (eds.), *Naturalistic Epistemology*, BSPS Vol. 100, Dordrecht: Reidel, 1987; »Selection Theory and the Sociology of Scientific Validity« in W. Callebaut and R. Pinxten (eds.), *Evolutionary Epistemology. A Multiparadigm Program*, Dordrecht: Reidel, 1987.

4. Cited by L. Couturat, *La logique de Leibniz*, Paris, 1901. Reprint Hildesheim 1961, on p. 149 (emphasis mine).

5. This classification makes use of Luhmann's analysis of the three dimensions of meaning - see N. Luhmann, »Sinn als Grundbegriff der Soziologie« in J. Habermas and N. Luhmann, *Theorie der Gesellschaft oder Sozialtechnologie. Was leistet die Systemforschung?*, Frankfurt am Main: Suhrkamp, 1971.

6. See the interesting description of this shift in P. Ramus, »Advertissements sur la reformation de l'Université de Paris«, 1562, in L. Cimber and F. Danjou (eds.) *Archives curieuses de l'histoire de France*, Première Série, Vol. 5, Paris 1835, p. 118-119. See for continuities of an oral culture in a Jesuit university up to the 1960s W.J. Ong, »Communications Media and the State of Theology«, *Cross Currents* **19** (1969), 462-480, p. 469-470.

7. See the remarkable analysis in E. Durkheim, *L'évolution pédagogique en France*, Paris: PUF, 1969, p. 220-233.

8. Ch. Thomasius, *Höchstnöthige Cautelen Welche ein Studiosus Juris, Der sich zu Erlernung der Rechts-Gelahrtheit ... vorbereiten will, zu beobachten hat*, Halle 1729, p. 379-381.

9. See L.W.B. Brockliss, *French Higher Education in the Seventeenth and Eighteenth Century. A Cultural History*, Oxford University Press, 1987, p. 172-173.

10. See A. Seifert, *Cognitio Historica. Die Geschichte als Namengeberin der frühneuzeitlichen Empirie*, Berlin: Duncker and Humblot, 1976.

11. See K. Morrison, »Stabilizing the Text: the Institutionalization of Knowledge in Historical and Philosophic Forms of Argument«, *Canadian Journal of Sociology* **12** (1987), 242-274.

12. See for the effects of the nearly complete availability of the important Greek texts in print in late 16th century and for bilingual folios as standard format A. Grafton, »Higher Criticism Ancient and Modern: The Lamentable Deaths of Hermes and the Sibyls« in A.C. Dionisotti et al. (eds.), *The Uses of Greek and Latin. Historical Essays*, London: The Warburg Institute, 1988, p. 159-160.

13. See A.M. Cartter, *An Assessment of Quality in Graduate Education*, Washington, District of Columbia: American Council on Education, 1966, p. 114-115; E. Gross and P.V. Grambsch, *University Goals and Academic Power*, Washington, District of Columbia: American Council on Education, 1968, p. 56-57. In early modern university towns one must also count the private libraries of the academics and should study the conditions of lending libraries and inter-library loans.

14. See R. Blum, *Bibliographia - Eine wort- und begriffsgeschichtliche Untersuchung*, Reprint from *Archiv für Geschichte des Buchwesens*, Vol. X, No. 4-5, 1969, Frankfurt am Main, 1969.

15. See N. Luhmann, »Theoretische und praktische Probleme der anwendungsbezogenen Sozialwissenschaften«, 1977, in *Soziologische Aufklärung 3*, Opladen: Westdeutscher Verlag, 1981.

16. See for the 16th century A. Grafton, »Teacher, Text and Pupil in the Renaissance Class-Room: A Case Study from a Parisian College«, *History of Universities* **1** (1981), 37-70, p. 48.

17. See W. Kühlmann, *Gelehrtenrepublik und Fürstenstaat. Entwicklung und Kritik des deutschen Späthumanismus in der Literatur des Barockzeitalters*, Tübingen: Niemeyer, 1982, p. 282ff.

18. See Seifert, *op. cit.*, 1976, note 10.

19. W. Temple, »An Essay upon the Ancient and Modern Learning« in *The Works*, Vol. I, London, 1720, p. 153.

20. J.G. Schlosser, »Ueber Pedanterie und Pedanten« in *Kleine Schriften*, 5. Theil, Basel, 1787, p. 261-262: »Meist sind ihr [der Wißbegierde] sogar die alten Kenntnisse, die schon durch mehrere Köpfe gegangen sind, und die in mehreren Händen liegen, die liebsten, weil sie mehrere Prüfungen ausgestanden haben, und folglich wahrscheinlich die richtigsten sind.«

21. Article »Polyhistorie« in J.H. Zedler, *Grosses vollständiges Universal-Lexikon*, Vol. 28, Leipzig, 1741, Column 1319: »Wer ... seine Kräfte und Arbeit in allzu vielen Dingen zerstreuet, der kan eben keine so grosse Geschicklichkeit in seiner Disciplin erlangen, folglich *dem andern nicht so dienen* ... Dahero wird man auch finden, daß die größten Polyhistores in der Profeßion, darinnen sie stehen, der Welt eben keine so grosse Dienste leisten, sondern eben weil sie Polyhistores sind, sich mit Allotrien beschäftigen.« (Emphasis mine). Compare the article »Polymathie«, Column 1320.

22. See the »Promemoria« from 1781 on the function and intellectual values of cameralism: » ... die ganze innere Staatsverwaltung zu dem Gegenstand seiner Betrachtung zu machen; und, wie der Publizist seiner Seite untersucht, was in Ansehen derselben gerecht ist, er seiner Seite aufzufinden, was in Ansehen derselben nützlich seyn dürfte.« (Cited in K.H. Osterloh, *Joseph von Sonnenfels und die österreichische Reformbewegung im Zeitalter des aufgeklärten Absolutismus*, Lübeck and Hamburg: Matthiesen, 1970, p. 245).

23. Cited by M. Stolleis, *Geschichte des öffentlichen Rechts in Deutschland. Vol. 1. Reichspublizistik und Polizeywissenschaft, 1600-1800*, Munich: Beck, 1988, p. 317: »Hochgelahrt und gemeinnützig als reine Synonymen.«

24. See W.J. Ong, *Ramus, Method and the Decay of Dialogue*, Cambridge, Massachusetts: Harvard University Press, 1958.

25. See especially N. Luhmann, *Soziale Systeme: Grundriß einer allgemeinen Theorie*, Frankfurt am Main: Suhrkamp, 1984.

26. See for a more detailed argument R. Stichweh, »Differenzierung des Wissenschaftssystems« in R. Mayntz et al. *Differenzierung und Verselbständigung: Zur Entwicklung gesellschaftlicher Teilsysteme*, Frankfurt am Main: Campus, 1988, p. 68-72, 101-106.

27. See for *multiple operationalism* and *triangulation* as two such internal differentiations M.B. Brewer and B.E. Collins (eds.), *Scientific Inquiry and the Social Sciences: A Volume in Honor of Donald T. Campbell*, San Francisco: Jossey-Bass, 1981.

28. See A. Falconnet, *Essai sur le barreau grec, romain et françois, et sur les moyens de donner du lustre à ce dernier*, Paris, 1773, p. 172-173.

29. See for an interesting early praise of *Galilei* and *Columbus* -as paradigms for sailing the unknown seas of learning - by a Jesuit scholar D. Bartoli, *Dell'Huomo di Lettere Difeso*, Firenze, 1645, esp. p. 58-71.

30. See Ong, *op. cit.*, 1958, note 24, p. 135 for Francisco de Vitoria.

31. See the interesting passage in Ch. Wolff, »Von den Regenten, die sich der Weltweisheit befleissigen, und von den Weltweisen, die das Regiment führen« (1730) in *Gesammelte kleine philosophische Schrifften*, pt. 6, Halle, 1740, p. 589: He says that up to now only the »art of measurement« (Meßkunst) has produced new inventions. But the »practical philosophy« (ausübende Weltweisheit), which is needed for a good government, may learn from this to search for an increase by new inventions. It may succeed in doing this by creating a connection among all the bits of knowledge already known (»wenn nur dasjenige, was wir bereits davon wissen, in einen Zusammenhang gebracht würde»).

32. See for different aspects R. Stichweh, *Zur Entstehung des modernen Systems wissenschaftlicher Disziplinen. Physik in Deutschland, 1740-1890*, Frankfurt am Main: Suhrkamp, 1984; »Die Autopoiesis der Wissenschaft« in D. Baecker et al. (eds.) *Theorie als Passion - Niklas Luhmann zum 60. Geburtstag*, Frankfurt am Main: Suhrkamp, 1987; »Technologie, Naturwissenschaft und die Struktur wissenschaftlicher Gemeinschaften. Wissenschaftliche Instrumente und die Entwicklung der Elektrizitätslehre«, *Kölner Zeitschrift für Soziologie und Sozialpsychologie* **40** (1988), 684-705; »Bildung, Individualität und die kulturelle Legitimation von Spezialisierung« in *Études sur la genèse du système moderne des sciences*, Presses Universitaires de Lille, 1990.

33. Brockliss, *op. cit.*, 1987, note 9, p. 189-192.

34. See, for example, C. Rollin, *De la manière d'enseigner et d'étudier les belles lettres*, 4 Vols., Paris, 1728-1731, Vol. I, p. 248-254, Vol. IV, p. 599; J.G. Fichte, »Deduzierter Plan einer zu Berlin zu errichtenden höheren Lehranstalt, die in gehöriger Verbindung mit einer Akademie der Wissenschaften stehe« in E. Anrich (ed.), *Die Idee der deutschen Universität*, Darmstadt: Wiss. Buchgesellschaft, 1956, p. 184-185.

35. See R. McRae, *The Problem of the Unity of Sciences, Bacon to Kant*, Toronto: University of Toronto Press, 1961, p. 128-129; see Seifert, *op. cit.*, 1976, note 10; Stichweh, *op. cit.*, 1984, note 32, ch. 1, 2 and 4.

36. See Stichweh, *op. cit.*, 1988, note 32.

37. See W. Lepenies, *Das Ende der Naturgeschichte*, Munich: Hanser, 1976; M. Guntau, *Die Genesis der Geologie als Wissenschaft*, Berlin: Akademie Verlag, 1984.

38. See S.F. Cannon, *Science in Culture - The Early Victorian Period*, Oxford University Press, 1978.

39. See Stichweh, *op. cit.*, 1984, note 32.

40. See Stichweh, *op. cit.*, 1987, note 32.

Science as a Self-Organizing System
Outline of a Theoretical Model

WOLFGANG KROHN, GÜNTER KÜPPERS
University of Bielefeld

1. Introduction - »Old« and »New« Systems Theory in the Sociology of Science

A renaissance of systems theory in the social sciences must have good reasons. At least there must be some hope to overcome the main deficit of the »old one«, i.e. the absence of mechanisms which could explain the macroscopic behavior of the system on the basis of its microscopic interactions.

In sociology systems theory is primarily associated with the Parsonian tradition, which through Robert Merton (1968) and his school (among others, Storer, Barber, Hagstrom) dominated the empirical sociology of science up to the beginning of the 1970s.[1] The main concern of the classical systems theory was the identification of *functions* and the related norms of action (functional imperatives) and role patterns. The functions were obtained (by Parsons) in a deductive-analytical way by means of a general functional theory of action (the four functions of the AGIL - scheme »Adaptation, Goal-Attainment, Integration, Latent pattern-maintenance«). The construction of functional systems was carried out by means of a formalistic technique of cross tabulation which was fed by the analytical self-application of the functions to themselves.[2] The relationship between the systems and their hyper-complex environment was defined by adaptive learning. The internal relation of the components was determined by categories of equilibrium and uni-directional evolution (adaptive upgrading) in »coping with increasingly wide rangings of exigencies«.[3] The participation of individuals in the functionally differentiated system was defined by their roles, whereby the competence to carry out these roles was again generated by functional systems.

The criticisms brought forward in the 1970s against the application of this systems theory in the sociology of science all took the form of an empirically and historically oriented protest against the ideal-type theory construct: real science has hardly anything in common with its ideal-type. Discrepancies were noted between (a) the functionalistic rationalism or cognitivism and the actual goals and methods of science;[4] (b) between the assumed role-patterns and models of gratification and the actual activities in industry and government research;[5] (c) the assumed evolutionary path of growth of objective knowledge and the actual historical development, which is more shaped by »dogmatism« and »revolution« than by adaptation;[6] (d) the assumed set of norms and that which is actually followed.[7]

If against this criticism systems theory shall be reintroduced into the discussion in the form of the theory of self-organizing systems, it is because the principle of self-organization has changed its theoretical value for empirical analysis. The main concern of the theory of self-organization is the modelling of the actual operation of real systems, not the functional description of analytically defined entities.

W. Krohn et al. (eds.), Selforganization. Portrait of a Scientific Revolution, 208–222.

Theoretical Entities versus Real Components

Parsons was predominantly inspired by the classical conception of system in the natural sciences. His idea was that »behind« the unanalyzable complexity of existing reality there lies a »pure world« of theoretical entities. »Real« behavior of the »real« world was attempted to be approximatively explained as superimposition of events in the theoretical world. Complexity was understood as piled up simplicity, not as an inherent property of the systems themselves. In contrast, the new systems theory attempts to define the specific identity of each system, not through theoretical entities, but through those forms of organization by which their self-organization is secured. Basically, self-organization is maintained by means of non-linear recursive interaction. Recursive processes can be intrinsically complex and, even if deterministic, unpredictable. This means, that a theorist who concentrates on the discovery of the modes of operation of complex systems must »leave it to the system« to form structures even of great divergence, so long as it continues to exist by maintaining its form of organization. The self-structuring of the system relieves the theorist from questionable *definitions* of structure.

Function versus Mechanism

The prominent position held by *functionalism* is counter-balanced by the importance of systemic *mechanisms* by which self-organization takes place. To be more exact, the analysis of the interdependency between function (and change in function) and structure (change in structure) plays the crucial role. The reason for this is that it is impossible to determine a complex non-linear system by means of functional analysis.[8]

Stable Equilibrium versus Dynamical Process

The concept of stable equilibrium which dominates in the classical systems theory and cybernetics is replaced by the dynamics of processes which determine - beyond thermodynamical equilibrium[9] - through structure building, phase transitions and instabilities, the unpredictable history of a system. The analysis of the dynamics of recursive processes has led to a theorem of great consequence: already in the case of deterministic systems, the sensitive dependency on boundary conditions can lead to great structural changes at the slightest perturbation. But, equally important, a high degree of perturbations can also be neutralized. From this follows that the actual behavior can only be predicted by theory slightly.

Adaptation versus Structuring the Environment

In classical systems theory the theorist observes from outside the relationship of a system to its environment and is from this Archimedean point in the position to evaluate the performance of the system. Modern systems theory shifts the theoretical position of the observer into the system and models all its operations from this perspective. Consequently, the notion of adaptation loses its meaning. For it is now the case that the referential frame for distinguishing a better or worse adapted behavior is no longer recognizable. The concept of adaptation must be replaced by that of the *structuring of the environment*, that is, by the attempt of a system to create according to its own evaluation favorable conditions for its continued existence.

These general comments on the difference between the classical and new systems theory will be exemplified, at least in part, in the following presentation for the system of science.

2. On the Construction of Science and Research

The essential point of our departure is a distinction between research and science. *Research* is the production of knowledge, *science* is concerned with the creation of conditions favorable to the continuation of research. The smallest social unit of research is the research group. Although membership in the group is more or less regulated on a formal basis by the respective institution, what is of particular interest to us are the processes of self-organizing the group as a social system and of self-regulating its performance. Analytically we distinguish between three different types of action carried out by the group in a distinct manner. The first type applies to the self-maintenance of the group: integrative activities. The second serves to achieve specific functional goals with respect to the production of knowledge: research activities. The third relates to the mobilization of external resources required for continuing the existence of the group: scientific activities.

Integrative Activities

Each group develops a »constitution« in which the rules governing its self-maintenance are laid down. We shall term this constitution *group matrix*. Its rules are neither formally defined nor unchangeable nor are they explicitly known to the members of the group. Such a matrix regulates in particular the coordination of the different interests and abilities of the members and provides for a commonly shared context of beliefs, attitudes and intention. In this context the specific »style« or »culture« of the group as well as its »self-image« are shaped. At the outset the members have individual ideas about cooperation. But to what degree they are consented by others and what they would mean in practice remains open. In coping with tasks, including the possibility of redefining them, the group does not only achieve problem solutions but also interactive definitions of group specific beliefs, attitudes and intentions. Theoretically speaking, this process is the formation of the specific boundary conditions out of its contingent initial conditions. They regulate the goal-oriented activities of a self-organizing social system. In the tradition of social-psychological group research, integrative activities which are directed toward the development and modification of this matrix of boundary conditions belong in the socio-emotive dimension.[10] This dimension is important also for research groups because each group must maintain its existence when faced with the attractiveness of other groups. The analysis of group formation has been the concern of manifold experiments and there is little doubt that cooperation in research is a typical example of group formation.

Research Activities

The task-oriented activities of research groups are directed toward the production of knowledge. As will be discussed in more detail in the following section, research activities are in a strict sense recursively interconnected. All activities directed toward a research goal have to be accepted or rejected by other members of the group as possible contributions for the solution of the common problem. Problem solving does not allow neutrality, but only action and reaction in a network of recursive moves. The further the interactive production of a piece of knowledge progresses the more this network is elaborated.

Scientific Activities

To secure its existence in its social environment the group must market its results and acquire new tasks. The groups' activities to mobilize resources for their self-maintenance (for instance, by gaining project assignments, participating in editorial boards, presenting experts and specialized reports or carrying out publicity activities) are only in part recursively networked because here scientists are not necessarily acting as group members and for group members.

This sequence of types of action does not presuppose any causal process. It is not necessary for a group to first secure its social integration before it can carry out research; a group can be formed on the basis of holding similar research interests and it is only subsequent to this that the problem of cooperation arises. The dynamic process of the group has an unlimited number of paths open to it from which, however, only very few lead to a stable solution, i.e. to the creation of rules which serve cooperation and can become part of the group matrix.

The group matrix provides the basis of self-organization. But this basis is not static. It is continually modified through goal-oriented research and science activities. On the other hand the group-matrix influences as a boundary condition the research process.

By integrating these three types of action, each research group performs three things with a certain routine: it secures the group members' willingness to cooperate (component up-keep), produces research results and mobilizes resources. In view of the obviously specific nature of the psychical and social boundary conditions, as also of the concrete research goals, it is only on the basis of detailed empirical observation that a statement can be made concerning the actual creation of a matrix.

3. Knowledge Production in the Research Group

Origin and existence of a research group as a social system depends on the production of new knowledge. The production of knowledge is social interaction (communication, cooperation), working with material objects (apparatus, instruments, substances, effects) and non-material entities (concepts, theories, methods). Research results are the work performed by the research group for the environment (other research groups and the scientific as well as non-scientific institutions).

Finding a general definition for the notion of »research result« is an extremely problematical task. In the philosophy of science different candidates have been selected repeatedly as representative contributers to scientific knowledge: ideas, laws, paradigms, structural cores, effects. They have mainly been derived from the reconstructive analysis of the generated body of knowledge and not from its generation. Alternatively we suggest a sociological definition that can be derived from the research process: a result of research is that which is considered by the research group to be suitable for preparation as *information to the outside*. The standard form for conveying information to the environment is the article in a scientific journal. Other variants are presentations in the form of patent specifications, monographs, lectures, and others. However, it is to be kept in mind that such contributions are not in themselves the research results of the group, but their stylization for an external purpose; this distinction will be of importance later.

Our concern here is solely with the product that research groups produce for themselves. For its description we have chosen the term »operation of knowledge«. It should suggest that it is a complex formation which has a structural and procedural side. The structure is formed from the combination of theoretical, methodical, empirical and interpretorial components; the procedure operates by the possibility of its repeated processing on reception of new entry values. For this reason, an article, which is a description of such a linkage, is always a complex unity comprised of the presentation of data, graphic illustrations, explanations, assumptions, and so forth. Not all but some of these components must have been newly researched. Its value, however, increases and decreases according to the quality of a coherent presentation. It must have a good gestalt to borrow Flecks description.[11] Therefore we say that when an *operation of knowledge* counts as being successful it has found an eigensolution, that is, it shows an inner stability of the theoretical, methodical, empirical and interpretorial components.[12]

The interpretation of scientific knowledge as an eigensolution of a research process is part of a constructivistic approach to epistemology. It joins the work done by Piaget, Bateson, Glasersfeld, Spencer-Brown and others.[13] The shared »central dogma« of a constructivistic theory of knowledge is a negative theorem: for the explanation of knowledge based on the theory of knowledge, the objectivity of reality offers no assistance, because any proof of a reference to reality falls back on constructions which already stabilize knowledge. For, there is neither an elementary level of contact to reality (as sensualism wished to see it) nor are there elementary unmanipulated objects (the view of realism) which are not influenced by theoretical decisions and conventions. From this it follows that knowledge must be a construction which can only be controlled by another construction. With the exception of Piaget's analysis on ontogenesis, which cannot easily be applied to the open processes of scientific knowledge acquisition, hardly any systematic investigations have focused on finding out of what these constructions are comprised.[14] Constructivism poses two systematic problems:

(a) How does the process of knowledge bind itself to the constructions on which it is based? What are the limiting conditions by which a construction once it has begun cannot take any arbitrary direction? In other words: What is coherence?
(b) If knowledge does not have reference in a given structured reality (or there is no way to know about it), how then is the contingency of experience constructed? How does knowledge become improbable if it is only based on constructions? In other words: what is surprise?

A constructivism that in the radical sense is non-realism, in that it does not take into account the experience of inconsistency and of unexpected events, would break down due to its own lack of structure. The task, therefore, is to show that, with an increase in the complexity of the operations of knowledge, there is an extension also in the unavailability of reality created by exactly these constructions.

The following diagram attempts to present in an ideal type the *structure* of knowledge production:

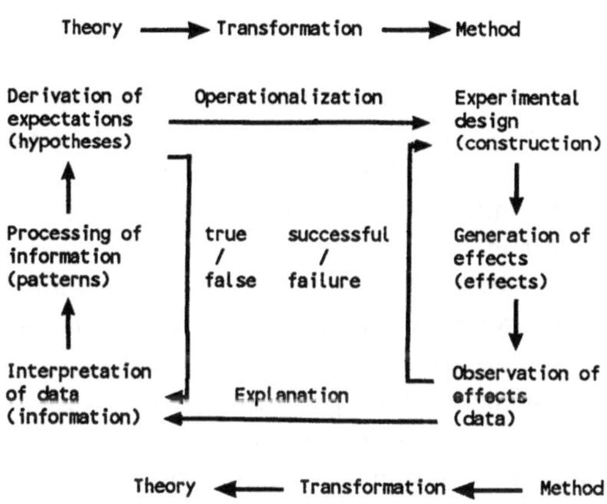

Diagram 1: Structure and dynamics of the »operation of knowledge«

The central idea behind this diagram is the assumption that all research of the empirical (if not all) sciences are comprised of two independent components. The left side of the diagram represents the *theoretical* part of recursion between the interpretation of data, the networking of interpretation to »patterns« or contexts[15], and the advancing of an expectation due to a distinction;[16] this part of the recursion operates on the basis of the valuation grid true/false. The right side of recursion represents the experimental or *operational* part of knowledge production. It proceeds along the grid success/failure via the stages of decision making for an experimental method, the carrying out of an experiment and the observation of the data delivered by the experiment. Taken separately, these two parts do not work recursively (experiments are theory-laden and theories are decision-laden[17]) Only if linked together they can operate in a way so that the output of one part is the input of the other.

These links are made by way of two transformations that bind the effective operation (right side) to that of the intellectual (the left side). We have named these transformations *operationalization* and *explanation*. At the beginning both transformations can be characterized as optional or loose couplings since no general rules exist for the transfer between the two differing valuation grids of both sub-operations. Instead, there is always a variety of options for the transfer from intellectual to effective actions, that is, for the operationalization of theoretical assumptions into an experimental design, on the one hand, and conversely, for the theoretical interpretation of experimental data (explanation) on the other. Neither are the individual stages in the two sub-operations coupled to each other in a fixed manner. Here also options exist which must be selected.

By decisions, taken in the progress of knowledge production, all links are successively changed into fixed couplings. These decisions are not completely arbitrary, because the spectrum of possibly accepted explanations and operationalizations depends on the commonly shared context of beliefs, attitudes, and intentions. In other words, they depend on the boundary conditions as specified in the group matrix. Although couplings which have already been selected once can be released and be replaced by others, at some stage all the loose couplings have been transformed into fixed ones and the production of knowledge becomes circular without further changes occurring. It has found its eigensolution »new knowledge«. The structure of this eigensolution is the research result, the description of which is its publication.

To have found an eigensolution for an operation of knowledge means to believe that the application of a certain method of investigation on any element of a set of objects yields similar data which support a general statement about a certain feature of this set, which implies that the application of a certain method on a *different* element yields similar data... The idea of knowledge as an eigensolution of an operationally closed combination between argumentative and experimental activities attempts to answer the initially posed question of how the construction of knowledge binds itself to its construction in a new way. The coherence of an eigensolution does not refer to an objectively given reality but follows from the operational closure of the construction. Still, different decisions on the selection of couplings may lead to different, equally valid eigensolutions. Between such different solutions no reasonable choice is possible unless a new operation of knowledge is constructed exactly upon the differences of the given solutions. But again, this frame of reference for explicitly relating different solutions to each other introduces new choices with respect to the coupling of operations and explanations. It does not reduce but enhances the dependance of knowledge on decisions. On the other hand, the internal restrictions imposed by each of the chosen couplings do not allow for any arbitrary construction of results. Only a few are suitable to mutually serve as inputs in an circular operation of knowledge.

4. Scientific activities

In contrast to research activities and group integration, scientific activities are initially not recursive since it is action in the environment of the research group. These activities are mainly concerned with the creation of conditions favorable to the survival of the group by mobilizing resources in the form of funding, legitimation and manpower. Seen from the research groups' view, central importance is assigned to the dissemination (marketing) of research results. Additionally, of relevance are activities as heads, experts, advisors, teachers, and publishers which give access to the infrastructure of institutions such as laboratories, committees, panels, journals, societies, and universities. These institutions form the *social system of science* as the research groups form the *social systems of research*.

The system of science and the research groups are coupled with each other in different ways, depending on the type of function of the institution. Decisive is that all these couplings are *cognitive feedback* for research. Research provides the system of science with new knowledge which in turn provides cognitive networks as orientation complexes for further development: generalized theories, research programmes, universes of discourse, focal promotion areas, contexts of relevance, and others. These cognitive networks are created in the institutional establishments of the science system: in journals in the form of citations, in curricula through texts and manuals, in research programmes through the coordination of perspectives, at public meetings via normative orientation. Research groups re-orient their research according to these cognitive networks. In turn the system of institutions safeguards research as an autonomous process of knowledge production.

5. On the Reformulation of Externalism

The importance of this cognitive feedback for the forming of scientific knowledge was first recognized by L. Fleck.[18] He assumed that research underlies an inner process of differentiation because every new knowledge is a special innovation which has as a consequence further specialization. According to Fleck, it is the so-called »exoteric circles« that carry on the counter tendencies like simplicification, understandability, systematization, and so forth. Research is cognitively dependent on these »popularizations« in the »exoteric circles«, the first of which are already formed by colleagues outside one's own narrow field of specialization. With this approach Fleck indicated a long unnoticed way out of the debate on the external versus internal determination of science, which we consider can be formulated even more plausibly with the help of the concept of self-organization.

The assumption of *cognitive* feedback of *social* action presupposes that the concepts »cognitive (rational)« and »social« are not dealt with at categorially distinct levels between which interaction is not possible. It was critical rationalism in particular that suggested an essential separation of categories between truth discourse and social behavior.[19] Totally opposed to this idea is Fleck's point of departure: »Cognition is the most socially conditioned activity of the humanity and knowledge is the social construction per se«.[20] Following on this idea, a different conceptual development would appear to provide a better basis for solving the problem of externalism.

Analytically separable aspects of action are *social interaction* (communication, cooperation) on the one hand and *working with material and non-material objects* (in science, amongst others, measurements, experiments, calculations) on the other. *Both* aspects of action can be cognitive or rational as well as more intuitive or emotional.[21] A further useful distinction can be made between the concepts »cognitive« and »rational« such that cognitive action is understood as action oriented towards problems and solutions while rational action is understood as justifications and explanations.

Against this background it is conceptually unproblematical that social interaction has retroactive effects on work with objects and vice versa. Anything to the contrary would be surprising. Such retroactive effects could consist of semantic analogies in everyday language, in constructed metaphors, as also in the reciprocal forming of knowledge interests, perception of problems, and ideals of explanations. These mutual influences can take a more cognitive-rational or a more intuitive-emotional course. And again, it can be expected and it is conceptually unproblematical that retroactive effects exist between cognition and emotion. They are for instance the concern of creativity research or research on prejudices.

Our starting point is based on the assumption therefore that all social interaction of scientists effects retroactively the way in which something is scientifically dealt with and vice versa. Taking such feedback to be self-evident and of an everyday occurrence, the question of concern to us narrows down to the structure building effect this feedback has on the research process.

6. Cognitive Feedback on the Research Process

We distinguish between six structurally different types of feedback on the research process. The underlying mechanism is the same throughout: the scientists of a research group seek to secure the conditions for the continuation of their research work. They continually strive to have this environment create favorable conditions for them by increasing their influence in the environment (e.g. through membership in publishing boards, programme committees, scientific foundations, self-administration bodies, and others). At the same time they orient their research towards the environmental conditions to be expected. This shaping of the environment necessitates the taking of an observer's perspective both towards the environment as also towards the groups' own field of research. The group must construct a frame of reference which, in turn, operates in a destabilizing manner. It makes the respective research group prone to disturbances from the environment and leads to changes in the research process. In the following diagram the six environmental loops (action in the environment including the retroactive effects released by them) have been summarized.

type of action	**functional goal**	**cognitive feedback**	**institutions**
recursive interaction	production of knowledge	cooperative formulation of research goals	research groups
informal (oral) communication/-cooperation	information	formation of self images and images from outside of the research process	workshops, conferences, congresses, institutions, laboratories
formal (written) communication	reputation	decentering of knowledge, modelling of research fields, construction of historical continuity (past)	reports, journals
science policy and research planning (forma- planning (formation of research programmes)	disposal of financial res- sources sources	formation of new research fields, integra- fields, integration of extra-scientific goals of knowledge, design of the future	hybrid communities, funding agencies agencies
teaching	recruiting of qualified manpower	unification and axiomatization of science	universities, medical schools etc.
practice	scientification of society	integration of social preferences, norms, criteria of relevance	science based industry, health system, education
public discourse	legitimation of science	integration of values	public media

Diagram 2: Cognitive feedback of sciencing on the research process and institutions as agencies of these feedback loops

In the following only three of the environmental loops presented in the diagram are to be discussed: a) informal communication, b) formal communication, and c) science policy and research planning.

Informal Communication

Informal communication includes all activities undertaken by scientists during their research for the purpose of bringing their work into relation to that of other groups. The standard way for this to occur is through the coordination of research processes in laboratories, most particularly in industrial and state institutions; then in »invisible colleges« which are maintained through telephone and computer link-ups, the dispatching of proposals, drafts and preprints, and by meetings at conferences; and finally on the basis of informal relations to members of non-scientific institutions. Since every research group is »disturbed« daily by all kinds of information, decisions have to be made concerning how this information is to be dealt with. Via these decisions perturbation by information can have a direct influence on the research process. This is the everyday externalism to which every research process is exposed. In spite of this on-line connection to the research process, this communication is not part of the recursive networking within the research group but part of the different external relations of its members. The good and bad messages cannot in fact be re-examined by all; they have to be either disregarded or used. The structuring of the relation between leadership and group spirit is based on this non-controllability of relevant information.

The analysis of cognitive feedback can only be carried out on the basis of laboratory studies because it concerns the development of knowledge previous to its publication. In general it can be said that disturbances occasion research groups to reflect upon their own activities as also to form an image of the activities of others. They are induced to plot a map of their immediate research environment and by stressing specific features (methodological aspects, integration of external goals, etc.) they attempt to form research in such a way that the hoped-for results fit into the niches of the environment as their map would suggest.

Formal Communication - The Construction of the Past

Every type of formal communication is a translation of research results into a *description* for others. For a general characterization of its cognitive feedback we use Piaget's notion of decentering.

Modelling of Given Knowledge

Each publication secures a place in the research landscape of a discipline, in that it indicates links to existing knowledge. This is done through the condensed summary of important results of others. Thus every publication performs a certain minimum of »historiographic« activity: it reconstructs its own history in its field of research. This does not happen in an indifferent way; the reconstruction is a stylization of the past development so that ongoing research seems to fit quite naturally. For other research groups it is information about the reception of their work. This obligation to model given knowledge is a corrective against the specialization of research and is the first aspect of decentering: one must perceive ones own knowledge from the »historical« perspective of an observer of the field in its general development.

Modelling of own research contributions

Due to the non-reproducibility of implicit knowledge in a descriptive language the printed presentation of a research result, the theoretical approach, methods and the observed facts are described in a pictorial way. This picture must be designed from the perspective of the outsider (the expected reader). Pictorial presentation means simplification and generalization and leads to an increase in coherence. By means of modelling, the unavoidable weak points in the relationship between theory, method, data and interpretation are ironed out in such a way that the burden of decision of the research process is minimal and the objectively enforced interpretation of the results is maximal. By this construction, a picture of the general meaning of the group's activities is formed which counteracts their own specialization.

A further aspect is to be stressed: the feedback on research from the making of »journals«. Journals, although vehicles of networking, are not a »mirror« of research. At the level of specialist journals, the editors, experts and publishing boards have a more influential position than that of authors. They lay down standards and criteria according to their own measures. In the case of journals which deal with topic-oriented volumes, or those with representative functions towards a scientific-oriented society, or else with commercial backgrounds, there is a reversal of the supply-demand relationship: journals demand contributions and a contribution seldom accidentally fits into the publication strategy of the controlling body. Although researchers try to meet the demand through publications on hand and by means of recombination techniques, the work becomes dependent on the journals. The publication is no longer only a modelling of ones own contribution with respect to the possible expectations of the intended recipients, but a modelling which is prescribed in its outline by the editorial policy.

The more extensive the networking of research is carried out by the scientific journal market, the more independently it structures itself and takes on systemic features. However, scientific publication activities - similar to the management in laboratories - are predominantly carried out by the scientist themselves. That means that scientists construct an environment for research with the aim to dominate and shape this for science, but the formed structures lead to cognitive perturbations of the research process.

Research Policy and Planning - The Construction of Future Continuity

The main instrument of research-policy is the research programme. Its institutional function in the system of science is to trigger innovations in the research process. That research, although its goal is to produce new knowledge, has to do with a problem of innovation has already been mentioned: research groups (like all social systems) are conservative.

Research programmes are means to counteract this conservatism. They are institutional perturbations in terms of their effect on research: physical and personnel resources for new subject matter on the one hand, the end of funding on the other. This institutional function is based on a cognitive performance which these programmes carry out for the system of science: programmes are a perspective construction for *future knowledge*. This modelling of the developmental potential of the present has an interesting symmetrical relation to the cognitive performance of publications: while publications connect current research to the past, programmes seek connections to the future based on extrapolations, partly on options. They mediate promises (of science) and expectations (on science).

Programmes are the mechanism of innovations. This mechanism is based on the positive feedback for outsider opinions which could not have been placed within the system.[22] When scientific outsiders make it known to politicians and foundations by way of memorandum and the like that the promotion of their field of research is necessary (e.g. in order to be on a par with international standards) or with adequate assistance it can have future use in the extra-scientific field (health, military, economy, etc.), then politicians and foundation members can initially as outsiders also mobilize votes in their party which result in an increase in support possibilities. With this result, scientific outsiders are successful in getting the attention of further colleagues, whereby externally the impression increases as to the absolute urgency of support. If, as a result, boards of authority are set up in which positively interested scientists and non-scientists reach an agreement with respect to their new ideas and present a plan of a support programme, all the conditions have been fulfilled through which, from a scientific (and perhaps also a political) outsider position, a new programme can be innovated. Therefore, there is normally mutual reinforcement (positive feedback) between the autonomy of science and control.

Programmes lead to institutional changes. Do they also lead to cognitive feedback for the research process - to a »programming of research« so to speak? In reaction to programmes, research groups propose projects in which their intentions of research and relevant findings are stylized with respect to the goals of the programme. Necessarily they integrate into the proposals »alien« problem definitions and relevance criteria, which gained access to the programme through the programme communities and now interpenetrate the actual research process. In these programme communities scientists and non-scientists (politicians, industrialists etc.) cooperate, especially if the programmes are not only inner-scientific ones.

Similar to the publication of research, the cognitive feedback of research programmes compels scientists to take an observer's position, both towards their field of research as also towards the environment. Research policy leads to a further decentering of the research area which allows the channelling of innovation into the process of research. Though, there are already fluctuations at the level of interacting researchers, it is especially through the activities within the boundaries of research policy that it is possible for outsiders to reinforce the perturbations to such a degree that instability finally causes the restructuring of the current research landscape.

7. Summary

In an extended analysis, not to be undertaken here, it could be considered how the system of science generates itself from the institutions, e.g., journals and programme committees created through scientific activities. It is clear that there is in turn interaction between these institutions by way of which science carries out its structuring performance towards the extra-scientific environment and its cognitive feedback on research.[23]

Our analysis started out by focusing on research groups because the functional primacy of the production of knowledge is rooted within them. But the research groups, influenced by the manifold system of institutions in science, are only »particles« that move in these »fields«, which they already produced. It is like the electron which follows its »lines of force« produced by its own electric field. Therefore, the internal self-organization of research groups, so essential as it is for the production of knowledge, is, in effect, dependent upon the external self-organization of science for research. This means that the development of science is not determined by research, but through the structuring performances which the science system can obligate in society.

Notes and References

1. T. Parsons, *The Social System*, New York: The Free Press, 1964; N.W. Storer, *The Social System of Science*, New York: Holt, Rinehart and Winston, 1970; B. Barber, *Science and the Social Order*, New York: Collier Books, 1962; W.O. Hagstrom, *The Scientific Community*, New York: Basic Books, 1965.

2. Paradigmatically presented in T. Parsons/G.M. Platt, *The American University*, Cambridge, Massachusetts: Harvard University Press, 1973.

3. T. Parsons/G.M. Platt, op. cit., note 2, 1973, p. 14.

4. P. Feyerabend, *Wider den Methodenzwang. Skizze einer anarchistischen Erkenntnistheorie*, Frankfurt am Main: Suhrkamp, 1976.

5. St. Cotgrove/St. Box, *Science, Industry and Society. Studies in the Sociology of Science*, London: Allen & Unwin, 1979.

6. Th.S. Kuhn, *The Structure of Scientific Revolutions*, Chicago: University of Chicago Press, 1962.

7. S.B. Barnes/R.G.A. Dolby, The Scientific Ethos. A Deviant Viewpoint, in *European Journal of Sociology*, **11** (1970), 3-25.

8. The simplest description (idealization) of complex systems is presented by H. von Foerster, »Principles of Self-Organization in a Socio-Managerial Context«, in H. Ulrich/G.J.B. Probst, (ed.), *Self-Organization and Management of Social Systems. Insights, Promises, Doubts, and Questions*, Berlin u.a.: Springer, 1984, 2-24 in the model of a non-trivial machine. It matches an entity to another on the grounds of its internal state and thereby changes its state on the grounds of its inner structure. Such a machine is not reconstructible from an analysis of the matching of entities (e.g. numbers) because of the hidden change in state. In the case of machines, the programme of which is not known because they were not built by us, i.e. natural and social systems, an empirical task follows to develop a model of its construction by means of the composition of functional analysis (investigation of the matching of input and output) with the investigation of the organization (mechanism of operating).

9. »Beyond thermodynamical equilibrium« means that the mechanical forces are balanced but the thermodynamical ones (gradients with respect to temperature, concentration, etc.) are not.

10. See R.F. Bales, *Personality and interpersonal Behavior*, New York: Holt, Rinehart and Winston, 1970, 5-8. In general this distinction is still considered to be useful.

11. L. Fleck, *Erfahrung und Tatsache. Gesammelte Aufsätze*, Frankfurt am Main: Suhrkamp, 1983, p.147-174.

12. Eigenvalues or eigensolutions describe certain parameters or states of a system which, depending on the boundary conditions, have been chosen from the number of theoretically possible parameters or states of a system. In contrast to classical systems where the boundary condition is independent of the state (by means of undetermined constants contained in the general solution, any boundary condition can be fulfilled), the boundary condition and state of the system in case of self-organized systems are closely connected with each other. It can be said, therefore, that the system forms its own boundary and the boundary determines the state of the system.

13. J. Piaget, *Die Entwicklung des Erkennens* (1950), 3 Vol., Stuttgart: Klett, 1972/73; G. Bateson, *Geist und Natur. Eine notwendige Einheit*, Frankfurt am Main: Suhrkamp, 1982; E. von Glasersfeld, *Wissen, Sprache und Wirklichkeit*, Braunschweig: Vieweg, 1987; G. Spencer-Brown, *Laws of Form* (1971), London: Allen & Unwin, 1979.

14. Exceptions here are the »radical conventionalism« of Ajdukiewicz (Ajdukiewicz, K., *The Scientific World Perspective and other Essays 1931-1963*, Dordrecht, Boston: Reidel, 1978) and perhaps the older tradition of the conventionalism of Duhem (Duhem, P., *The Aim and Structure of Physical Theory*, New York: Atheneum, 1962) and Poincare (Poincare, H., *Science and Hypothesis*, New York: Dover, 1952). Here, however, a reference to social interaction was missing. They only reconstructed theories and world images. The more recent laboratory studies exhaust themselves in arguments against their rivals in the field of theory of knowledge - rationalists and positivists.

15. Following on G. Bateson, *op. cit.*, 1982, note 13.

16. As expressed by G. Spencer-Brown, *op. cit.*, 1971, note 13.

17. N.R. Hanson, *Perception and Discovery. An Introduction to Scientific Inquiry*, San Francisco: Freeman, Cooper & Co., 1969; K. Knorr-Cetina, *Die Fabrikation von Erkenntnis. Zur Anthropologie der Naturwissenschaft*, Frankfurt am Main: Suhrkamp, 1984.

18. L. Fleck, *Entstehung und Entwicklung einer wissenschaftlichen Tatsache*, Frankfurt am Main: Suhrkamp, 1980, p. 152ff.

19. K. R. Popper, *The Logic of Discovery*, New York: Harper & Row, 1976; I. Lakatos, »History of Science and its Rational Reconstruction«, in *Boston Studies in the Philosophy of Science*, VIII, 91-136. L. Laudan, *Progress and its Problems*, London, Routledge & Kegan Paul, 1977.

20. See Fleck, *op. cit.*, 1980, note 18, p. 58.

21. The complementary concepts are therefore: The complement class of rational action is ir-rational action regardless of whether performed against a subject or object. Ir-rational action is, according to the everyday use of the word, for example, emotional or intuitive action. Social action is oriented towards subjects capable of interaction and communication regardless of whether rational or emotional. The complement class of social action is action directed toward objects. One could argue whether the demarcation line between the class of complements is essential or gradual. This would not be just a debate on concepts but an empirical one (e.g. on the rationality of emotional strategies, communication with animals, amongst others).

22. See M. Maruyama's »deviation amplifying mutual causal processes«, in «The Second Cybernetics. Deviation-Amplifiying Mutual Processes«, in W. Buckley (ed.), *Modern Systems Research for the Behavioral Scientist*, 2. ed., Chicago: Aldine, 1969, 304-316.

23. W. Krohn/G. Küppers, *Die Selbstorganisation der Wissenschaft*, Frankfurt am Main: Suhrkamp, 1989.

Actor-Networks vs. Science as a Self-Organizing System: A Comparative View of Two Constructivist Approaches

HELGA NOWOTNY
University of Vienna

Largely unnoticed by mainstream social studies of science, concepts of self-organization and self-referentiality of systems or of autopoiesis have quietly invaded a number of scientific disciplines. They range from cybernetics, one of the main breeding grounds for these concepts under the pioneering influence of Heinz v. Foerster, to work conducted by Manfred Eigen (and later also by Peter Schuster) on molecular self-organization. Hermann Haken's work on »synergetics« as part of a theory of the laser has to be mentioned as well as that of Ilya Prigogine on the thermodynamics of irreversible processes. Other »adumbrationist« heritages can be located in modern ecology (connected with the name of B. Holling) or, in a historical perspective, can be located even in the natural philosophy of Schelling.[1] There exists also a wider range of practical applications that these concepts have found. They span organizational theory and family therapy; principles of self-organization are apparent in the structure and programs of new social movements. Their popular appeal has led to a rise in sales figures of popularizing books highlighting, for instance, principles like »networkthinking«.[2] While some authors have celebrated the theory of self-organization as a scientific revolution already[3], others have emphasized the fresh and stimulating perspectives that the redefinition of the system/environment boundary has brought to a wide range of problems in epistemology.[4] Especially the work of Humberto Maturana, originally conducted within biology and neurophysiology of cognition, has become the focus for a boom in what could be called »cognitive constructivism«.

It is not the purpose of this contribution to assess the epistemological mono- or inter-disciplinary implications that reconceptualizations revolving around theories of self-organization may bring with them, nor to pursue the question under what social and cognitive conditions concepts gain acceptability in fields that are otherwise literally »worlds apart«. Other contributions in this volume provide at least partial answers. Rather, I want to address a theory of self-organization of science, such as it has been proposed by Krohn and Küppers in this volume, in the context of historical developments as well as in a contemporary, comparative perspective: (i) where does such a theory stand with regard to the oppositionary dualism that has characterized the sociology of knowledge, separating »social constructions« from »scientific facts«, but also »the social« from »nature« or »objects« (things, technological artifacts) and (ii) compare it with another constructivist approach whose foremost and outspoken proponent is Bruno Latour. My thesis is that much of sociology of science and also of sociology of knowledge is heir to a general predicament of social science when it comes to conceptualizing and accessing the world of »nature« and its objects as well as those of the material world, notably in the form of technology[5], and that social constructivism has to be seen as a rather limited case of constructivism. While a theory of science based upon principles of self-organization does not entirely escape the traps of previous theories of science,

W. Krohn et al. (eds.), Selforganization. Portrait of a Scientific Revolution, 223–239.

notably either wanting to be a meta-theory or of being based upon a notion of science modelled after the leading discipline of the day, it offers a wider range for including other empirical as well as theoretical accounts, with some interesting points of convergence with alternative approaches.

Until recently, the social sciences could either ignore the natural world as falling outside their sphere of competence or they were facing it only indirectly. If »nature« entered the social science discourse at all, it was either in a hermeneutical sense, that is, through the meaning social actors attached to it, or it was seen as a »resource« to be exploited, commodified or treated as a free good. In a wider sense, the inherited dualism between »nature« and »society« (or »culture») ran very deep not only through philosophy but through all of the social sciences, although this relationship undoubtedly underwent some drastic reversals in the last centuries. In a more narrow sense, nature, the natural environment, as well as technological artifacts were assumed to be »real«, with their existence grounded in a »reality« outside and different from that which constituted society and social life. Interactions between the two kinds of reality were largely modelled in a causal way, imitating the prevalent methodology of the natural sciences. The question was what kinds of »influence«, »impact«, or »consequences« could be detected that the transformation of nature effected largely through science and technology for the social world. For understandable reasons technological artifacts and the genesis as well as possible forms of control were given special attention. The natural environment was seen as being transformed through human intervention and »responding«, but the nature of this interaction remained until recently couched in rather superficial, if not simplistic or crude accounts. The intellectual objects and constructs discovered and constructed by the natural sciences - the force of gravity, the periodic system, microbes, or the logical structure of mathematics - came only to the fore when the sociology of knowledge of the early 1920s put them on the agenda - and decided, for the time being, not to meddle with them.[6]

Such a conceptual agnosticism based upon a profound opposition between nature and society found its legitimation in the division of labour among the sciences and in the historical process of constructing their own disciplinary identities.[7] Was it not an even longer-term process of differentiation through which »society« emerged from an anthropomorphic and organic mixture with nature and the cosmos, which then, when society had finally established itself as a separate entity, saw in turn the natural sciences emerging from the social matrix? The later-born social sciences, children of the Enlightenment and spurned during the Industrial Revolution, were passionately concerned with the problems of the newly emerging society and its order: the political order, manifest in new demands for rights and representation; the economic order largely operating through the market; and the social order in all its many-faceted dimensions of societal division of labour, solidarity and interpersonal relations. While the natural order intermittently figured as »ideal«, metaphorically or ideologically interpreted, to which the actual deficiencies of the social world could be compared, its real place, however, was »outside« of society. In this society-centered, nature-exclusive view that came to dominate the social sciences, a tiny enclave was at last established within the sociology of science.

It was science and technology that had become visible as the major intermediary links between nature and society; powerful forces of transformation that impinged upon society, that were finally coming under more systematic observation and empirical analysis from a social science perspective. »Science« was »society« in its dealings with »nature«, with technology as its more utilitarian branch. It is therefore not surprising that what was considered open to observation was initially the institutional structures and norms operating inside science and the organizational behaviour of scientists themselves. Science as a system meant essentially the social support structure enabling the study of the natural world and the production of knowledge it entailed. Within the structural-functionalist paradigm as well as within the early sociology of knowledge, the care of natural science knowledge, intellectual objects and constructs that constituted nature remained eclipsed. Yet, from its very beginning, there existed an alternative approach to a sociology of knowledge and of science limiting itself to a more narrow social perspective. Nourished by a positivistic philosophy of science, it culminated in the programmatic appeal of a »science of science«: the best that the scientific method had to offer was to be applied to the institutions that produced science, to their products, and procedures. Almost in a mirror-image-like fashion, the social institutions were to be treated as though they were part of nature. For Derek de Solla Price, for instance, the quantitative method would enable the study of science like a gas under the varying conditions of temperature and pressure.[8]

Only within the last decades a slow, somewhat erratic convergence between these opposite views was initiated. It was triggered largely, but not exclusively, through developments that took place within epistemology and that helped to advocate constructivism. As Freudenthal and Löwy see it, it was specifically the antipositivistic and antirealistic shift within the philosophy of knowledge and of science, linked to names like Quine, Kuhn, and Feyerabend, that drove home the point that theories, and propositions in general, are underdetermined by empirical reality, so that (even) scientific knowledge is not simply mirroring reality. This insight opened the way to arguments that (even) scientific knowledge and a fortiori all other knowledge may be codetermined by social factors.[9] On the side of sociology, the rise of ethnomethodology with both its participant-observatory and constructivist bias as well as the integration of theoretical notions like that of »symbolic capital« and the structure of the »scientific field« as advanced by Pierre Bourdieu opened the way for sociologists of science to enter the laboratories and to take a closer look at the »manufacture of scientific facts«.[10] A renaissance of sociology of scientific knowledge occurred in the wake of these and related developments[11] to prove - be it through a strong or a weak program - its assumptions of the determining influence of social factors on the epistemological core and structure of knowledge. Equally influential became theoretical and empirical points of view directly or indirectly derived from Bourdieu who had from very early on emphasized the sociological dimensions of power and of scientific struggle that were taken to culminate in the power to define science and scientific reality[12], hence implicating the nature and social production of scientific knowledge itself. The stage was thereby set for what sociologists of science would now be looking for: social interests, the definition of scientific reality, the construction of texts and facts, the struggles for reputation as well as the ultimate prize that would mercilessly replace the old-fashioned goal of »truth«: success. The clamorous - and successful - entry of laboratory studies into the field of social studies of science was overshadowed by a somewhat artificial controversy about relativism and realism, while other issues received much less attention than they deserved.

Prominent among the unresolved issues and vigorously coming to the fore in both approaches to be examined here, is the problem of which language and concepts to use. In what language - meaning what kinds of concepts and dominant explanatory mechanisms that are embedded in them and hold them together - should sociology of science speak? Can it escape the imitation, conscious or not, of the scientific language used by its subjects and their way of seeing and interpreting the world? If sociologists of science following scientists in action have an open mind and approach their objects of study without preconceptions, as Latour urges us to do[13], will they not inevitably end up becoming either (mere) interpreters or adopt the presumably more powerful conceptual inventory to which they have now gained access in the hope of becoming more powerful themselves by association? That these problems are by no means problems of social scientists only following natural scientists or the development of science but pervade all other attempts to account for what scientists do and through what mechanisms science progresses is illustrated by attempts like the most recent one to recast scientific development as an evolutionary process. David Hull, reconstructing a history of taxonomy covering almost 2,500 years but examining the last 40 years in minute detail, wants to account for scientific progress in terms of mechanisms based entirely on the selection of concepts in close analogy (if not isomorphism) to the mechanisms of selection that drive biological evolution.[14] In these and similar attempts[15] it becomes obvious that a very fine line has to be threaded between merely drawing analogies, fruitful as they may be, and of hypothesizing isomorphic relationships that are extremely difficult to prove. The other danger comes from »converting« completely into the language and the conceptual way of seeing and interpreting the world in the light of one single discipline.

That a danger of »going native«, as the anthropologists call it, was inherent in laboratory studies was in evidence from their very beginning. In fact, it was Jonas Salk, the director of the Salk Institute for Biological Studies in whose institute the first study of this kind was carried out, who stated it, presumably unintentionally, in his introduction to the book that resulted from his opening the lab to those he saw as being »nonscientists«. After admitting honestly to the usual prejudice scientists hold against outsiders when they approach science, he praises the presumed achievements reached by them:

»In this book the authors demonstrate what they call the ›social construction‹ of science by the use of honest and valid examples of laboratory science. This in itself is an achievement for they are, in a sense, laymen to laboratory science and are not expected to grasp the fundamentals, but merely expected to comprehend only that which is easiest to understand, such as the superficial aspects of laboratory life.«

He then recounts his »amazement« when realizing »how ›scientific‹ a study of science could be when viewed by an outsider who felt impelled to *imitate* the scientific approach he observed (italics, H.N.). The authors' tools and concepts are crude and qualitative, but their will to understand scientific work is consistent with the scientific ethos ... for me the most interesting part of the work and of its outcome, is that Bruno Latour, a philosopher-sociologist, began a sociological study of biology and along the way came to see sociology *biologically* (italics in original). His own style of thought was transformed by our concepts and ways of thinking about organisms, order, information, mutations etc. Curiously, instead of sociologists studying biologists, who in turn are studying life processes - in a sort of infinite regression - here are sociologists coming to recognize that their work is only a subset of our own kind of scientific activity, which in turn is only a subset of life in the process of organization.«[16]

Latour has followed this route with consequence, remaining faithful to his own program. Recently, if I interpret him correctly, he has advocated with eloquence that sociologists of science should merely act as the representatives through which »nonhumans«, that is, the objects of the natural world or technological or intellectual artifacts, such as electrons, gravity, microbes, »speak«.[17] Looking at his own writing, however, it is also obvious that it cannot be sufficient for a student of scientists in action to merely speak like a biologist, high-energy physicist or analytical chemist. Rather, Latour himself, when coining new phrases or introducing a whole new set of terminology like enrolment, inscriptions, translation or interessement, is much closer to the language of (French) literary criticism and semiotics than to the world of nonhuman entities like gravity, electrons or microbes or to the language that scientists make them speak.

Latour is extreme in his demands for »going native« and radical in his social constructivism. Since only humans speak, language is the final arbiter of all scientific struggles (»Language is a faculty whose exact distribution is the very outcome of our fiercest struggles»). Thus, humans necessarily become »spokespersons«, »who talk, represent and interpret non-humans«. In Latour's »actor-network« approach, nonhumans have become »party to all our disputes«, they are seen as being »actors like all others - open or closed, active or passive, wild or domesticated, far or close depending on their (and our) negotiating strength«. But they also need to be endowed with some kind of language through instruments. »Microbes, electrons, gravity do the talking and writing, not Pasteur, Millikan or Newton«.[18] But since most of us, I would like to add, cannot hear or listen to them without microphones, microbes and someone sitting in an interpreters' booth, the problem is merely relegated to another level. We still need, it seems, interpreters and translations even at the most elementary level, namely to listen to what is being said and what goes on in the original language.

The problems of representation raised by the questions of language (whose language is to be used) and which concepts (powerful in different degrees) are not only, as W. Shrum correctly notes, problems common to the study of all institutions that are »successful« and that therefore can raise the questions of what words should be used to describe and account for what scientists do and why they are so good at convincing others of their value.[19] Nor is it only a problem of representation in the sense of analyzing the assumptions under which representation occurs. For representation occurs at least on two levels: that of standing in for someone else, which raises questions described above of who speaks for whom in what and whose language, what effects this has and how much »distortion« and of what kind occurs on the way. But there is also the deeper level of representing concepts in images and words. This brings us back to the question raised at the beginning of this paper, whether in fact social studies of science have succeeded in overcoming the »dividing line« that separates the natural world from the social. If Latour proposes to abolish, for the purpose of accounting for the power of science, all distinctions that normally are made between humans and nonhumans; if he - boldly and correctly - claims that microbes can alter human relations and therefore ought to be included in social studies of science as fully legitimized members to be studied and to forge alliances with, this is not only a radical and original insight, but one that points to the direction in which the social sciences in general ought to move by incorporating at least a part of the nonsocial world. But if one looks closer, the radicalism falls far short of its self-proclaimed aims. It is one thing to propose to draw nonhumans into human games of competition or struggles for power. But is quite another - and it is a pitiful spectacle - to see them take on anthropomorphic qualities in this process. Latour speaks of the »politics of things« and it seems that humans are able to impose, or at least extend, *their* politics on them. But what is gained by it? If we cannot be sure when the scientists' nonhuman allies, entities like gravity, electrons, or microbes, speak physics or biology and when they speak sociology, political science, or literary criticism, we have either succeeded in building another tower of Babel, or social constructivism, despite its claims to the contrary, has simply turned back upon itself. It has become grounded then exclusively in *social* constructions, the most social (and flexible) of which is language itself. In doing so, nonhumans have become fully colonized: they act and even feel exactly like humans; they have been transformed into anthropomorphic puppets. To a lesser extent, these limitations of social constructivism hold also when applied to technological systems and artifacts. While making plausible and identifying research sites for the »thick descriptions« of social influences on technological development, relatively little can be learned about the material or technological qualities of the artifacts, although they might also contribute to their success or hardness. The »interpretative flexibility« of artifacts, particularly in times of controversies, clearly takes precedence over other constructions. It elucidates the process of closure, stabilizing the socioeconomic or political interests of the relevant social actors, but technological artifacts remain a distorted and humanized junior partner throughout the process.[20]

Social constructivism as evidence in the actor-network approach and in similar attempts to pursue the social construction of technology has made many inroads into the territory of »nature« and »hard technology« after having crossed the boundary that prevented the older sociology of knowledge and of science from addressing such issues. It has focussed upon the constructivist interpretations that render plausible a certain degree of flexibility and of indeterminacy that reside in artifacts and theories, and it has equally plausibly demonstrated the social nature of science in its agonistic and power aspects. But by edging its way a bit further into the realm of nature and technology, by filling interstices that are malleable and open to social influence and interests, and by showing the gradual process of »closure« that occurs on the way, it also remained profoundly bound by its own means of interpretation, that is, by the *social* nature of its *constructions*. When nature and scientific knowledge become thoroughly anthropomorphized, when nonhumans are made to behave exactly like humans, the social construction of science becomes a construction modelled entirely after its constructors, in this case the social scientists. But the point about constructivism is that claims built upon constructions can only be compared and assessed against other constructions. This is a point well underscored by Latour, who proposes that controversies will tell or that deliberate »trials« set up between rival constructions will allow us to find out who is the winner. In these trials, each side musters all the allies it can, human and nonhuman alike, including their relative political strength and strategies as well as the forces of nature that have become a part of each network. The trial will tell us which side is »stronger«, »harder«, more »resistant«, or »robust«. But since all that can be said about a network remains couched in social terms only, we *learn little about qualities, processes, or mechanisms that confer »robustness« or »hardness« to some objects but not to others*. It all depends, *Latour concludes stoically*, on the *concrete historical situation*.

Let us now turn to the other constructivist account of science, constructed as a self-organizing system, and see how it fares when confronted with the same kinds of problems that have plagued social constructivism and the actor-network approach. Seen from an actor-network perspective, a theory of science as a self-organizing system can already boast a great comparative advantage: it is a theory whose proponents have linked up with relatively powerful allies. They include a concept of »system« that has been empirically enriched in its operationalization and empirical content in fields ranging from thermodynamics to lasers and the neurophysiology of the brain. Moreover, there are relatively precise indications for the mechanisms that separate each system from its environment, and claims are made that these mechanisms apply to *any* system, regardless of its location in a specific discipline, in nature, or in society. From its general perspective, no distinctions between systems are made. Social systems are therefore only one out of many possible and actual constructions. But epistemologically speaking, a theory of science as a self-organizing system is constructivist in a radical way, as Spencer Brown, v. Glasersfeld, and others have maintained.[21] Before proceeding further and mustering the relative strengths of both claims and the allies that both sides have lined up, let me give a highly simplified description of the main principles of self-organization as Niklas Luhmann, among others, has developed before turning to the theory of science as self-organizing system proposed by Krohn and Küppers.[22]

Contrary to the old system theoretical view, according to which the system was seen as adapting more or less mechanically to its environment, the new system theory assumes that the system itself defines and differentiates its own boundaries, thereby constituting everything as »environment« that is beyond this boundary. The relation between system and the environment is organized by a set of boundary conditions that constrain the arbitrariness of the structures of the system and expose it to evolutionary selection. »The world« is always seen and interpreted through the systemic reference, that is, through the interface that differentiates the system from its environment. In the case of a social system, the interface can become an instrument of reflexivity for the system itself, thereby increasing its internal complexity. The elements of a system can either be substantive units (e.g. individuals in the case of a social system), or »operations«, which are said to be self-referential or autopoietic, as the case may be.[23] They can only be produced and reproduced with the help of a network of similar operations. There exists a difference in complexity between system and environment insofar as each environment is always more complex than the system itself (for the system). This makes it necessary for the system to reduce environmental complexity, but such reduction can only take place within a system and through reference to the system and its environment. While there are various possibilities for effecting such reductions of complexity, the most successful has been the evolutionary. It has basically been achieved through an increase in the system's own complexity, which in turn facilitates more heterogeneous relations and operations between system and environment. Self-referential and autopoietic systems are inherently »restless« and disposed to reproduce, while the environment functions both as precondition for the possibility of reproduction and as constraint: the system is supported by the environment and perturbed, but it is never forced to adapt (as the old system theory had it). The system produces and reproduces its own structures in order to facilitate the continuation of the self-organizing process - or it ceases to exist. Perturbations that come from the environment allow the system to restructure itself.

The relationship between system and its environment is not determined by simple input/output mechanisms but by structural coupling; a system is coupled to its environment not through operations but only through perturbations. It is on the basis of its eigendynamics that the system decides how to cope with perturbations. Either they are »taken in« and then become part of the system's order by increasing it, or they are not taken in and hence dissipated. In any case, it is the system with its eigendynamics that regulates how to operate under the impact of perturbations. Another tenet is that the complexity of the world must have found an adequate representation in the system that can be actualized by »making sense«, as Luhmann puts it. In a more general way, this means that as long as the boundary conditions define the states of the system and vice versa, information about the environment is also coupled with the eigendynamics of the system. It follows that the system has its own distinctions as to what constitutes information and what not: information is a purely system-internal property. Information is not transferred from the environment to the system (the environment may at best hold »data»), for in order to constitute »information« some kind of distinction has to be made that allows the separation of »this«, but »not that«, that is, a boundary must be drawn so that self-observation becomes possible.[24] For Luhmann it follows from these assumptions that each concrete subsystem of society is based upon a binary code (»the distinction which matters») that constitutes the world view through which the system distinguishes between »possible« and »notpossible«. The binary code performs a number of useful functions: it integrates the two opposites; it produces paradoxes, and so forth, but it also facilitates further differentiation of the system along the lines indicated by the code. According to Luhmann, the codes of the various societal subsystems are badly integrated among themselves and cannot claim any validity outside their own subsystem. Nevertheless, subsystems »act« upon each other, at least indirectly. They can »export« distinctions that then can become relevant environments for other subsystems, provided that they are recognized as information there.

Science, according to Luhmann, constitutes such a societal subsystem. Its binary code, which provides the line for differentiation, is that of true/false. The code specializes in communications of a special kind: those that refer selectively to nonsubjective experience and the production or acquisition of new knowledge. The evolution of science consists essentially in the increase of the ability of the system to dissolve and recombine knowledge: the higher the rate at which this happens and the more visible, the more powerful the consequences. Due to its equally highly developed structures for producing resonance, the dissolution and recombination potential is very great indeed, but whenever the world begins to solidify, it can become transparent again through the operation of a new dissolution and recombination. This is one reason why society as a whole has not taken over the scientific world view: science is useless for societal orientation. But science does have an impact. What it really »exports« is twofold: it is a general awareness of selection with regard to undetermined (unknown/untried) recombinant possibilities and it is technology as a recombination that has already taken place and therefore has taken on a definite form. Only a tiny fraction of what is scientifically possible is realized - the rest is economically, legally, and politically not feasible. But the pressure waves of contingency that emanate from the scientific subsystem are increasingly felt in other systems that may, or may not, treat them as perturbation.

For Krohn and Küppers, the emphasis is not upon science as a *social* subsystem.[25] Far more ambitiously, they have set out to develop a *theory of science* as a self-organizing system. Towards this end, they have to begin with the »act of creation« of every self-organizer. In the famous words of Spencer Brown, the act of creation consists in »make a distinction«. Their distinction, the boundary that they impute to science as a self-organizing system, runs right through what is normally considered to constitute scientific practice. It is the boundary that separates the core of scientific activity, namely the production of scientific knowledge - which they call »research-acting« (emphasizing the action which always contains the decision of doing research) - from the rest - that is, a variety of the tasks, functions, and activities that scientists normally also engage in, submitting articles to journals or refereeing them, writing research proposals and applying for funds, going to conferences or waiting in the corridors of ministries, and so forth are all activities grouped together under the name of »science-acting«, that is, acting into and thereby structuring the »environment« to the »system« constituted by research in the sense described above. As with all other system/environment relations, the system structures its environment first and above all in order to maintain itself, that is, to organize its survival. Its (evolutionary) success in doing so depends on the kind and complexities of the internal structures it has been able to build up. In the production of scientific knowledge, the typical »operations« that occur are based upon cooperative structures among scientists. They are organized in such a way that typically recursive and operationally closed interactions take place. For Krohn and Küppers, the research group is the basic acting unit of the system. It is built by recursive interactions among the members of the group that thereby also maintains its identity and boundaries through mechanisms of inclusion/exclusion.

The main point, diverging from common sense and conventional theories of science, is that it is not the nature or content of knowledge that matters per se, in the sense of, for instance, certain kinds of knowledge production being able to attract more funds. Recall that it is the system that structures its environment through its self-organizing mechanisms. Following its own criteria of evaluation, the system attempts to create an environment that holds the favourable conditions for its continued existence. This means, among other things, all that research groups really want is to be able to continue to do research (and this as unperturbed as possible). Their »science-acting« has the sole purpose of serving this end. Alas, the environment does not permit this without producing some kind of - essentially unpredictable - perturbations. How a system reacts to a perturbation is exclusively defined in terms provided by the system itself. Perturbations do not lead to adaptations, but they regulate the eigensolutions that the system has found at a previous stage. If the perturbation is sufficiently disturbing (in terms specified by the system), then the system will find a new eigensolution and alter its structure accordingly. The theoretical term for this process is »order through fluctuation«.[26] This is basically also the way in which systems evolve over time.

The epistemological core of this model for a theory of science as self-organizing system is thus put inside the system constitution »research«. According to Krohn and Küppers, the »operation« of scientific knowledge is a complex set of formations and procedures with a structural and a procedural side. The structure is formed from the combination of theoretical, methodological, and empirical, as well as interpretational components, while the procedural side operates through the repeated processing of new entry values that have been obtained. Furthermore, a distinction is made between an argumentative component based upon a true/false distinction and an effective action component based upon success/failure. The »operation of knowledge« produced then has obtained an inner stability of its theoretical, methodological, empirical, and interpretational components. The »operation of knowledge« contains transformations (the operationalization of hypotheses in experimental methods and the explanation of data as theoretically interpretable information) that move the components from being initially »loose« couplings to becoming gradually »fixed«. The operation thus transformed becomes a »result« grounded in scientific conviction. Put in more technical language, »the operation of knowledge applied to its description (knowledge) leaves the description invariant.« It has reached a stationary state that can be called »successful« insofar as it has found an eigensolution (for a certain complex unity, similar to Fleck's gestalt).

It is with regard to these operations taking place inside the research system and their transformation into »fixed couplings« that resemblances to Bruno Latour's approach come to mind again. He points repeatedly to the process of »science in the making« that differs fundamentally from science »made«, that is, the textbook version of scientific facts and results that have become well established. It is the process of the »hardening« of results, of the initial claims that scientists make becoming more »robust« in the sense that it becomes more difficult and »costly« for any critic or opponent to dismantle them. The hardening of results is not so much visible in the research process itself, at least Latour gives no criteria or indication how it could be observed while still in the making. For him, the revelation of hardness comes only in those dramatic moments or episodes when results are indeed challenged, when controversies erupt, and when opponents are engaged in the fierce struggles that eventually lead to victory only for one side and defeat for the other. It is in these battles when the »moment of truth« comes that both opponents reveal their relative strength: it becomes apparent that the networks include not only social or political links, not only careful and skilled strategies, hierarchies and positions, but consist also of the forces of nature and of things. The one combattant that proves to be more resistent to the attacks mounted by the opponent wins. »Reality« writes B. Latour, »is what resists. Let us call trial whatever tests this resistance. The favourite gesture of realists, thumping a table, is a trial of a sort. The table resists their fists. They are right in saying that it is solid and real. Nevertheless, it would be absurd to conclude from this trial by fist that hacking the table with a sledge-hammer will have the same result. Another trial will define another reality relation and hence another reality ...«.[27] But where he keeps insisting on trials followed by more trials, Krohn and Küppers have the more elegant solution.

The concepts of »robustness«, resistance, »stabilization«, or transformation from »loose« into »fixed« couplings, testify to varying degrees to the constructivist nature of knowledge production. The construction of reality is a process in which reality is becoming increasingly stabilized. But what does constructivism mean in the context of the two approaches? For Krohn and Küppers, the theory of self-organization is based on the shared central dogma of a constructivist theory of knowledge: it is a negative theorem. The objectivity of reality offers no assistance whatsoever in deciding between claims or constructions, because any proof or reference to reality falls back upon constructions that already stabilize knowledge (about reality). Since a self-organizing system is by definition operationally closed, there can be no contact with reality, neither at an elementary level nor are there any elementary unmanipulated objects (as realism holds) that are not already influenced by theoretical decisions and conventions. Knowledge as a construction can only be controlled by other constructions.

The main difference between the two approaches towards constructivism lies in my opinion in the following. For Krohn and Küppers, the construction of scientific knowledge is seen as a complex operation culminating in the (repeated) finding of eigensolutions of the knowledge-producing system. Therefore, research results have to satisfy a number of preconditions or criteria; the work of a research group is oriented towards the production of such a product. »The construction problems to be overcome in the process result not only from the conditions of stability, of theoretical consistency and from the reliability of the experimental arguments, but more from the *improbability* (italics, H.N.) that couplings occur which allow a fixed arrangement between both sides, that is, the argumentative based on t/f and the experimental based on s/f which are usable for further research.«[28]

Eigensolutions therefore can be interpreted as truths or knowledge of reality under the conditions that they do *not* provide information *about* reality (italics, H.N.), but a description of an operation in this reality, which is in itself closed, intellectually stable, and effective. Moreover, the finding of an eigensolution or its setting up does not exclude the existence of others. Krohn and Küppers stress that alternatives and different routes are always open to the research group and its rivals; they do not follow any logical sequence or procedure in the path of development of knowledge and can be modified until the last moment. Operations of knowledge and hence, the finding of eigensolutions do depend, however, upon the specified conditions of cooperation in their cognitive-emotional matrix, that is, upon the social interactions that prevail in the research group. The group defines for itself its own mode of working. Hence it is free to decide upon the modification of its operation of knowledge. There is no coordination between the process of group integration and the operation of knowledge in the sense that an increasingly fixed social interaction would result in an increasingly stable operation of knowledge. It all depends upon the actual working of the group and its finding of a product that has an eigensolution. In this sense, group integration acts as boundary condition for the operation of knowledge.

In Krohn and Küppers' view therefore, the research group as the knowledge producing system (or agent in Latour's terminology) does not depend upon finding any »heterogenous allies« that are nonhuman, while Latour's scientist, who is portrayed as a lonely fighter, needs such assistance. For Latour, the power of social constructions is encompassing since it penetrates defiantly the world of nonhumans and succeeds in involving them in human politics. Things and nature are made part of a kind of *Realpolitik* in which everything becomes enmeshed in heterogeneous networks and alliances. All is done in order to make claims more resistant, but resistance can only be revealed in real social trials. The seamless web of forces, as W. Shrum has called it, consisting of the social, political, economic, technical, and rhetorical, leads to a pan-social construction in which Gods, demons, plants, viruses and electrons freely engage with humans and vice versa, entangled and conscripted as they have become to the games that humans as scientists play among themselves. Success comes to those who have succeeded in finding the more powerful allies, the more resistant and costly claims and constructions. But in the end, this is a social and political world that has been extended to the realm of nature. It is the *social* constructions - success, costs, victory/or defeat, struggle, and so forth that determine how science works and how scientific knowledge is produced. Latour has bravely opened the door for nonhuman agents to enter, but when they do, they have been thoroughly socialized and anthropomorphized. In the end, they behave like us. Realism may have been defeated, but the victory of the relativism of social conductivism is obtained at a very high cost.

Krohn and Küppers' constructivism is a social one. It is social in the sense that it is based upon the social mechanisms of group integration (essentially a Fleckian notion) that provide the boundary conditions for the »operation of knowledge«. The latter never becomes arbitrarily social since it exhibits a distinct structure of its own. Although it follows its own dynamics, it remains tied, however, to the social group (in this case the research group) that produces it.

It is a constructivism that has been derived and adopted from a (relatively) successful construction: the theory of self-organization that has become stabilized in different scientific fields including biological accounts of the working of the human brain. By radically closing access to reality and by pitting construction against construction, it resembles only superficially, however, one network of allies having to confront another network of allies. Rather, the resistance of successful constructions is based upon their having found an eigensolution: a product resulting from the operation of knowledge that has to satisfy certain specified criteria, both of an argumentative and experimental (effective) kind. But the emerging eigensolution as a construction that has gained a certain amount of stability as well as coherence becomes in the ultimate moment, that is, when it reaches its stability, independent of its producers and of the conditions of production, although retrospectively it has to be seen as having arisen from the social matrix that specifies the interactions of the research group. It thus bears an imprinting of the social conditions of its production, but at the same time has evolved into a product in its own right. The construction is one that has detached itself from the social matrix out of which it was born, while still bearing the marks of its social parentage. It is a constructivism in which social interactions (of a cooperative and not an agonistic nature, moreover) serve as preconditions in the sense that the recurrent interaction of members of a research group are needed for knowledge constructions to evolve at all. Once they have found their eigensolution, however, they cease to be social constructions other than by origin. The marks of their origin remain with them, however.

In a recent paper, which is impressive through the clarity of its observations and argumentation, Terry Shinn has shown how the research products of a French research laboratory in experimental physics, that is, the publications written by the members of this lab, differ systematically in results, and that a relationship exists between these results and the social positions that their authors occupy in the laboratory.[29] Shinn calls this the »social hierarchy of research results« which he compares with results emerging from an equally existing, but more latently operating, cognitive hierarchy. Although a closer analysis of T. Shinn's results would lead too far in the present context, it is nevertheless this kind of empirical, laboratory-based research that is very illuminating for our discussion. Shinn's findings lend support to the *dual nature* of research results that have found their eigensolution in the sense just described: they clearly bear the marks of their social origins. In this case of the hierarchical social position that researchers occupy in the French lab, while at the same time they obey specified criteria of an argumentative and effective (experimental) kind that qualify them as having found their eigensolution. This is all the more striking since Shinn's results have not been obtained with »testing« constructivisms in mind, nor have they anything to do with notions of self-organization. His paper makes one wonder, moreover, how the social hierarchy that manifestly operates in the lab and in the research results alike, has obtained *its* high degree of stability. This opens a whole new set of questions about social interactions and how eigensolutions found in the operations of knowledge might in turn serve to back up and stabilize social interactions. Another track which cannot be pursued here, however, is the role that Peter Galison assigns to stabilization occurring both at the level of the experimental situation and at the level of the scientific community.[30]

The purpose of the comparison between the two constructivist approaches has not been to stage a trial of relative strength, although, stimulated by Bruno Latours proclamations, I could not help but occasionally apply his concepts to what I was doing. Of course, a theory of science as self-organizing system could be recast in Latour's frame of reference as having been more successful, since Krohn and Küppers had the more powerful concepts and the help of *all* systems on their side. In doing so, Latour's approach would, of course, thereby become vindicated. But I think that this is too easy a way out for all concerned.

The comparison has shown, I conclude, that *social* constructivism by remaining too closely bound to the social world and its concepts (struggles, success, victory, allies, power, etc.) has limited itself - perhaps unnecessarily - even when it has entered the realm of nature, of technological artifacts and nonhuman entities - all of whom have been invited or conscripted to join the human struggles and games. To put it for once also into the language of selforganizing systems: Latour's mark of distinction has been to make no distinction. But in doing so, he anthropomorphized everything and the price is high. Moreover, he refuses to make any prediction whatsoever, relegating us to a world consisting of single events (the trials or controversies through which the winners are determined) that can only be analyzed retrospectively. Constructivism, as it underlies Krohn and Küppers' approach, avoids limitations that block social constructivism. It contains, as I have tried to show, especially in connection with empirical data, interesting new avenues that might help us collectively to overcome some of the persistent and old problems that the social sciences have inherited from the century-old separation prevailing between nature and society. In further pursuing hints about the dual nature of research results, we might come to understand better how the social world is represented inside nature and vice versa, or how constructions operate through and upon other constructions.

With regard to the second issue raised initially, which conceptual language a sociology of science or a theory of science is to use if it wants to avoid the trap of either becoming completely integrated into what it studied or of merely transferring concepts from a specific discipline to ways of studying it through metaphor or analogy, it remains to be seen to what extent a theory of science as a self-organizing system will fare. So far, attempts at constructing theories of science have either been meta-theories (and failed) or have taken the leading discipline of the day as the model after which all other sciences have been construed. It is unclear at the moment to what extent the epistemological foundations upon which the theory of science as a self-organizing system rest, namely the biological and neurophysiological descriptions and explanations of how the human brain operates (as theorized by Maturana), will provide a stable basis for further knowledge to build upon. It is partly couched in the language of a leading discipline that has relevant things to say. How far this language will carry remains also to be seen. It may well turn out that multi-linguism will be the selective criteria for future theories and social studies of science, including abilities like those of code-switching.[31] But it also may turn out that a multi-levelled kind of multi-linguism will be needed: constructivism grounded in neurophysiological theories of cognition may serve well for explaining perception and even operational descriptions of perception. As social beings, we are still bound by the physiological and neurological structure and functions of the human brain to certain ways of perceiving the world. Nevertheless, new structures and complexities enter on the level of the social system. They may again be cast in terms of self-organizing principles, but new levels of complexity in describing and explanation would need to be added and so on ... On all these levels, the processes of self-organization may operate by operational closure, giving rise to structures that regulate their own criteria for information, relevance, resonance, and so forth, and thereby set (internal) standards, while at the same time deviations from these standards are minimized.[32]

To obtain coherence between the different levels of complexity may also exact its price. Success, also for the system of science, is defined in degrees. First of all and basically, a system is successful if it is able to maintain itself. It is then successful in increasing its eigen-complexity, which is not simply a quantitative or qualitative increase, but rather manifests itself in generating more and heterogeneous kinds of relations to its environment as well as its capacity to structure a more complex environment. This means that within the system, it works more selectively; its structures and components become more contingent: it evolves.

The system of science, to recall Luhmann's view, has been successful with reference to society not by providing a world view or by bestowing upon society the scientific method. Neither has been taken over by other social subsystems. Rather, science excels in the ability of dissolution and recombination. Technology is based upon »successful« recombinations that have hardened since they have already taken place. New recombinations, made of theoretical knowledge, are in store with an almost indefinite potential for recombination. It is the technological products and the promises of future products based upon theoretical knowledge and their recombination that make science successful beyond its own confines.

Yet the question remains as to the conditions that enabled such successful operations. Operational closure with complex internal structures in the sense of having obtained a relatively high degree of autonomy of the research system would perhaps be one possible answer. But the relationship of science as a system or as a societal subsystem to its environment remains a tricky one. For there remains quite a bit of anticipation built into evolution. Science enables society to structure its environment in the most encompassing sense, but it depends on this environment for its own survival. If a theory of science as self-organizing system enables us to see old problems in a new light, it may well become a potentially powerful ally to be adopted with considerable pay-off by social studies of science.

Notes and References

1. W. Krohn, G. Küppers, R. Paslack, »Selbstorganisation - Zur Genese und Entwicklung einer wissenschaftlichen Revolution«, in S.J. Schmidt (ed.), *Der Diskurs des Radikalen Konstruktivismus*, Frankfurt am Main: Suhrkamp, 1987.

2. See, for instance, the books by F. Vester, *Denken, Lernen, Vergessen*, Munich: dtv, 1978; *Phänomen Streß*, Munich: dtv, 1976; *Unsere Welt - ein vernetztes System*, Munich: dtv, 1978.

3. Krohn, Küppers, Paslack, *op.cit.*, 1987, note 1.

4. G. Roth, »Die Theorie H.R. Maturanas und die Notwendigkeit ihrer Weiterentwicklung«, in Schmidt (ed.), *op.cit.*, 1987, note 1.

5. For a recent summary of literature on technology and technological systems, see B. Joerges (ed.), *Technik im Alltag*, Frankfurt am Main: Suhrkamp, 1988.

6. V. Meja, N. Stehr, *Der Streit um die Wissenssoziologie*, 2 Bd., Frankfurt am Main: Suhrkamp, 1982.

7. L. Graham, W. Lepenies, P. Weingart (eds.), *Functions and Uses of Disciplinary Histories*, Dordrecht: D. Reidel Publishing Co., Yearbook in the Sociology of the Sciences, 1983.

8. D.J. De Solla Price, *Little Science, Big Science*, New York: Columbia University Press, 1963.

9. G. Freudenthal, I. Löwy, »Ludwik Fleck's Roles in Society: A Case Study Using Ben-David's Paradigm for a Sociology of Knowledge«, *Social Studies of Science* **18** (1988), 625-51.

10. K. Knorr-Cetina, *The Manufacture of Knowledge*, Oxford: Pergamon Press, 1981.

11. H.M. Collins, »The Sociology of Scientific Knowledge: Studies of Contemporary Science«, *Annual Review of Sociology* **IX** (1983), 265-285.

12. P. Bourdieu, »The Specificity of the Scientific Field and the Social Conditions of the Progress of Reason«, *Social Science Information* **14**, 6 (1975), 19-47.

13. B. Latour, *Science in Action*, Milton Keynes: Open University Press, 1987.

14. D. Hull, *Science as a Process. An Evolutionary Account of the Social and Conceptual Development of Science*, Chicago: University of Chicago Press, 1988.

15. S. Toulmin, *Human Understanding, Vol. 1*, Oxford: Clarendon Press, 1972, was already advocating an evolutionary approach to the development of scientific concepts.

16. J. Salk, »Introduction«, in B. Latour, S. Woolgar, *Laboratory Life. The Social Construction of Scientific Facts*, London: Sage Publications, 1979. See also H. Nowotny, »Leben im Labor und Draußen: Wissenschaft ohne Wissen?«, *Soziale Welt* **33**, 2 (1982), 208-220. J. Salk's »Introduction« was, as it turns out, actually written by B. Latour himself. Bruno Latour, written communication, 18-02-89. This does not detract from the argument presented here, however, on the contrary.

17. B. Latour, *Clothing the Naked Truth. Guiding through some Objections Raised against our Sociology of Science*, Paris: Ecole des Mines de Paris, 1988. If not otherwise indicated, all references and quotations to Latour's work refer to this source.

18. Latour, *op. cit.*, note 17.

19. W. Shrum, »Review Essay: The Labyrinth of Science«, *American Journal of Sociology* **94**, 2 (Sept. 1988), 396-403.

20. M. Callon, J. Law, A. Rip (eds.), *Mapping the Dynamics of Science and Technology: Sociology of Science in the Real World*, London: Macmillan, 1986. W.E. Bijker, T.P. Hughes, T. Pinch (eds.), *The Social Construction of Technological Systems: New Directions in the Sociology and History of Technology*, Cambridge, Massachusetts: MIT Press, 1987.

21. Schmidt (ed.), *op.cit.* 1987, note 1.

22. N. Luhmann, *Ökologische Kommunikation*, Opladen: Westdeutscher Verlag, 1986.

23. For differences, see this volume.

24. Luhmann, *op.cit.*, 1986, note 22.

25. W. Krohn, G. Küppers, »Science as a Self-Organizing System - Outline of a Theoretical Model«, this volume.

26. I. Prigogine, »Order through Fluctuations. Self-Organization and Social Systems«, in E. Jantsch, C.H. Weddington (eds.), *Evolution and Consciousness. Human Systems in Transition*, Reading, Massachusetts, 1976, p. 93-126.

27. Latour, *op.cit.*, 1980, note 17.

28. Krohn, Küppers, *op.cit.*, note 25.

29. T. Shinn, »Hiérarchies des chercheurs et formes de recherches«, *Actes de la recherche en sciences sociales* **74** (Sept. 1988), 2-22.

30. P. Galison, *How Experiments End*, Chicago: The University of Chicago Press, 1987.

31. Terry Shinn, for instance, states at the beginning of his paper that the vocabulary he used to describe the research results of the lab he observed had a three-fold origin: some terms are part of the language used by the physicists studied; others come from the philosophy of science and others have been borrowed from the sociology of science or the cognitive sciences. T. Shinn, *op.cit.*, 1988, note 29.

32. G. Grössing, »Kybernetik der Wahrnehmung - über fundamentale Naturgesetze, Seefahrer und das Irrationale«, *Delfin* **VIII** (1987), 10-22.

(3) The scientific theory of self-organization as a tool for the legitimation of NSM demands, or as a means to their self-understanding: my observation here is that, since the 1980s, the self-reflection of the NSM and the motivation of their sociopolitical program has increasingly had recourse to the scientific concept of self-organization; whereby the reception of this concept is concomitant with a growing need for consistent worldimages of a »new thinking«, and is in part, strongly influenced by »New Age« ideas and the new esoteric trend. (In this context it is to be noted that this agreement with science with a view to the legitimation of social movements is in no way something new: the early Labour Movement already made use of Darwin's ideas.)

1. Self-Organization as the Principle of Organization of the NSM

Self-organization relates here not only to particular NSM groups, such as citizens initiatives and self-help groups, but also to the cooperative interlacing of such groups to social movements: as, for example, to the citizens initiative movement, the self-help movement, the ecology, peace, and alternative movement, amongst others. The organization of the NSM, that of a manybranched and mainly informally structured network of loosely binding initiative groups, is for the traditional sociology of organization a considerable problem to solve in terms of classification: the NSM neither allows itself to be defined as a formal organization along the lines of a business enterprise nor as a normative-legal organization along the lines of a voluntary association with political, religious, or similar world-view orientations; their multiplex form necessitates placing them, as Neidhardt phrased it, »somewhere between interaction and organization, between small group and large society. To this extent, there is no clear-cut place for them in just microsociology, or in just macrosociology.«[10] It is not to be decided here whether the NSM in its *totality* constitutes its own societal subsystem, as is occasionally maintained; for it is difficult to deal with the NSM by way of functional description in the way it appears possible to do so for political and economic systems by having recourse to generalized media of communication such as power and money.[11] Instead, it will be referred to simply in terms of an NSM »network« that clearly shows up certain systemic features: at least there are sufficient recursive interactions between the networked initiative groups to allow one to speak of the existence of an autonomous »alternative culture«. The difficulties organization theory encounters in an attempt to arrange the NSM systematically can possibly be overcome by shifting the focus of the analytical perspective; that is, by focusing on the *self-organizing* character of the NSM. Viewed in terms of organization theory, with the NSM we are dealing with a rather improbable case in that the constitutive force of informal processes extends beyond the dynamics of individual groups and leads to the build-up of a complex network of initiative groups.

Should one now consider the formation of social movements under the aspect of their *self*-organization, then social structural problem definitions such as economic crises or the crisis of parliamentarianism would appear to be of secondary importance in attempts to explicate the *emergence* of such movements. Resorting to the assumed

»objective contexts« fails to recognize that the social situation to which social movements respond is largely co-created by them; described by Klaus Japp as the »objective« environment of a social movement that appears as its policy arena which »it itself played a fundamental role in creating.«[12] Correspondingly, the formation of social movements is not determined from outside, but, at most, provides an *impulse*.

The self-organization of social movements means that they themselves determine their structures and changes. This becomes clear when the methodical starting point focuses on the individual NSM group below the network level and analyses them as social action systems in which the different individual actors' needs and interests are expressed, their knowledge and skills are shared, as also in part, their time and financial resources are invested. This, however, does not just occur in an arbitrary manner. A prerequisite for such investment is the existence of a joint goal of action that provides those involved with identification: be it a protest, (e.g. the prevention of a municipal construction project), be it the actualization of a desired project (e.g. the setting-up of a youth center), be it the self-help projects for the unemployed, for parents, addicts and so forth, or be it the setting-up of small urban or rural communes or the founding of a self-managed collective. Without a common action goal of this kind, the interactions do not result in the solidarization necessary for group-building. However, should it exist, a group dynamical process begins, about which, at this point, only the following observation is to be made: as is the case with all collective action, in like manner, the action of the NSM groups necessitates regulation; the NSM group as an action system must create its own rules of behavior while conforming to certain boundary conditions (an example of which would be norms related to criminal law). The self-organization of NSM groups (and their networks) are to be understood in terms of their own dynamics *and* their interactive connectedness with the societal environment. Only then can it be shown that the self-organization of groups is the requirement for their autonomy, which »internally« is displayed in the formation of its group-specific norms and description of reality, and »externally » is shown in the ability to reach agreement on mutual boundary conditions for reciprocal exchange vis-à-vis other societal groupings and institutions. For, as time goes by, the more specific and coherent the groupinternal interactions become, the clearer the group's own rules of selection are outlined and, consequently, the more clearly defined are the connection rules for further communication, as also rules governing group inclusion and exclusion. In short: a group-specific, socio-cultural »actionmatrix« gradually crystallizes from this, as it were a specific actionsemantic that affects the cognitive, normative, emotional, and strategic behavior of the group members.[13]

The self-referential »semantic condensation« of NSM groups, especially those from the »alternative scene«, shows itself not only in the fact that action and living within the group is in accordance with criteria of a »better life«, but also that these criteria are thematized *as* alternative. In other words, alternative life-style is lived experience and is, at the same time, the subject of discourse, experiment, demonstration, and so forth. The actualization of alternative life is experienced *as* alternative everyday living and working and is propagated as progressive in terms of evolution. It is essential for NSM groups not only to maintain the processes of self-organization; this applies to all self-organized groups, such as family and research groups. It is, moreover, of particular importance for NSM groups to maintain *reflexivity* in the course of self-organizational processes, more so than it is for families, research groups, associations, and so forth, which as recognized institutions need not defend their *form* of organization. In this context then, that would explain the justification and publicity work on the part of the NSM groups.

So long as the NSM groups succeed in creating a mainly unbureaucratic, antiauthoritarian, and understanding-oriented internal world, then only so long are they also able to remain as institutions for *self*-representation of citizens' interests, *self*-help groups, and *self*-administered cooperatives. There is, indeed, always the »danger« that in the course of the legalization of initiatives and as a result of claiming »government cash«, traditional forms of work allotment, bureaucratization, and professionalization would take root again. Attempts to ward off this transmutation of »movement energy« into »establishment energy«, to turn to Josef Huber's formulation, remains at all times precarious, and it is to be expected that parts of the NSM will abandon their »self-organization culture«.[14] The maintenance of self-organizational potential means, then, that the group members must have access to the ability to intercept such erosion tendencies within the group and, at the same time, to secure a minimum of cohesion. The latter is only then possible, however, if at least a fundamental consensus about values remains opinio communis[15], to which all group innovatory processes can be referred.

That the NSM groups stress the importance of the maintenance of their self-organizing interior structures outlines the significant distinction between them and the earlier type of social movements. And even though it may apply that self-organization was important for the formation of *all* »progressive« social movements of modern times as, for example, in the Labor Movement of the 19th century, the NSM is actually the first to reflect and propagate explicitly self-organization. In doing so, self-organization is applied and defined as a principle of social organization. For the NSM, self-organization is *not only* an organizational necessity to be employed as a strategic tool to achieve other political or social goals, although it is *also* that, it is more an end in itself, that is, it is an essential part of their cultural identity and message. In the light of this, the NSM groups would appear to be the descendants of the political and literary societies and clubs of the 18th century or also, in the case of self-help groups, of the organizations of the workers' self-help in the 19th century.[16] However, one must bear in mind that the self-organization of »enlightened« and »cultured« men in the 18th century occurred under the specific conditions (though »enlightened») of absolutism and under the banner of early civic society, whereas today's groups step into a mature democratic tradition and avail of the boundary conditions of a constitutional state which in the past had to be fought for. Likewise, in contrast to soviet republican models (in Germany), here it is to be observed that the understanding of democracy, definition of motivation, legal situation, available technical means of communication, public structures and so forth, have changed considerably, whereby not only the model conceptions have been affected, but the chances for the realization and success of self-organized forms of life and work have also changed.

NSM groups not only organize themselves, whether it be in a nonlegally binding or associated form, but they also *cultivate* self-organization, and by so doing, aim at creating something like an alternative »*self-organization*culture«, which for its part, should prepare the way for the gradual transformation of the existing society into a type of »self-organization society« for which self-determining willingness for social experimentation is characteristic. In short, for the NSM, self- organization is an *eigenvalue,* that is, it is the basis for a new form of collective living and working, a new »paradigm of life-style«[17]; or in other words, the organizational medium is at the same time here its »message«. It is true, certainly, that structural continuity or partially analogous patterns are to be found between the NSM and earlier social movements The differences between them, however, are on the one hand the institutional and life-world framework conditions, the issues and conflict definitions, and on the other the emphatic interpretation of social self-organization as »eigenvalue«, which, seen in the light of movement history, amounts to a qualitative step forward in (self-)reflection and sets the NSM up as favored candidates for an »ideological« reciprocal action with the new scientific concepts of self-organizing systems.

At this point, a short look at *movements at an international level* should show that my presentation oriented toward the NSM in West Germany also holds true, grosso modo, for other industrial nations in the West.

A comparison with the USA, France, and Great Britain[18] clearly shows that since World War II the sociocultural situation in all of these countries has been characterized by two opposite development trends: on the one hand, by a greater degree of *individualization* in the face of a relatively higher standard of living, better education, and increased mobility, and on the other hand, by an increasing *homogenization* of the individual lifeconditions determined by urbanization, mass culture, and bureaucratic-legal standardization.[19] The resulting shifts in the classical lines of conflict provide a feeding ground for greater interest in citizens' self-organization. From this emerges the developing »postindustrial« manifestations of alternative culture that, although pertinent to the international NSM scene, are nowhere articulated in such a fundamental way as in West Germany; also particular to the Federal Republic is the formation of a relatively tightly-knit NSM net with »semantically condensed« main currents. The reasons for this are to be found in the different sociopolitical opportunities for development in the respective countries. In this context, Brand[20] referred to the differences in the mobilization force of the traditional left-right axis in the degree of social state regulations, in the different access to the political system, and in the prevailing political culture.

While the clear-cut separation of centralized and corporate forms of mediation of interests and decentral-autonomously organized movements promotes the anti-Establishment development of the NSM in the Federal Republic, the NSM in the *United States* can look back to a long tradition of political self-organization of concerned groups in which there has always been community self-help and the pluralistic exertion of influence onto a relatively open state apparatus in the form of »pressure group politics«. However, the absence of a securely anchored social state, alongside the domination of property-ownership individualism results in the relative lack of form and mobilization strength of the NSM.[21]

The NSM in *France* share with West Germany the stronger anti-Establishment fillip against what is considered to be an authoritarian state; however, with a revival of the political left since the mid-1970s, the impulse of decentrally organized units has yielded to the partial tying to, and with this the transfer of its movement energy to the socialist parties and trade unions. Here a greater degree of autonomy can only be attained by the regionalistic protest groups and the anti-nuke protest movement.[22]

In contrast to this, the political system in *Great Britain* is characterized by a high degree of accessibility; on the other hand, the Labour Party draws to itself much of the system critical protest. The traditional strong left-right polarization subdues the radical mood of the counter-culture protest and narrows the scope for self-organized social experiment.[23] In spite of these differences, there are obvious analogous efforts being made in the countries mentioned that aim at the utilization and extension of existing possibilities of self-organized forms of self-help, of codetermination, and of collective living.

2. Self-Organization in the Political Program of the NSM.

Self-organization is not only practiced »internally«, but it is also articulated as one of the main issues of concern vis-à-vis different policy fields of the state. The paradigm of self-organization, in the form it is articulated by the NSM groups, is neither compatible with the hierarchical and centralized structures of power and organization nor with the primacy of the functional differentiation of society into service institutions, diverse publics, and private lifeworlds. Certain needs and value patterns traditionally assigned to the private sphere, for example, emotions, spontaneity, solidarity, and so forth, should be included via the NSM into political, economic, and sociocultural life. These postulates extend beyond, though at the same time adhere to the classical demands for greater social justice, equal rights, and so forth, and underline what is new about the NSM program.[24] In line with these principles, the area of formal social relations in politics and economy should be transformed, at least partially, according to the model of informal social relations, in like manner to the way that is already lived in the NSM groups. Generally, it is to be seen as a new definition of the relationship between the private sphere and the institutional sphere (or »lifeworld« and »system« as defined by Habermas).

This section concerns itself mainly with the NSM in the Federal Republic thereby omitting the features particular to movements in other countries where their political efforts are in part less polarizing (e.g. USA), in part equally rigorous (e.g. France), for the reason that they are to a wide extent dependent upon the degree of inflexibility of the respective government administration in each country. Furthermore, my focus is primarily on the dominant constitution conformist and »reformist« groupings within the West German NSM. The aims of these groupings are twofold: on the one hand, tasks now assigned to the state and social sphere (examples of such would be education, culture, and certain forms of psychosocial care) should be returned to the private sphere, thus the autonomy of the private life-style can once again be increased; on the other hand, there should be greater scope for citizens to have the right of codetermination in politics through the introduction of basic democratic elements. In the political arena this means, on the one hand, the struggle for new scope for freer, self-determined activities - either in the form of the relative enclosure of the private sphere towards the social state bureaucracy or as the redirection of work into cooperative forms of production. Groups less interested in this are the more »escapist« initiatives who withdraw into their »niches« in society, as is the case with the majority of small rural communes, but is of more concern to those self-help groups and self-administered collectives who themselves win this scope by taking the offensive and who want to have a transforming impact on society. On the other hand, the political conflict has to do with new forms of political participation or an »optimal decentralization« of political planning and decision-making structures; this constitutes the field in which most citizens' initiatives are active.

On both fronts, the criticism held jointly by all West German groups is functionalism - be it in terms of bureaucracy or in terms of industrialization - in the name of a totally experienced life. The primary goals are to maintain the natural basis of existence, the private sphere, political influence, and the historically grown (e.g. the aged areas threatened by urban renewal); in short, the NSM groups work towards the maintenance but also the spreading of »self-organized« natural and social lifespace. In juxtaposition, an international comparison[25] shows that criticism forwarded by the NSM in all the Western industrialized countries focuses on the trias of industrial modernization (civilization criticism), economic growth, and central government-bureaucratic interventionism; everywhere there is the felt urgency to redefine societal progress and social lines of conflict. The NSM in all of these countries show themselves to be at least latent carriers and catalysts of structural change on a political, social, and cultural level. Thus, while traditional patterns of authority are continually being challenged, there is a persistent striving for the creation of a new collective identity. However, the chances of participation and success are dependent upon the opportunity structures and cultural tradition existing in the respective countries. Accordingly, »German movements ... have still the more fundamental approach, the French are intensive and shortwinded, the British breathe the spirit of ›fair play‹, the Swedish are reformist and integrative, and the Americans are mainly pragmatic-instrumental«.[26] To be sure, the West German example can be used to pinpoint with paradigmatic precision the ubiquitous political issues of the NSM.

Therefore, the area of main concern in the following is the second line of conflict: the struggle for more political participation based on the West German example. The political demands articulated by the citizens' initiatives for more participation and decentralization complement each other to the extent that participation means cooperation »from the bottom upwards« and decentralization means the shifting of competencies »from the top downwards«: The goal in both cases is a progressive representation of citizens' interests or citizens' political participation at a regional and local level, not only in the forms of policy decision but also in terms of content, most especially in the field of infrastructure and social policy decision making (covering energy supply right across the board to medical care). Also articulated are demands for decentralization and participation in economy and technology policy making.

Relatively »realistic« postulates for optimal state authority and for the self-organization of politically and socially mature citizens can be found among the »real political« advocates of the ecology-peace movement, most especially among the cohorts of the eco-liberals and the undogmatic eco-socialists. Should one follow the description of »eco-socialist essentials« as presented by Scherer and Vilmar[27], such moderate-radical demands for self-organization manifest themselves as in:

1. the socio-ethical demand for the creation of an ecologically responsible society that lives and works collectively with common interests and goals; among other things, this would necessitate the disintegration of patriarchy (this demand is formulated by parts of the alternative culture, the small urban commune movement, and the new women's movement);

2. the demand for greater accentuation of the democratic principle, which would lead to the recognition of the parliamentary-democratic principle of representation and thus, to a strengthening in basic democratic and plebiscitarian mechanisms for decision-making (the main advocates of this are parts of diverse citizens' initiatives and the »Green Party»);

3. the demand for an »ecology-oriented economic democracy« based on democratic outline planning and employee control without thereby the wish to remove the market economy mechanism, for an equal share in codetermination for employees, an ecology-based cycle economy along the lines of the recycling principle, a restriction on economic growth, and the removal of industrial conurbation areas in favor of the production of ecologically amicable energy and goods at a decentral level, concomitant with a strengthening in the »informal« sectors of a collective economy in line with the Gorz and Huber »dual-economy« concept[28] (this is demanded mainly by the ecology and anti-nuke movement, as well as some alternative project initiatives);

4. the demand for a »reorganization of the social state« by means of a filling out of the state »social security« network in the form of decentralized and self-organized »voluntary associations« covering such areas as youth, the aged, medical care, educational, and cultural programs, and so forth (speaking out for this are most especially the self-help initiatives, youth centers, and squatters groups); and finally

5. the demand articulated by the peace movement for a principle of a policy of nonviolence, which aims at the gradual removal of nuclear and conventional arms, as well as the removal of the image of the threatening enemy juxtaposed to the increase in the citizens' ability for »social defence«.[29]

To summarize this admittedly somewhat accentuated presentation: »decentralization« and »participation« are fundamental principles of the NSM program, in like manner to the way »self-organization« and »networking« are the fundamental concepts of the NSM organization. Both conceptualizations are closely related and provide the framework for the anticipated new society. Their vision, however, creates the need for justification, and for this reason, science is increasingly being called upon for help; most especially when it develops models of order that allow an understanding of the world according to self-organizing principles.

3. The Relation of the NSM and Self-Organization Research

The NSM assign the greatest importance to securing scientifically, and as their vision of the world, their programmatic »issues« and internal projects both in science and as their vision of the world, not least because they are working on the development of a »new style of thinking«, an all-round »paradigm-change«, and also even an »alternative science«.[30] It is not surprising, therefore, that their »chief thinkers« are on the lookout for philosophical and scientific approaches that would not only provide a base for their own self-understanding but also furnish them with legitimation toward the societal environment. That such efforts were already being made during the early phase of the ecology movement is in evidence however, whereas at that time, primarily models of ecological balance were being sought, anchored in the basic assumptions of the classical systems theory, the focus is now also on the concepts of self-organizing systems that have already gained access to the conception world of the NSM: under this influence, the natural and social world appears as a »networked system« of loosely coupled autonomous systems, and thus is similar to the network structure of the NSM.[31] The knowledge that natural systems beyond the state of equilibrium can reorganize themselves by means of fluctuation, furnishes the argument that in the societal sphere also, self-organizing systems and network structures obtain a higher degree of flexibility and evolutionary force than relatively rigid organizations, structured according to a bureaucratic pattern.

Clearly, the expectations placed on science are, then as now, of a traditional nature: to be instrumental in the provision of »counter-expertise« (against the building of a nuclear power plant, for instance) or to develop alternative types of technology. However, the theories of self-organization are steadily gaining ground due to the attempts on the part of prominent members of the »Green Party« to inaugurate ecology as a new »steering science«, and in so doing, to bring to bear the »network« or the »environment-related« idea of ecology[32] alongside the exactitude model of physics. Within the concept of self-organizing systems, it would appear that understanding and explaining nature, objective knowledge of science, and a moral-practical striving for autonomy are no longer noncombinable, and furthermore, that a union between the divided domains of morals, ethics, and science now seems to have entered the realm of the possible. Under the banner of self-organization, a synthesis that broaches also cosmological dimensions is to be imported to the classical ideas of »permanent peace« and »harmony with nature« - something along the lines of a »biocybernetic world machine«.[33] To give but one example: W.-D. and C. Hasenclever write (1982) with reference to F. Vester's cybernetic of self-organizing system: »Cybernetics shows us that the stability and the developmental potential of a system ... is all the greater the more functioning subsystems it has with the greatest degree of independence,« and therefore, they advise politicians to provide »the respective institutions and bodies with a high degree of self-administration ... here the well-known demand for decentralization articulated by the ecology movement is justifiable on cybernetic grounds.«[34] Other authors from the NSM scene refer to, for instance, Lovelock's »Gaia« hypothesis[35], Jantsch[36], Bateson[37], or to the holistic-oriented New-Age prophets such as Capra and Ferguson, who for their part are greatly influenced by Prigogine, Bohm, Varela, and others.

That such a fusion between scientific concepts and normative-political expectations should immediately provoke a critical ideology critique speaks for itself, but it is not my intention to go into it further at this point.[38] My main concern here is the statement that in the Federal Republic the reception of scientific theories of self-organization on the part of NSM has been continually increasing since the 1980s.[39] This situation mainly came about due to the scientific popularization of the concept of self-organization among a broad spectrum of the public, and on the other hand, as a result of the increasing density of network links between the individual NSM, there was a greater need to do more systematic and reflective groundwork on the NSM program. The willingness of the NSM to be on the lookout for new, generalizable scientific concepts was influenced to a large extent by what is termed »New Age Philosophy«: it was here already during the 1970s that there was explicit use of self-organization theories. Thus, for the USA, the land of origin of »New Age«, it was particularly favorable that already at an early stage, specific forms of New-Age ideas were linked to specific forms of alternative life-style (e.g., the small rural communes).[40] On the other hand, however, in view of the importation of strong spiritual and esoteric elements of New-Age thinking into the reception of self-organizational research, the relationship between the numerous more politically pragmatic NSM groups and the adherents to the view of »New Age« was then, as now, at variance.[41] Some of the »founding fathers«, such as Prigogine, Haken, and Maturana, offer themselves as agents of the theory of self-organization by popularizing their concepts and propagating them as instrumental in overcoming the mechanistic worldview.

Generally, it would appear that the theory of self-organization and the NSM alternative culture, both of which have had independent, but parallel developments to date, are now entering the phase of reciprocal interaction; that is, the epistemic affinities are mutually perceived, and can from now on be drawn upon for the legitimation of movement policy on the one hand, and for science policy on the other. At all events, there seems to exist a mutual interest on both sides in that the one side uses the other respectively »catalytically«, as it were, to further its own interests. The enormous planning and management problems of highly complex societies, so emphatically stressed by the NSM, as well as the NSMs' »alternative« humanizing perspective based on social self-organization, inspires many self-organization researchers to have aspirations for societal policy measures. For science this would mean, on the one hand, a speedy diffusion of the concept of self-organization throughout new (practice-sensitive) disciplinary fields of application, such as sociology, social psychiatry, and business management, and on the other hand, an increase in state funding for research as a result of the greater interest shown by a general public for the practical application of self-organization models. Thus, the »catalytic« usage of the NSM by the theorists of self-organization has to do more with research policy measures and is less of a conceptional nature: the sociocultural self-organization idea of the NSM is not in the position as such to affect theory construction, but, at most, to extend the sphere of the »intended applications« of the scientific concept of self-organization and, in doing so, display itself as a prime example of a sociological self-organization theory and as such lead to its advancement.[42]

The question can be raised whether the mutual reinforcement between the NSM and the scientific self-organization theory is only a contingent occurrence. It was the aim of this study therefore, to show that there is reason to assume that the process of mutual interaction is of a systematic nature in view of the existence of relatively strong structure analogies between both sides. Finally, this connection is to be seen within a general perspective. In both cases we are dealing with social systems: the system of science within which there are distinctly defined disciplines comprised of many research groups on the one hand, and on the other hand, the network of the NSM groups. In both cases concern is with the carrying through, and stabilization of internally generated »deviations« (the theory of self-organization/the vision of a new society) that would secure chances for innovation. The realization of innovation is problematic for social systems in view of the conservatism of social systems, which, for their part, are operationally closed. Therefore, innovations internal to the system are not possible without the mobilization of support from the system's environment. It is only then, by means of such »feedback loops«, that processes of mutual »deviation amplifying«[43] can be set into motion via which »the neutralization« of the innovation is intercepted as a purely local monosystemic fluctuation and the fundamental structure-conservatism of the individual systems is severed, thus, between them, allowing the formation of qualitatively new structure couplings. This mechanism applies to processes of innovation in all social systems. Focus here, however, was solely on the reciprocal interaction between two social systems. It is only when this field is extended onto other systems (political-legal, economic system, etc.) that it is possible to say whether society is on the threshold to progressive change under the banner of self-organization.

Notes and References

1. On the historical genesis of the self-organization concept, see W. Krohn, G. Küppers, R. Paslack, »Selbstorganisation - Zur Genese und Entwicklung einer wissenschaftlichen Revolution« in S.J. Schmidt (ed.), *Der Diskurs des Radikalen Konstruktivismus*, Frankfurt am Main: Suhrkamp, 1987, p. 441 ff., particularly from p. 446 onwards.

2. As discussed in H. Haken, *Synergetics - An Introduction*, Berlin, Heidelberg, New York: Springer, 1977; or in I. Prigogine, *Vom Sein zum Werden*, Munich: Piper, 1979; and for a somewhat more tentative account, see M. Eigen and R. Winkler, *Das Spiel - Naturgesetze steuern den Zufall*, Munich: Piper, 1975.

3. See, with regard to the successful penetration of the quantum theory, the now already classical study from P. Forman, »Weimar Culture: Causality and Quantum Theory, 1918-1927; Adaptation by German Physicians and Mathematicians to a Hostile Intellectual Environment«, in *Historical Studies in the Physical Sciences* **3** (1971), 1 ff; also T.S. Kuhn, *The Copernican Revolution*, Cambridge, Massachusetts: Harvard University Press, 1957.

4. The interviews in question, as yet unpublished, were carried out in the context of a research project on the history of self-organization (to which the author also contributed) at the Center for Science Studies, University of Bielefeld.

5. For an account of the adherence to the inner-scientific tradition in the earlier works of v. Foerster, Prigogine, Eigen, Haken and Holling on self-organization, see: Krohn et al., *op. cit.*, 1987, note 1, 448 ff.

6. Thus, for example, Eigen stated: »This coincidence of two events in different areas is indeed notable ...«, *op. cit.*, note 4.

7. For example, M. Haken: »It could be that they (the »founding fathers«; R. P.) suddenly discovered a certain reference which catalytically had a further effect . . .«, *op. cit.*, note 4. It is obvious that such »memories« have limited statement value in view of the virtually unavoidable inclusion of self-stylizations and smoothing-out of the historical process.

8. Especially interesting here is M. Ferguson, *Die sanfte Verschwörung*, Basel: Sphinx 1982; see also F. Capra, *Wendezeit. Bausteine für ein neues Weltbild*, Bern, Munich, Vienna: Scherz, 1983.

9. Particularly I. Prigogine and I. Stengers, *Order out of Chaos. Man's new Dialogue with Nature*, London: Heinemann, 1980.

10. See F. Neidhardt, »Einige Ideen zu einer allgemeinen Theorie sozialer Bewegungen«, in W. Hradil (ed.), *Sozialstrukturen im Umbruch. Karl Bolte zum 60.Geburtstag*, Opladen: Leske & Budrich, 1985, p. 193.

11. At least when based on the criteria of the Luhmannian systems theory; see, for example, N. Luhmann, *Soziale Systeme*, Frankfurt am Main: Suhrkamp, 1984.

12. K. Japp, »Selbsterzeugung oder Fremdverschulden - Thesen zum Rationalismus in den Theorien sozialer Bewegungen«, *Soziale Welt*, **35** (1984), 326.

13. See W. Krohn, G. Küppers, *Die Selbstorganisation der Wissenschaft*, Report 33, University of Bielefeld, 1987; here the process of interactive group-building is reconstructed in the case of research groups.

14. See J. Huber's study on the Berlin self-help model: *Die neuen Helfer*, Munich, Zurich: Piper, 1987, especially p. 76 ff.

15. Of significance here is the type of »postmaterialistic« value ethic first put forward by R. Inglehart in *The Silent Revolution: Changing Values and Styles Among Western Publics*, Princeton: University Press, 1977.

16. The comparisons drawn by J. Huber between the »Berlin Model« of self-help groups on the one hand, and the »Elberfeld Model« of 1850, resp. the history of the development of the workers' social aid on the other hand are instructive here; in Huber, *op. cit.*, 1987, note 14.

17. At any rate, this is how J. Raschke sees it: *Soziale Bewegungen - Ein historischsystematischer Grundriß*, Frankfurt am Main, New York: Campus, 1985, for example, p. 387 ff.

18. See K.-W. Brand (ed.), *Neue Soziale Bewegungen in Westeuropa und in den USA - Ein internationaler Vergleich*, Frankfurt am Main, New York: Campus, 1985.

19. Brand, *op. cit.*, 1985, note 18, p. 314.

20. Brand, *op. cit.*, 1985, note 18, p. 319.

21. H. Kitschelt, »Zur Dynamik neuer sozialer Bewegungen in den USA«, in Brand, *op. cit.*, 1985, note 18, p. 248-305.

22. C. Leggewie, »Propheten ohne Macht. Die neuen sozialen Bewegungen in Frankreich zwischen Resignation und Fremdbestimmung«, in Brand, *op. cit.*, 1985, note 18, p. 83-139.

23. D. Murphy, »Von Aldermaston nach Greenham Common. Politischer Protest und neue soziale Bewegungen in Großbritannien«, in Brand, *op. cit.*, note 18, 1985, p. 83-139.

24. See W. Nelles, »Neue soziale Bewegungen und alte Politik«, in P. Grottian and W. Nelles (eds.), *Großstadt und neue soziale Bewegungen*, Basel: Birkhäuser, 1983, p. 83 ff.

25. See Brand, *op. cit.*, 1985, note 18.

26. Brand, *op. cit.*, 1985, note 18, p. 333.

27. K.-J. Scherer, F. Vilmar, *Ökosozialismus? - Rot-grüne Bündnispolitik*, Berlin: Verlag Europäische Perspektiven, 1985, p.48 ff.; or from the same (eds.), *Ein alternatives Sozialismuskonzept: Perspektiven des Ökosozialismus*, Berlin: Stattbuch-Verlag, 1984 p. 21 ff.

28. For example A. Gorz, *Abschied vom Proletariat*, Reinbek (Hamburg): Rowohlt, 1983; or J. Huber, *Die Regenbogen-Gesellschaft*, Frankfurt am Main: Fischer, 1985.

29. See, in particular, Th. Ebert, *Gewaltfreier Aufstand. Alternative zum Bürgerkrieg*, Frankfurt am Main: Fischer, 1980.

30. See, for example, the reader edited by G. Böhme, E. Schramm, *Soziale Naturwissenschaft*, Frankfurt am Main: Fischer, 1985; or G. Michelsen, H. Siebert, *Ökologie lernen*, Frankfurt am Main: Fischer, 1985.

31. Particularly of influence here were the works of F. Vester, most especially *Unsere Welt - ein vernetztes System*, Stuttgart: Deutsche Verlags-Anstalt, 1978.

32. See L. Trepl, *Geschichte der Ökologie*, Frankfurt am Main: Athenäum, 1987, p. 22. See also M. Maren-Grisebach, *Philosophie der Grünen*, Munich, Vienna: Olzog, 1982.

33. F. Vester, *Neuland des Denkens*, Stuttgart: Deutsche Verlags-Anstalt, 1980.

34. W.-D. Hasenclever, C. Hasenclever, *Grüne Zeiten. Politik für eine lebenswerte Zukunft*, Munich: Kösel, 1982, p. 144 f.

35. J.E. Lovelock, *Gaia - A new look at life on Earth*, Oxford: University Press, 1979.

36. In particular, E. Jantsch, *Die Selbstorganisation des Universums*, Munich, Vienna: Hanser, 1979.

37. G. Bateson, *Steps to an Ecology of Mind*, Los Angeles: Chandler Publishing, 1972.

38. One only has to read the critical assessments of »New Age« in the journal *Wechselwirkung*, August 1985.

39. Very informative here are the interviews carried out with the West German »Green Party« (for examplee, the P. Kelly, L. Beckmann, and R. Bahro interviews) reported in Ch. Spretnak's book *Die Grünen*, Munich: Goldmann, 1985.

40. See, for example, the report based on the experiences of an avantgarde of the American counter-movement, W.I. Thompson, under the heading »Gaia-Politik« in the *taz*, December 6th, 1986, p. 15 ff.

41. Criticism ignited mainly due to the partly naive or partly blasé attitude of many of the »New Age« supporters toward the question of political power, the reasons for which are mainly ideologically based.

42. This tendency is in evidence in, for instance, A. Touraine, *The Voice and Eye. An Analysis of Social Movements*, Cambridge: University Press, 1981.

43. M. Maruyama, »The Second Cybernetics: Deviation-Amplifying Mutual Causal Processes«, *American Scientist* **51** (1963), 164-179.

Person Index

Aristotle 91, 159, 162
Ashby 118, 119
Atlan 195
Baldwin 81
Barber 118, 208
Barnes 221
Barsch 125
Barthes 155, 156
Bartlett 188, 191
Bateson 5, 6, 130, 131, 212, 250
Baudrillard 99
Ben-Eli 151
Berger 89
Bergson 183
Bernard 175
Bohm 250
Bourdieu 99, 225
Brand 5, 245
Brunswik 73
Campbell 73, 81
Cannon 175
Capra 5, 250
Carroll 87
Chekhov 154, 159-161, 166, 167
Cohen 99
Conant 118
Culler 156
Darwin 3, 80, 174-177, 242
de Man 155
de Shazer 131, 139
Dell 131
Derrida 155, 159, 167
Descartes 71
Dewey 81
Douglas 96
Dreyfus 93
Driesch 183
Duguit 89
Durkheim 96, 114, 117, 122, 205
Easterbrook 104
Eco 155, 248
Eichenbaum 158
Eigen 3-6, 181, 223, 240
Einstein 21
Fechner 174-178
Ferguson 250
Feyerabend 225
Fichte 147
Fleck 96, 212, 215, 216, 232, 234
Foerster 2-5, 130, 132, 171, 181, 195, 223
Foucault 88, 90-96, 98
Freud 158, 175-177
Freudenthal 225
Galilei 81, 203
Galison 236
Gibson 182
Glasersfeld 78, 181, 212, 229
Goethe 147
Goolishian 129
Gordon 90
Gorz 248
Guntern 130
Gunther 87
Habermas 88-96, 246
Hagstrom 208
Haken 3, 4, 6, 181, 184, 186, 191, 223, 240, 250
Haley 130
Hasenclever 249
Hayek 90
Hegel 176
Hejl 144, 145
Helmholtz 171, 172, 177, 178, 191
Hempel 58, 59
Hering 177, 191
Holling 223
Holton 2
Huber 244, 248
Hull 226
Humboldt 204
Hunt 125
Husserl 81,
Illich 5
Jackson 130
Jakobson 155, 158
James 2, 75, 81
Jantsch 5, 6, 130, 250
Japp 243
Jhering 99, 102
Kant 1, 65, 66, 91, 92, 95, 172, 173, 177, 204
Keeney 131
Kelsen 89
Kennedy 90
Kepler 60
Klopstock 150
Klüver 145
Köhler 177, 183, 184, 186, 189, 191
Krohn 150, 178, 223, 229, 231-234, 236
Kuhn 52, 58, 157, 225
Küppers 150, 178, 223, 229, 231-234, 236
Latour 223, 226-229, 232-234, 236

Leibniz 197
Lewin 187
Lilly 5
Lord 163
Lorenz 3, 173
Lotman 155
Lovelock 250
Löwy 225
Luckmann 89
Luhmann 87-91, 93, 95, 96, 102, 115, 118, 132, 134, 143, 151, 199, 202, 229-231, 237
Mach 172, 177, 178
MacKie 74
Mannheim 64
Marshall 98
Maruyama 130
Maturana 7, 46-50, 66, 69, 78, 79, 131-133, 137, 143, 171, 181, 223, 236, 241, 250
Mayr 38
McCulloch 120
Melville 165
Mendlovitz 6
Merton 87, 208
Moritz 147
Müller 170-172, 177
Neidhardt 242
Newton 60, 130, 203, 227
Nietzsche 75, 158
Ockham 154, 167
Oken 173-175, 177
Oppenheim 58, 59
Parry 163
Parsons 208, 209
Paslack 150
Pasteur 227
Peirce 176, 177
Piaget 78, 212, 213, 218
Pizzorno 88
Plato 64
Polanyi 55
Popper 81, 90
Pound 102
Pribram 5
Prigogine 3-6, 8, 130, 181, 223, 240, 241, 250
Probst 151
Quine 225
Rabinow 93
Rescher 81
Rickert 61
Riffaterre 155
Rogers 129
Roth 143, 144
Saleri 6
Salk 6, 226
Saussure 154, 156-159
Schelling 54, 173, 223
Scherer 248
Schiller 147, 148
Schlosser 200
Schlözer 201
Schumacher 192
Schuster 223
Schütz 70
Shannon 2
Shinn 235
Shrum 227, 234
Smith 2
Solla Price 225
Sonnenfels 201
Spencer 5, 68, 176, 212, 229, 231
Spencer-Brown 5, 68, 212, 213, 229, 231
Storer 208
Tarski 91
Tart 5
Temple 200
Thomasius 197
Thompson 6
Varela 6, 77, 78, 143, 171, 250
Vester 249
Vilmar 248
Watt 2
Watts 5
Watzlawick 130
Weber 75, 88, 90, 91
Weizsäcker 60-63
Wertheimer 184, 192
Williamson 104
Windelband 60
Windscheid 100
Winkler 5
Winter 105, 159
Wittgenstein 154
Zedler 200

Subject Index

absolutism 244

acceptability 13, 14, 17, 24, 137, 223

action 13-17, 19, 21, 24, 25, 27-34, 73, 88-90, 93-97, 100, 114, 116-124, 134-136, 138, 144-147,150, 164, 171, 174, 187, 202, 208, 210, 211, 215-217, 226, 227, 231, 232, 243, 245
communicational 134, 135
norms of 208
program of 128
social 27, 134, 202, 215, 243
types of 145, 210, 211

activity 14-17, 41-44, 182-184, 198-200, 215-217
goal-oriented
integrative
neural
of cells 44
of mind 44
(s. literary; research; science)

actor 4, 88-90, 93, 95, 96, 98, 105, 115, 145-147, 187, 223, 224, 227-229, 243
collective 88, 105
(s. network)

actuality 44, 45, 47, 48

adaptation 3, 7, 133, 198, 208-210

aesthetics 131, 144, 145

aesthetization 150

AGIL 208

agnosticism 224

alchemy 200

analysis
deconstructive 102
economic 102, 104
empirical 208, 225
functional 209, 221
metatheoretical 8
policy 102, 103
reconstructive 212
sociological 241
structural 51

artificial-intelligence 42

astrology 200

attitudes 129, 138, 210, 214

authenticity 157

authority 87, 88, 92, 99-102, 104-106, 131, 164, 220, 247, 248
epistemic 92, 99, 100, 102, 104-106

autocatalysis 3, 185, 188

autonomization 95, 99, 147, 149

autonomy 89, 96, 98, 99, 103, 105, 121, 147, 182, 184, 195, 201, 202, 220, 237, 243, 246, 247, 249
epistemic 89, 98, 105, 106
(s. order; system)

autopoiesis 1, 36, 40, 46, 87-91, 93-95, 97, 98, 102, 132, 195, 196, 199, 201, 202, 223, 240
social 91, 94, 95

autopoietic
network 47, 48
organization 41, 46
structures 70
system 46, 75, 78, 94, 98, 143, 195, 202, 204, 230

behavior 4-7, 37, 38, 43, 45, 52, 57-59, 79, 101, 114-121, 123-125, 128, 129, 133, 139, 163, 174, 181, 184, 187, 208-210, 216, 243
collective 52
(s. eigen-)

behaviorism 90, 191

beliefs 22, 27, 210, 214

bifurcation 3, 4, 189

»Bildung« 154

binary codes 75

Biological Computer Laboratory 3

biology 4-6, 13, 20, 36-38, 41, 42, 44, 46, 52, 53, 55, 78, 79, 133, 191, 223, 227, 228
cellular 37
evolutionary 38
of cognition 46, 191
molecular 36, 53
neuro- 37

black-box 80

bodyhood 28, 29, 32

brain 7, 37, 38, 42-49, 64, 67, 70, 72, 78, 171, 177, 183, 184, 186, 229, 235, 236, 237
mind-brain problem 42

»Brussaville«-project 6

bureaucracy 161, 245-247, 249

causality 3, 104, 130, 151
lineal 130, 151

chaos 1, 3-5, 176
chaotic phases 190
chaos research 3, 4
deterministic 1

chemistry 4, 37, 38, 41, 51, 52, 54, 56, 178
bio- 37, 81
physical 4, 51

circle 40, 46, 47, 66, 68
communicative
exoteric 215
self-referring 66, 68

circularity 65, 92-94, 130, 203, 204

closure 7, 47, 65, 66, 68, 76, 79, 81, 95, 119, 131, 145, 202, 204, 214, 228, 229, 237
operational 67, 76, 119, 131, 133, 202, 204, 214, 232, 233, 237, 251
semantic 47
systemic 7

co-determination 105, 106

cognition 2, 7, 13, 34, 45-47, 64-66, 68-74, 76-78, 81, 89, 91-96, 98, 101, 102, 105, 132, 149, 176, 181, 182, 184, 191, 203, 204, 216, 223, 237, 260, 282
(s. biology)

cognitive 12-14, 22-30, 43-47, 69-71, 101, 102, 115-117, 146-149, 176-178, 184-189, 215-221
abilities 13
acts 70
conditions 223
domain 12-14, 16, 19, 22-30, 32, 33, 131, 148
functions 43
instruments 64, 81, 97
(s. constructivism; hierarchy; operation; system)

cognitivism 191, 208

coherence 6, 100, 101, 213, 214, 219, 235, 237
social 102

collectivism 89, 90, 96

commitment 26, 33

communication 15, 44, 70, 78-80, 87, 88, 94-99, 104, 130-136, 139-151, 162-165, 216-218, 242-244
formal 217, 218
informal 217
internal 88, 146
legal 97, 104
(s. system)

complex 2-7, 40-44, 53-59, 182-184, 208, 209, 230-233
biomolecules 41
phenomena 53-56, 58, 59
(s. complexity; order; system)

complexity 6, 7, 37, 38, 43, 48, 53-55, 57, 61, 77, 87, 118, 130, 134, 143, 150, 151, 181, 182, 185, 209, 213, 230, 231, 237
analytical 181, 182
-generating 53, 61
levels of 48, 237
of a phenomenon 54
social 134
(s. eigen-)

components (systemic) 14, 37-42, 44, 48, 55, 98, 114-125, 133-135, 143, 144, 146, 198, 208, 209, 211, 212, 213, 232, 237

(s. dynamics)

computation of computation 94

conceptual
change (in science) 52
frame 114
re-orientation 54
solution 1, 114

conditions 54-61, 92-94, 143-145, 183-186, 209-211, 213-216, 232-235, 243-245
boundary 38, 44, 54-59, 61, 183, 184, 209-211, 214, 229, 230, 234, 243, 244
creation of 210, 215
environmental 56, 150, 216
favorable 136, 210, 216
framework- 145
initial 4, 55, 57, 59, 144, 183, 190, 210
limiting 213
of possibility 155, 159
selection- 55
stimulus- 184, 185, 191
subsidiary 60
sufficient 58
(s. cognitive; equilibrium)

connectivity 38, 201

consensus 87, 91, 92, 122, 132, 200, 244
consensual relations 15
(s. coordination)

consequentialism 103

conservatism 219, 251

construction 71-77, 88-91, 94-102, 210-214, 216-219, 226-236
social 89-91, 94, 216, 226, 228, 229, 234
(s. reality; world)

constructivism 64-66, 69, 70, 73-75, 77, 78, 80, 81, 89, 91, 93-97, 99, 130, 132, 213, 223, 225, 227-229, 233-237
cognitive 223
radical 94, 99, 132
social 223, 227-229, 236

contingency 57, 59, 60, 91, 93, 94, 147, 210, 213, 231, 237, 241, 251

conversation 16, 132, 137
monovalence/polyvalence 146

cooperation 3, 131, 138, 210-212, 216, 217, 234, 247

cooperatives 244

coordination 3, 79, 114, 151, 210, 215, 217, 234
consensual 97

Copernican revolution 130

cosmology 173

couplings
fixed 214, 232
loose 214
strong 200
structural 241

creativity 29, 31, 121, 149, 216

culture 16, 23, 31, 33, 89, 148, 155, 163, 164, 197, 210, 224, 242, 244-248, 250, 260, 264-266
alternative 242, 245, 248, 250
mass- 245
oral 205
political 245
self-organization- 244, 245
(s. laboratory)

curiosity 17, 200

cybernetics 4, 5, 56, 128, 130, 132, 209, 223, 249
first order 101
of observing 132
second order 73, 98

decentralization 247-249

decision-making 103, 105, 119, 123, 247

deconstruction 156

democracy 244, 248

»Denkkollektiv« 96

determinism 19, 21, 22, 131, 133, 183, 284

(s. structure; system)

development 2-4, 6, 36, 42-44, 51, 55, 59, 61, 78, 81, 93, 128-130, 132, 144, 146, 150, 161, 163, 170, 171, 173, 174, 176, 177, 183-185, 191, 195, 200-202, 208, 210, 215, 216, 218, 226, 228, 234, 240, 241, 245, 249
historical 128, 208
of science 51, 201, 226

deviation amplifying 251

»différance« 167

differentiation 37, 69, 78, 79, 95, 98, 100, 114, 117, 122, 123, 144, 145, 148-151, 161, 172, 173-176, 196, 200, 204, 215, 224, 230, 231, 246

discontinuity 68, 201

discourse 14, 30-32, 87-102, 132-139, 143-150, 224, 243
legal 87, 89, 97-102, 104
rational 91, 92, 94
super- 101
universes of 215
(s. literary; scientific)

dissipation (s. structure) 3

distinction 16, 22, 24, 40, 41, 48, 52, 65-77, 79-82, 94, 122, 132, 133, 145, 171, 175, 198, 210, 212, 213, 216, 230-232, 236, 244
of distinguishing 65, 66
system-environment 66

dogmatism 208

dualism
Cartesian 5
transcendental 20

dynamics 3, 6, 7, 13-15, 22, 29, 32, 55-57, 98, 115-119, 121, 123, 129, 130, 135, 151, 158, 174, 177, 183, 209, 213, 235, 240, 242, 243
inner 177, 183
interpersonal 130
intrapersonal 130
of components 116
system- 55
(s. eigen-; state; theory; thermodynamics)

ecology 5, 6, 38, 223, 241, 242, 248-250

eigen-
complexity 237
dynamics 230
fluctuations 3
solution 212, 214, 221, 232-235
time 151
value 73, 245

élan vital 183

emergence 36-40, 42, 44, 47, 48, 64, 70, 74, 94, 136, 144, 146-149, 151, 181, 242
emergentist view 38
emergent properties 36-40,

emergentism 36-40, 48, 49
holistic 49
physicalist 49

emotions 14, 15, 17, 21, 24, 33, 149, 164, 216

encyclopedia 110, 197, 199, 200

environment 5-8, 66-73, 76-80, 123-128, 146-150, 208-212, 215-221, 229-232, 249-251

entelechie 183

epistème 91-96, 98

epistemological 6, 11, 12, 46, 47, 64-67, 88-94, 130, 131, 223-225, 262
conditions 74
problems 47, 134, 143
questions 74
shift 103

epistemology 1, 46, 67, 75, 78, 80-84, 87-93, 95, 96, 99, 102, 131, 132, 149, 177, 212, 223, 225, 264, 267
bioepistemology 132
legal 88, 90, 91, 96, 102
social 89, 92, 93, 95, 96, 99

equifinality 183

equilibrium 3-5, 7, 53, 54, 59, 183, 184, 191, 201, 208, 209, 249
final 183
material 53
non- 7, 53, 54, 59
stationary 183
(s. state; thermodynamics)

ethics 27, 28, 131, 175, 249

ethnomethodology 225

evolution 5, 6, 38, 73, 80, 95, 98, 105, 116, 155, 158, 159, 167, 174, 176, 196, 202, 208, 226, 231, 237, 240, 241, 243
co-evolution 95, 241
legal 105
organizational 116
social 80, 98, 159

evolutionism 240

experiences 17-33, 45, 116, 121, 123, 129
spiritual 20, 27

externalism 215, 216, 218

family 4, 129-132, 136-140, 159-161, 223, 244, 263
(s. theory)

feedback 2, 117, 130, 188, 215-221, 251
positive 220
self-regulative 130

feeling 5, 45, 148, 156, 176

flexibility 117, 198, 228, 249
interpretative 228

fluctuation 2-4, 34, 130, 184, 185, 188, 189, 220, 232, 249, 251, 265
(s. eigen-; order)

functional imperatives 208

functionalism 209, 247

»Gaia«-hypothesis 250

Gestalt 4-6, 176, 177, 181, 184, 186, 187, 189, 191, 212, 232
-psychology 5, 181, 186, 191
-quality 182
-theory 4, 181-184, 191

Green Party 248. 249

group 1-3, 6, 7, 114, 115, 120-124, 129-131, 210-212, 214-221, 232-235, 241-251
egalitarian 120
group research 210
initiative- 242
-integration 215, 234
-matrix 210, 211, 214
pressure- 245
research groups 115, 210-212, 215-221, 232-235, 244, 251
self-help- 242, 244, 247
social 1, 7, 146, 235
squatters- 248
(s. initiatives)

Hempel-Oppenheim model 58, 59

hermeneutics 103

hermetism 200

heterarchy 120, 124
(s. system)

heteronomy 89, 98, 105, 106

hierarchy 33, 41, 65, 92, 117-124, 145, 161, 182, 233, 235
cognitive 235
-formation 117
logical 65
organizational 119
social 235
temporary 122, 123
(s. organization; regulation; system)

holism 89, 125
(s. medicine)

homeostasis 130, 131, 175

humanities 1, 36, 51, 52, 59-61

idealism 54, 64, 76, 77, 80, 81, 89, 90, 173
German 54, 173

identity 42, 81, 88, 92, 98, 147, 148, 151, 197, 202, 209, 232, 244, 247
collective 88, 257

ideology 73, 162, 250

inclusion/exclusion 74, 124, 232, 243

indeterminacy 90, 228

individualization 95, 114, 117, 135, 245

Industrial Revolution 224

industrialization 54, 247

infinite regression 92, 93, 227

information 2, 4, 7, 59, 61, 64, 67, 69, 71, 78-80, 94, 96, 97, 101, 106, 117, 131, 163, 164, 182, 196, 212, 213, 217, 218, 227, 230-232, 234, 237, 265

informationally closed 184

initiatives 242, 244, 247, 248
alternative project- 248
citizen's 242
self-help- 248
(s. group)

innovation 2, 65, 105, 121, 203, 215, 219, 220, 251

intention 90, 93, 96-98, 146, 151, 178, 210, 214, 220, 250

interaction 3, 7, 13-15, 39-46, 79, 114-116, 119-123, 133-138, 144-148, 208, 209, 212-216, 232-236, 242, 243, 250, 251
change 39
modes of 39-44
recursive 232, 242
social 114, 134, 212, 216, 234-236

interconnection 95

interference 89, 95, 100, 101

interpretation 87, 103, 143, 149, 154, 155, 166, 172, 212-214, 219, 229, 245

intersubjectivity 66, 80, 91-94

invisible colleges 217

isomorphism 184, 226
psychophysical 184

journals 215, 219, 220, 231

jurisprudence 99, 103-105
rational 103
sociological 99, 103-105

Kantian revolution 95

knowledge 12, 13, 23-27, 31-41, 64-70, 72-84, 89-91, 101-105, 117-119, 136, 137, 170-172, 195-204, 210-219, 223-225, 231-237, 262-268
absolute 117, 119, 121, 122
empirical 36
foundation of 65
nomological 103
ordering 198-200, 203
production 204, 212-215, 232, 233
social 102, 105
theories of 66, 203
transcendental 34
(s. scientific)

laboratory 3, 4, 89, 100, 157, 204, 215, 217-219, 225, 226, 235
studies 218, 226

language 12, 15-17, 20, 28-34, 42-44, 46, 47, 64, 72, 79, 81, 87, 88, 93, 104, 132, 154-160, 162, 164, 167, 170, 172, 182, 197, 198, 200, 202-204, 216, 218, 226-228, 232, 236
and literature 154, 155, 158
constitution of 16
metalanguage 156
ordinary 88

languaging 12, 15, 16, 26-28, 132, 137

laser 3, 4, 223

law (natural and juridical) 4, 5, 7, 28, 36-38, 53-60, 87-92, 95-106, 160-164, 170-172, 212, 243

biological 37, 42, 49
Boyle's law 60
Charles' law 60
evolutionary 49
fundamental 54, 58
general 58, 60
historical 38
Kepler's law 60
Müller's law 171, 172, 177
Newton's law 60
natural 28, 55-60
of figure formation 186
physical 37, 55, 176
special 60

learning 3, 7, 33, 71, 128, 129, 196-199, 201, 203, 204, 208
adaptive 208
principles 128

legislature 106

legitimation 7, 149, 204, 215, 217, 224, 242, 249, 250

life 12-15, 27-30, 32-34, 98-100, 148-150, 173-177, 181-183, 243-247
alternative 243, 250
basis of 40
daily 12-15, 17, 21, 30, 33, 150
everyday 99, 148, 181, 182
social 27, 98, 224
spiritual 27
(s. system)

linguistics 64, 167

literarity 157

literary
activities 143-145, 147, 150, 151
circles 158
criticism 146, 151, 227, 228
discourse 144, 148-15o
fiction 155
field 151
form 158
modern 144, 147
modes of composition 163
norms 150
production 144, 151
socialization 143
societies 244
studies 143, 144
text 143, 146, 154, 155, 158, 196
works 147, 151
(s. system)

literature 104, 132, 143, 145-148, 150, 151, 154-159, 204
and education 148
pornographic 148

macro-phenomena 90

management 219, 250

mathematics 4, 61, 65, 72, 81, 224

Maxwellian demon 199, 202

meaning 16, 43, 44, 48, 57, 64, 71, 72, 76, 88, 89, 96, 97, 99, 104, 130, 132, 134, 135, 147, 149, 156-160, 166, 167, 172, 181, 210, 219, 224, 226

mechanism 2, 7, 18-23, 31, 32, 38, 55, 72, 128, 147, 149, 208, 209, 220, 226, 229-234, 248, 251
monetary 149
systemic 209

medicine 5, 128, 129, 203
holistic 5

membership (member) 19, 24, 88, 96, 114-117, 120-123, 125, 130, 132, 134-138, 145, 149, 172, 210, 211, 216, 218, 220, 228, 232, 235, 243, 244, 249

methodological individualism 88-90, 96

microscopic vs. macroscopic 3, 36, 184, 185, 208

monadism 101

morals 75, 91, 178, 249

motivation 146, 242, 244

movement 15, 61, 71, 90, 99, 102, 105, 156-160, 172-175, 223, 240-251
alternative 242
anti-nuke- 246, 248

citizens initiative- 242
ecology- 249, 250
Labour- 149, 224, 242, 246
new women's 248
new social movements (NSM) 240-251
peace- 242, 248, 249
self-help- 242
urban commune- 248
(s. group; initiatives)

Neo-Kantianism 60, 73, 178

network 14-16, 43-49, 94-98, 103, 104, 134, 183-185, 227-230, 234, 235, 248-251, 260, 261
communicational 134
neural 37, 43, 44, 48, 185
recursive 68
(s. autopoietic; system)

networking 213, 218, 219, 240, 241, 248
transdisciplinary 240, 241

neural assemblies (s. network) 44

neurology 42

neurophysiology 78, 81, 223, 229

New Age 5, 241, 242, 250

non-linearity (s. process) 150

objectivism 66, 130, 137

objectivity 24, 26, 131, 133, 137, 138, 149, 172, 212, 233
objectivity-in-parenthesis 133, 137

observation 7, 19-22, 52, 53, 68-75, 77-80, 87, 91, 99, 100, 183, 203, 204, 211-214, 225, 235, 242, 243, 265, 266
of observation 73, 79
recursive 73, 75, 79
self- 7, 91, 230
(s. system)

observer 12-33, 41, 46-48, 68-80, 116, 131-137, 188, 195, 210, 216-218, 263, 265
and observing 12, 13,
external 116, 188

ontogenesis 6, 213

ontological
domain 47
gaps 36, 37
representation 67
status 12, 46, 47

ontology 12, 67

openness 3, 95, 104, 145
(s. system)

operational 7, 13-15, 17-29, 31-34, 55, 56, 67, 76, 104, 119, 131, 133, 202, 204, 213, 214, 232, 233, 237, 251, 265
capacity 13
consensuality 23
domain 13-15, 19, 21, 31, 32
effectiveness 17, 32
realization 29
validation 21
(s. closure; conditions)

operationality 25, 30, 32, 137

operationalization 213, 214, 229, 232

operations (s. operational)
circular 214
cognitive 76, 99, 101
communicational
elementary 96, 202
external 14
intellectual 197
legal 89
recursive 26, 29, 31, 69, 75

order 2-9, 12, 18, 30, 40, 41, 43, 45, 46, 48, 53, 56-58, 61, 65, 66, 68, 71-73, 76, 78, 79, 91-93, 97, 98, 101, 119, 120, 128, 132, 134, 135, 143, 144, 147, 156, 172, 181-185, 188-191, 195-199, 203, 220, 224, 227, 230-232, 234, 240, 241, 249
autonomous 79, 181, 184
causal 172
complex 184, 240
economic 224
formation 181, 182, 184, 185, 188, 189, 191
from noise 2
internal 4, 195
in time 172

legal 97
natural 224
order/disorder 2, 128, 195, 196, 205
social 224
through fluctuation 3, 232

organism 14, 37, 38, 47, 56, 66, 128, 133, 134, 173-175, 183

organization 1-9, 36-43, 46-48, 52-54, 59-61, 114-133, 143-145, 148-151, 170-188, 195-199, 233-237, 240-251
economic 105
formal 114, 149, 199, 242
functional 42, 48
governmental 106
hierarchical 117
of life (living beeings) 40, 41
of the brain 48
(s. autopoietic; self-organization)

organizational change 115, 116
(s. evolution)

ortho-model 149, 150

paradigm 2, 3, 51-54, 57-61, 95, 128, 130, 195, 199, 203, 204, 212, 225, 240, 241, 245-247, 249
shift 51-53, 57-60
»superparadigm« 240

paradox 38, 45, 46, 67, 68, 74, 75, 93, 94, 118, 182

participation 12, 19, 29, 31, 80, 105, 119-123, 137, 154, 208, 247, 248

perception 7, 37, 42, 45, 72, 75, 78, 79, 89, 91, 134, 172, 176, 182-186, 191, 198, 237, 241

perpetuum mobile 55

personality 88, 99, 100, 128, 129, 158

perturbation 2, 3, 72, 97, 209, 218-220, 230-232
by information 218
cognitive 219
critical 3
environmental 2, 3, 97

phase transition 186, 188

philology 198, 200

philosophy 2, 12, 46, 51, 52, 54, 64, 73, 92, 102, 223-225, 250
natural (»Naturphilosophie«) 54, 173, 178, 223
of science 12, 51, 52, 212, 225
pre-Socratic 2
positivistic 255
transcendental 73

physicalism 5, 41-44

physicotheology 200

physics 4, 5, 36-38, 41, 42, 44, 51-54, 56, 58-60, 64, 81, 129, 130, 155, 183, 203, 228, 235, 249
of becoming/of beeing 53
quantum- 64, 81
solid-state- 37

physiology 38, 171, 183

pluralism 74

poetics 144, 147-149, 151, 164, 167

political planning 118, 247

polycontextuality 74

popularization 6, 215, 250

population 197

post-modernity 156

post-structuralism 87-90, 155

pragmatics 91, 131
universal 91

pragmatism 81

praxis of living 18, 19, 21, 25-28, 30-32

predictability 133, 138

primates 15

principle
  basic- 183
  dualistic 55
  methodological 188
  Nirvana- 175, 264
  of transitivity 120

problem solving 121, 211

process 1-7, 37-48, 51-60, 77-79, 89-104, 114-120, 130-134, 144-151, 172-177, 181-186, 199-204, 209-220, 226-234, 240-244
  chemical 56, 144
  cortical 43
  dynamical 209, 243
  evolutionary 3, 38, 226
  genetic and epigenetic 38
  historical 53, 61, 224
  irreversible 3, 59, 223
  mental 42, 44
  nonlinear 191
  psychic 78, 128
  recursive 209
  self-organizing 40, 116, 144, 173, 174, 184, 230
  spatio-temporal 43

psychoanalysis 73, 128, 129

psychology 5, 37, 61, 78, 79, 104, 128, 175, 176, 178, 181, 182, 186, 187, 191
  perceptual 182
  social 187
  (s. Gestalt)

psychopathology 128, 129, 136

rationalism 208, 215
  critical 251

rationality 87, 91, 103, 261, 266
  communicative 87, 91

re-entry 68

realism 64, 88-91, 99, 100, 212, 213, 226, 233, 234
  anti- 89

reality 19-31, 41-47, 67-71, 75-81, 87-106, 131-137, 148-150, 170-172, 209-214, 224-226, 233-235
  autonomous 89
  construction of 89-91, 94, 233
  external 91, 100, 170, 171
  juridical 105
  legal 99, 100, 105
  objective and independent 21
  outside 94, 101
  physical 60
  realities-in-parenthesis 132
  scientific 225, 226
  social 70, 88, 89, 91, 97-99, 102, 103, 106
  theory-dependent 99
  therapeutic 132
  transcendental 133
  universal 20
  (s. world)

recursion 15, 59, 185, 213
  indefinite 59

recursiveness 155

recursivity 68, 72, 74

reductionism 20, 37, 40, 41, 48, 181, 182, 184
  reductionist view 37, 38

redundancy 69

reference 7, 19, 22, 26, 28, 31, 33, 60, 66, 68, 70, 75, 78, 92-95, 102, 135, 138, 143, 146, 148, 171, 177, 178, 188, 204, 212-214, 216, 230, 233, 236, 237, 249
  external 70
  hetero- 95
  self- 7, 66, 68, 70, 75, 92-95, 143, 148, 171, 177, 178, 204
  synreferentiality 145

reflection 12-14, 21, 22, 29-33, 64, 67, 68, 89, 90, 144, 199-202, 242, 245
  (s. self-)

reflexivity 151, 230, 244

regularities 71, 93, 94, 129, 176

regulation 114-119, 121-125, 143, 145, 146, 148, 163, 175, 243, 245
hierarchical 121
(s. self)

reinforcement 220, 241, 251
cross-catalytical 241

relativism 81, 100, 226, 234

religion 105, 147

representationalism 97

research 1, 3, 4, 6-9, 24, 25, 32, 42, 52, 54, 56, 61, 64, 78, 81, 94, 101, 105, 115, 124, 130, 138, 139, 154, 163, 181, 182, 188, 191, 199, 204, 208, 210-221, 228, 231-237, 240, 244, 249-251, 261
-activities 210, 211, 215
-interests 211
-planning 217
-policy 219, 220, 250
-process 211, 212, 216-220, 233
-programme 1, 215, 219, 220
(s. group)

responsibility 27, 28

reversible figures 182, 186

Romanticism 156

Russian Formalists 154, 157, 158

science 1-7, 22-34, 51-54, 57-62, 79-84, 99-104, 155-158, 194-204, 210-212, 215-217, 223-229, 236-242, 248-251
-activities 211
alternative 249
brain- 42
cameral 201
classifications of 198
cognitive 78, 81
empirical 54, 64
natural 4, 7, 51, 52, 54, 59-62, 128, 181, 203, 209, 224, 225
poetry of 25
-policy 217, 250
positivist 104
social 1, 7, 36, 37, 100-105, 191, 208, 223-225, 228, 236
structural 61

scientific
community 24, 29, 89, 136, 236
constructs 100-102
controversies 102
disciplines 36, 54, 181, 223, 240
discourse 99, 100, 157
explanations 12, 14, 16-26, 28, 30-34
facts 100, 223, 225, 232
inquiry 31, 33, 102, 104, 128
investigation 52
knowledge 23, 24, 27, 31-33, 79, 104, 105, 136, 199, 212-215, 225, 229, 231-234
method 12, 203, 225, 237
-oriented 219
outsiders 220
process 51, 100, 204
publication 219
revolution 1, 6, 58, 167, 178, 221, 223, 240
solution 136
truth 100, 105
validity 12, 205
(s. reality; theory)

selection 3-5, 55, 57, 69, 71, 80, 97, 100, 105, 119, 121, 134, 143, 147, 175, 177, 196, 201, 214, 226, 229, 231, 243, 266

selectivity 77, 114, 119, 120, 202, 203

self-
actualizing 129
administration 216, 244, 249
aggregation 40
amplification 185
cognition 46
consciousness 119, 156, 173
control 7
description 46, 47, 88, 241
experiment 48
explicative 171
expression 148
indication 77
interaction 46
legitimating 147
maintenance 40, 210, 211
movement 175
optimalization 144, 51
organization (s. culture; society; system; structure; process) 1-9, 36, 37, 40, 43, 52-54,

59-61, 89, 114-117, 123, 128, 130, 132, 143-145, 148, 150, 151, 155-157, 170-178, 181, 183-188, 191, 195, 196, 199, 201, 203, 208, 209, 211, 215, 223, 229-233, 235, 237, 240-251, 266
preservation 7, 8
reference (s. reference)
referential 1, 65, 81, 93-95, 98, 143-145, 149, 150, 155, 170, 171, 177, 181, 201, 202, 204, 230, 243
referentiality 87, 92-94, 131, 143, 145, 147-149, 151, 177, 178, 195, 203, 223
reflection 30, 89, 144, 199, 201, 242, 245
regulation 114-119, 122, 123, 130, 145, 266
reproduction 3, 97, 98, 101
specification 43
structuring 209
(s. observation)

semantic condensation 243, 245

semiology 158

sensation 45, 170-172, 177

simultaneity 2, 70, 71, 101

social change 125

social experiment 246

socialization 96, 143

society 2, 16, 54, 70, 78-81, 87-91, 93, 95-101, 106, 145-149, 161, 162, 197, 199, 204, 215, 217, 219, 224, 225, 229-231, 236-237, 240-242, 244--248, 250, 251
bourgeois 2
civic 244
feudal 147
fragmentation of 95
self-organization- 245

sociology 36, 64, 67, 73, 78, 79, 89-91, 101, 102, 114, 208, 223-228, 236, 242, 250
legal 101
macro- 242
micro- 242
of knowledge 64, 73, 91, 223-225, 228
of organization 242
of science 67, 208, 223, 224, 226, 236

solidarity 2, 114, 224, 246

solipsism 97, 101

spontaneity 22, 156, 174, 176, 246

spontaneously change 21

stability 3, 4, 71, 122, 130, 131, 143, 147, 150, 170, 174-177, 181, 185, 190, 212, 232, 233, 235, 249
bi-stability 192
instability 3, 209
multistability (s. state) 185, 186, 188, 191

stable solution 211

state
attractor state 188, 190
change of 23
dynamics of 29
equilibrium 53
internal 145, 221
mental 42-48, 97, 182
multistable 184
physical 177
psychical 177
stable 185
transformed 39

state (political)
authoritarian 246
constitutional 244
social 245, 247, 248

stimulus pattern 184, 185

stochastics 184

structuralism 87, 90, 93, 155

structure 1-3, 20-24, 39-43, 51-59, 69-80, 130-136, 169-171, 188-190, 198-202, 212-214, 223-225, 229-234, 243-249, 260
and function 43
client's 138
definitions of 209
-determinism 131, 133
dissipative 1, 49, 240
explanatory 58, 59

family- 136, 140
generative 20
innovative 8
molecular 59
physical 39, 56
self-organizing- 54
symmetric(al) 190, 191
therapist's 138
(s. autopoietic; couplings)

subjectivity 148, 149

substantialism 49

Surrealism 156

symmetry breaking 186-188

symptomatology 158

synergetics 1, 3, 6, 186, 191, 223, 240

system
autonomous 133, 249
biological 3, 37, 38, 46-48, 145, 183, 191
brain-mind 43
circular 46
closed 67, 145, 183
cognitive 43, 44, 47, 69-71, 81, 149, 170, 171, 176, 177, 178, 181, 182, 184-186, 188, 189, 191
communication- 79
complex 6, 38, 42, 55-57, 209
deterministic 209
developing 195
differentiated 80, 122, 208
eco- 53
functional 155, 199, 208
genre- 151
heterarchical 120-124,
hierarchical 119-121, 124
inanimate 38
juridical 149, 150
knowing 66, 67, 70
legal 87, 105
limbic 43, 44
literary 143-151
living 12-14, 19, 21, 24, 27-30, 38, 40, 41, 46, 53, 56, 59, 78, 133, 143, 145, 170
nervous 14, 29, 45, 47, 170
networked 249
non-linear 182
observing 47, 69, 205
open 53, 130, 131, 174
problem- 135, 136
psycological 78
recursive 7
self-referential 1, 81, 143, 144
self-organized 170, 171, 176
social 7, 72, 74, 79, 80, 86, 91, 96-98, 114-117, 122-125, 131-137, 143-151, 195, 196, 199, 201, 202, 204, 210, 212, 215, 219, 229, 230, 237, 251
therapy- 131, 132, 137
value 117, 120
visual 42, 182-184
(s. autopoietic)

systems theory 3, 5, 46, 66, 67, 72, 114, 128, 130, 132, 135, 143, 154, 155, 199, 208-210, 249
classical 155, 208, 209, 249
sociological 132, 143

technology 2, 31, 139, 167, 223-225, 228, 229, 231, 237, 248, 249

theoretical 57-61, 117, 118, 125, 130-132, 139, 140, 145-151, 181-184, 212-214, 224, 225, 232, 233
decisions 212, 233
entities 209
frame 57
ideas 60
interpretation 214
model 125, 183, 208
rearrangement 57
structure 57, 76

theory (s. theoretical)
automata- 3
clinical 128, 132, 136
consensus- 91
critical 87-91
dialectical 76
dynamics 57
evolutionary 196
field- 4, 183, 184, 186, 187, 191
functional 208
game- 5, 61
Grand Theory 87, 89
information- 2, 61
legal 87, 89, 91, 96, 101
non-metaphysical 87

of gases 60
of signs 172
of truth 91, 106
phlogiston- 53
quantum- 41
self-explanatory 46
social science 100
socio-legal 87
sociological 78, 103, 133, 196
therapy- 131
(s. Gestalt; knowledge; systems theory)

therapy 4, 128-140, 223
behavior- 129
-evaluation 136
family- 4, 129-132, 138-140, 223
psychosocial 138
psycho- 128-132, 139
systemic 128, 130-133, 139
(s. theory)

thermodynamics
linear 183

thing-in-itself 172, 177, 178

transcendentalism 66, 92
communicative 92

truths 28, 100, 199, 202, 234
(s. scientific)

undecidability 74

university 1, 3, 5, 36, 106, 114, 125-128, 139, - 154, 167-170, 197-201, 208, 240, 251

variety 53, 74, 105, 118, 119, 173, 174, 214, 231
behavioral 119
requisite 118, 119

viability 104, 136, 137

»Weltanschauung« 1

»Weltgeist« 96

world 2, 5, 6, 20, 21, 31-34, 36, 37, 47-49, 64-73, 76-80, 87-90, 94-99, 101-105, 118, 124, 129, 130, 133, 137, 139, 145, 146, 149, 154, 155, 159, 160, 167, 170-174, 176, 178, 182, 183, 196, 199, 201, 202, 209, 223-228, 230, 231, 234, 236, 237, 240-242, 244, 245, 249
-construction 80
external 64-69, 71, 76, 79, 171-173
-image 241
legal 87, 101, 105
-machine 249
-making 99
material 223
-model 146
objective 26, 45, 172
outside 94, 97
perceptual 182, 183
phenomenal 45, 47-49, 124
physical 45, 176
psychical 95, 176, 178
real 47, 48, 53, 70, 87, 97, 99, 199
social 99, 224, 236, 249
-view 51, 97, 230, 231, 237, 240, 242
(s. reality)

»Zeitgeist« 128

Sociology of the Sciences

1. E. Mendelsohn, P. Weingart and R. Whitley (eds.): *The Social Production of Scientific Knowledge.* 1977 ISBN Hb 90-277-0775-8; Pb 90-277-0776-6
2. W. Krohn, E.T. Layton, Jr. and P. Weingart (eds.): *The Dynamics of Science and Technology.* Social Values, Technical Norms and Scientific Criteria in the Development of Knowledge. 1978 ISBN Hb 90-277-0880-0; Pb 90-277-0881-9
3. H. Nowotny and H. Rose (eds.): *Counter-Movements in the Sciences.* The Sociology of the Alternatives to Big Science. 1979 ISBN Hb 90-277-0971-8; Pb 90-277-0972-6
4. K.D. Knorr, R. Krohn and R. Whitley (eds.): *The Social Process of Scientific Investigation.* 1980 (1981) ISBN Hb 90-277-1174-7; Pb 90-277-1175-5
5. E. Mendelsohn and Y. Elkana (eds.): *Sciences and Cultures.* Anthropological and Historical Studies of the Sciences. 1981 ISBN Hb 90-277-1234-4; Pb 90-277-1235-2
6. N. Elias, H. Martins and R. Whitley (eds.): *Scientific Establishments and Hierarchies.* 1982 ISBN Hb 90-277-1322-7; Pb 90-277-1323-5
7. L. Graham, W. Lepenies and P. Weingart (eds.): *Functions and Uses of Disciplinary Histories.* 1983 ISBN Hb 90-277-1520-3; Pb 90-277-1521-1
8. E. Mendelsohn and H. Nowotny (eds.): *Nineteen Eighty Four: Science between Utopia and Dystopia.* 1984 ISBN Hb 90-277-1719-2; Pb 90-277-1721-4
9. T. Shinn and R. Whitley (eds.): *Expository Science.* Forms and Functions of Popularisation. 1985 ISBN Hb 90-277-1831-8; Pb 90-277-1832-6
10. G. Böhme and N. Stehr (eds.): *The Knowledge Society.* The Growing Impact of Scientific Knowledge on Social Relations. 1986 ISBN Hb 90-277-2305-2; Pb 90-277-2306-0
11. S. Blume, J. Bunders, L. Leydesdorff and R. Whitley (eds.): *The Social Direction of the Public Sciences.* Causes and Consequences of Co-operation between Scientists and Non-scientific Groups. 1987 ISBN Hb 90-277-2381-8; Pb 90-277-2382-6
12. E. Mendelsohn, M.R. Smith and P. Weingart (eds.): *Science, Technology and the Military.* 2 vols. 1988 ISBN Vol, 12/1 90-277-2780-5; Vol. 12/2 90-277-2783-X
13. S. Fuller, M. de Mey, T. Shinn and S. Woolgar (eds.): *The Cognitive Turn.* Sociological and Psychological Perspectives on Science. 1989 ISBN 0-7923-0306-7
14. W. Krohn, G. Küppers and H. Nowotny (eds.): *Selforganization.* Portrait of a Scientific Revolution. 1990 ISBN 0-7923-0830-1

KLUWER ACADEMIC PUBLISHERS – DORDRECHT / BOSTON / LONDON

MIX
Papier aus verantwortungsvollen Quellen
Paper from responsible sources
FSC® C105338

If you have any concerns about our products,
you can contact us on
ProductSafety@springernature.com

In case Publisher is established outside the EU,
the EU authorized representative is:
Springer Nature Customer Service Center GmbH
Europaplatz 3, 69115 Heidelberg, Germany

Printed by Libri Plureos GmbH
in Hamburg, Germany